Analysis on Lie Groups
An Introduction

The subject of analysis on Lie groups comprises an eclectic group of topics which can be treated from many different perspectives. This self-contained text concentrates on the perspective of analysis to the topics and methods of non-commutative harmonic analysis, assuming only elementary knowledge of linear algebra and basic differential calculus.

The author avoids non-essential technical discussion and instead describes in detail many interesting examples, including formulae which have not previously appeared in book form. Topics covered include the Haar measure and invariant integration, spherical harmonics, Fourier analysis and the heat equation, the Poisson kernel, the Laplace equation and harmonic functions.

Perfect for advanced undergraduates and graduates in geometric analysis, harmonic analysis and representation theory, the tools developed will also be useful for specialists in stochastic calculation and statistics. With numerous exercises and worked examples, the text is ideal for a graduate course on analysis on Lie groups.

JACQUES FARAUT is Professeur émérite in the Institut de Mathématiques de Jussieu at the Université Pierre et Marie Curie in Paris.

CAMBRIDGE STUDIES IN ADVANCED MATHEMATICS

Editorial Board:

B. Bollobas, W. Fulton, A. Katok, F. Kirwan, P. Sarnak, B. Simon, B. Totaro

All the titles listed below can be obtained from good booksellers or from Cambridge University Press. For a complete series listing visit: http://www.cambridge.org/series/sSeries.asp?code=CSAM

Already published

Analysis on Lie Groups
An Introduction

JACQUES FARAUT

CAMBRIDGE
UNIVERSITY PRESS

CAMBRIDGE
UNIVERSITY PRESS

University Printing House, Cambridge CB2 8BS, United Kingdom

Cambridge University Press is part of the University of Cambridge.

It furthers the University's mission by disseminating knowledge in the pursuit of education, learning and research at the highest international levels of excellence.

www.cambridge.org
Information on this title: www.cambridge.org/9780521719308

© J. Faraut 2008

First published 2008
(French edition published by Calvage et Mounet 2006)

A catalogue record for this publication is available from the British Library

Library of Congress Cataloguing in Publication data

Faraut, Jacques, 1940–
Analysis on Lie groups: an introduction / Jacques Faraut.
p.   cm. – (Cambridge studies in advanced mathematics)
Includes bibliographical references and index.
ISBN 978-0-521-71930-8 (hardback)
1. Lie groups.   2. Lie algebras.   I. Title.
QA387.F37   2008
512'.482–dc22   2007053046

ISBN 978-0-521-71930-8 Hardback

Contents

Preface

This book stems from notes of a master's course given at the *Université Pierre et Marie Curie*. This is an introduction to the theory of Lie groups and to the study of their representations, with applications to analysis. In this introductory text we do not present the general theory of Lie groups, which assumes a knowledge of differential manifolds. We restrict ourself to linear Lie groups, that is groups of matrices. The tools used to study these groups come mainly from linear algebra and differential calculus. A linear Lie group is defined as a closed subgroup of the linear group $GL(n, \mathbb{R})$. The exponential map makes it possible to associate to a linear Lie group its Lie algebra, which is a subalgebra of the algebra of square matrices $M(n, \mathbb{R})$ endowed with the bracket $[X, Y] = XY - YX$. Then one can show that every linear Lie group is a manifold embedded in the finite dimensional vector space $M(n, \mathbb{R})$. This is an advantage of the definition we give of a linear Lie group, but it is worth noticing that, according to this definition, not every Lie subalgebra of $M(n, \mathbb{R})$ is the Lie algebra of a linear Lie group, that is a closed subgroup of $GL(n, \mathbb{R})$. The Haar measure of a linear Lie group is built in terms of differential forms, and these are used to establish several integration formulae, linking geometry and analysis. The basic properties of irreducible representations of compact groups, that is the Peter–Weyl Theory, are first presented in a general setting, then described explicitly in the case of the simplest non-commutative compact Lie groups: the special unitary group $SU(2)$, and the special orthogonal group $SO(3)$, then further in the case of the unitary groups $U(n)$. The topics in analysis we present are centred on a basic object: the Laplace operator. Fourier analysis on a compact linear Lie group provides a diagonalisation of the Laplace operator, and the Fourier method is in particular a natural method for solving the Cauchy problem for the heat equation on the group $SU(2)$. Similarly, analysis on the sphere in $\mathbb{R}^n$ uses the spherical harmonic decomposition and makes clear the interaction which exists between the orthogonal group $O(n)$ and Fourier analysis on $\mathbb{R}^n$, shown for instance by the Bochner–Hecke relations and potential theory, while expanding a harmonic function as a series of harmonic homogeneous polynomials. Questions of the same nature arise as one considers the action of the orthogonal group $O(n)$ on the space $Sym(n, \mathbb{R})$ of real symmetric matrices, or the action of the unitary group $U(n)$ on the space $Herm(n, \mathbb{C})$ of Hermitian matrices. The formula for the radial part of the Laplace operator plays an important role; in particular, it leads to evaluation of integral orbitals via solution of the Cauchy problem for the heat equation on the space $Herm(n, \mathbb{C})$ of Hermitian matrices. To study the irreducible representations of the unitary group $U(n)$ we start from the highest weight theorem. This is a special case of the Weyl theory of irreducible representations of compact Lie groups. The characters of the irreducible representations of the unitary group are expressed in terms of Schur functions, for which we establish some combinatoric properties. These make it possible to write explicit Fourier expansions of some central functions, and also general Taylor expansions for holomorphic functions of a matrix argument.

The invariant analysis topics we are dealing with in this book illustrate how Lie groups are involved in many fields: matrix analysis, Fourier analysis, complex analysis, mathematical physics.

Each chapter is followed by numerous exercises. Some topics which are not treated in the text are introduced as problems. For example, in Chapter 7, we present the construction of an equivariant isomorphism between the space of polynomials in two variables, homogeneous of degree 2ℓ, and the space of harmonic polynomials in three variables, homogeneous of degree ℓ.

Many books deal with the theory of Lie groups. We cite several of them in the bibliography. We were inspired at several points by the presentation given by J. Hilgert and K.-H. Neeb in their book *Lie-Gruppen und Lie-Algebren*, and we have included elegant arguments from the book by R. Mneimné and F. Testard, *Introduction à la Théorie des Groupes de Lie Classiques*.

We thank Rached Mneimné, Hervé Sabourin, and Valerio Toledano for reading and commenting on preliminary versions of this text, and giving us valuable advice for improving it.

1

The linear group

The linear group $GL(n, \mathbb{R})$ is the group of invertible real $n \times n$ matrices. After some topological preliminaries we present some subgroups of the linear group which play an important role in geometry and analysis. We establish the polar and Gram decompositions, which will be useful for proving some topological properties of these groups.

1.1 Topological groups

A topological group is a group equipped with a topology such that the maps

$$(x, y) \mapsto xy, \quad G \times G \to G,$$
$$x \mapsto x^{-1}, \quad G \to G,$$

are continuous. This amounts to saying that the map

$$(x, y) \mapsto xy^{-1}, \quad G \times G \to G$$

is continuous.

A topological group is Haussdorff if $\{e\}$ is closed (e is the identity element of G).

Let H be a subgroup of a topological group G. If H is open then H is also closed. In fact, if $g \notin H$, gH is a neighbourhood of g contained in H^c, therefore H^c is open.

Let G_0 be the connected component of e in G (one says the *identity component*). Then G_0 is a normal subgroup of G. In fact, if $g \in G_0$, then $g^{-1}G_0$ is connected and contains e, hence $g^{-1}G_0 \subset G_0$, and G_0 is a subgroup of G. Furthermore, if $g \in G$, then gG_0g^{-1} is connected and contains e, hence $gG_0g^{-1} \subset G_0$, and G_0 is a normal subgroup.

Proposition 1.1.1 *Let V be a connected neighbourhood of e in a topological group G, then*

$$\bigcup_{n=1}^{\infty} V^n = G_0,$$

where G_0 denotes the identity component of G.

Hence a connected topological group is generated by any neighbourhood of the identity element.

Proof. In fact, if V is a neighbourhood of e, then the increasing union $U = \bigcup_{n=1}^{\infty} V^n$ is an open set since V^{n+1} is a neighbourhood of each point of V^n.

If V is connected then U is connected as well since it is a union of connected sets, all of which contain e. Therefore $U \subset G_0$. Let $W = V \cap V^{-1}$, then

$$U' = \bigcup_{n=1}^{\infty} W^n$$

is an open subgroup of G, hence closed. Since $U' \subset G_0$, because $U' \subset U$, then $U' = G_0$, and therefore $U = G_0$. $\qquad\square$

The topology of a topological group is determined by the set $\mathcal{V}$ of the neighbourhoods of e. This set has the following properties.

(a) If $V \in \mathcal{V}$, there exists V_1 and $V_2 \in \mathcal{V}$ such that $V_1 \cdot V_2 \subset V$.
(b) If $V \in \mathcal{V}$, then $V^{-1} \in \mathcal{V}$.
(c) Let $V \in \mathcal{V}$ and $g \in G$, then $gVg^{-1} \in \mathcal{V}$.

Conversely, if G is a group and if $\mathcal{V}$ is a family of non-empty subsets of G with the following properties:

every subset of G which contains a subset of $\mathcal{V}$ belongs to $\mathcal{V}$,
any finite intersection of subsets of $\mathcal{V}$ belongs to $\mathcal{V}$,

(i.e. $\mathcal{V}$ is a filter), and also properties (a), (b), and (c), then there exists a unique topology for which G is a topological group such that $\mathcal{V}$ is the family of the neighbourhoods of e.

The neighbourhoods of an element $g \in G$ are the subsets gV ($V \in \mathcal{V}$).

1.2 The group $GL(n, \mathbb{R})$

Let $M(n, \mathbb{R})$ denote the algebra of $n \times n$ matrices with entries in $\mathbb{R}$, and $GL(n, \mathbb{R})$ the group of invertible matrices in $M(n, \mathbb{R})$, which is called the *linear group.* We will consider this group from the viewpoints of topology and differential calculus.

We consider on $\mathbb{R}^n$ the Euclidean norm

$$\|x\| = \sqrt{x_1^2 + \cdots + x_n^2}$$

associated to the Euclidean inner product

$$(x|y) = x_1 y_1 + \cdots + x_n y_n,$$

and on $M(n, \mathbb{R})$ the norm

$$\|A\| = \sup_{x \in \mathbb{R}^n, \|x\| \leq 1} \|Ax\|.$$

Let us recall that, on a finite dimensional vector space, all the norms are equivalent. Note that the norm we consider on $M(n, \mathbb{R})$ is an algebra norm:

$$\|AB\| \leq \|A\|\|B\|.$$

One can check that the product on $M(n, \mathbb{R})$ is a continuous map.

Proposition 1.2.1 *The group $GL(n, \mathbb{R})$ is open in $M(n, \mathbb{R})$. The map $g \mapsto g^{-1}$, from $GL(n, \mathbb{R})$ onto itself, is continuous.*

Proof. One can prove this proposition using Cramer's formulae. In fact,

$$GL(n, \mathbb{R}) = \{g \in M(n, \mathbb{R}) \mid \det(g) \neq 0\},$$

and

$$g^{-1} = \frac{1}{\det g}\tilde{g},$$

where $\tilde{g}$ denotes the cofactor matrix whose entries are polynomials in the entries of g. We will give another proof which holds if, instead of $M(n, \mathbb{R})$, one considers any Banach algebra, possibly infinite dimensional.

(a) *Let $M \in M(n, \mathbb{R})$. If $\|M\| < 1$, then $I + M$ is invertible and*

$$\|(I + M)^{-1}\| \leq \frac{1}{1 - \|M\|}.$$

In fact, the series $\sum_{k=0}^{\infty}(-1)^k M^k$ converges in norm and its sum is equal to $(I + M)^{-1}$:

$$(I + M)^{-1} = \sum_{k=0}^{\infty}(-1)^k M^k.$$

Furthermore,

$$\|(I+M)^{-1}\| \le \sum_{k=0}^{\infty} \|M^k\| \le \sum_{k=0}^{\infty} \|M\|^k = \frac{1}{1-\|M\|}.$$

(b) *Let A be an invertible matrix. If B is a matrix such that*

$$\|B-A\| < \|A^{-1}\|^{-1},$$

then B is invertible, and if $\|B-A\| \le \varepsilon < \|A^{-1}\|^{-1}$,

$$\|B^{-1}-A^{-1}\| \le \frac{\|A^{-1}\|^2 \varepsilon}{1-\|A^{-1}\|\varepsilon}.$$

One can write

$$B = A(I + A^{-1}(B-A)),$$

and one applies (a) to $M = A^{-1}(B-A)$. Note that $\|M\| \le \|A^{-1}\|\varepsilon$. Therefore, if $\varepsilon < \|A^{-1}\|^{-1}$, then $I+M$ is invertible and

$$B^{-1} = (I+M)^{-1}A^{-1}, \quad \|B^{-1}\| \le \frac{\|A^{-1}\|}{1-\|A^{-1}\|\varepsilon}.$$

Furthermore,

$$B^{-1}-A^{-1} = B^{-1}(A-B)A^{-1},$$

hence

$$\|B^{-1}-A^{-1}\| \le \frac{\|A^{-1}\|^2 \varepsilon}{1-\varepsilon\|A^{-1}\|}. \qquad \square$$

Therefore we can state the following.

Theorem 1.2.2 *The group $GL(n,\mathbb{R})$, equipped with the topology inherited from $M(n,\mathbb{R})$, is a topological group.*

From now on $GL(n,\mathbb{R})$ will denote this topological group.

Proposition 1.2.3 *The subsets*

$$\{g \in GL(n,\mathbb{R}) \mid \|g\| \le C, \ \|g^{-1}\| \le C\},$$

where C is a constant, are compact, and every compact subset of $GL(n,\mathbb{R})$ is contained in a subset of that form.

Proof. Let us show that the subset

$$Q = \{g \in GL(n,\mathbb{R}) \mid \|g\| \le C, \ \|g^{-1}\| \le C\}$$

is compact. Let (g_n) be a sequence of elements in Q. Since a closed ball in $M(n, \mathbb{R})$ is compact, it is possible to extract from the sequence (g_n) a subsequence (g_{n_k}) which converges to an element g in $M(n, \mathbb{R})$ with $\|g\| \leq C$. Since $\|g_{n_k}^{-1}\| \leq 1$, it is possible to extract from the sequence $(g_{n_k}^{-1})$ a subsequence which converges to an element h in $M(n, \mathbb{R})$ with $\|h\| \leq C$. Furthermore, for every n, $g_n g_n^{-1} = I$, hence $gh = I$ or $h = g^{-1}$, $g \in GL(n, \mathbb{R})$, and $g \in E$. $\qquad\square$

Note that the group $GL(n, \mathbb{R})$ is equal to the increasing sequence of the compact subsets

$$Q_k = \left\{ g \in GL(n, \mathbb{R}) \mid \|g\| \leq k, \ |\det g| \geq \frac{1}{k} \right\} \quad (k \in \mathbb{N}^*).$$

1.3 Examples of subgroups of $GL(n, \mathbb{R})$

(a) Let $SL(n, \mathbb{R})$ denote the *special linear group* defined by

$$SL(n, \mathbb{R}) = \{g \in GL(n, \mathbb{R}) \mid \det g = 1\}.$$

It is a closed subgroup of $GL(n, \mathbb{R})$ which is normal because it is the kernel of the continuous group morphism

$$\det : \ GL(n, \mathbb{R}) \to \mathbb{R}^*.$$

(b) Let $O(n)$ denote the *orthogonal group* defined by

$$O(n) = \{g \in GL(n, \mathbb{R}) \mid \forall x \in \mathbb{R}^n, \ \|gx\| = \|x\|\}.$$

By polarising one can show that $g \in O(n)$ if and only if

$$\forall x, y \in \mathbb{R}^n, \quad (gx|gy) = (x|y),$$

and this can be written

$$g^T g = I, \quad \text{or } g^{-1} = g^T,$$

where g^T denotes the transposed matrix of g. Therefore, if $g \in O(n)$, then $\det g = \pm 1$.

The rows of $g \in O(n)$ are orthogonal unit vectors, and the same holds for the columns. The subgroup $O(n)$ is a compact subgroup of $GL(n, \mathbb{R})$. This follows from Proposition 1.2.3. In fact, for g in $O(n)$,

$$\|g\| = 1, \quad \|g^{-1}\| = 1.$$

The *special orthogonal group* $SO(n)$ is the subgroup of orthogonal matrices with determinant equal to one:

$$SO(n) = O(n) \cap SL(n, \mathbb{R}).$$

(c) More generally, let us consider a non-degenerate bilinear form b on $\mathbb{R}^n$, and the subgroup $O(b)$ defined by

$$O(b) = \{g \in GL(n, \mathbb{R}) \mid \forall x, y \in \mathbb{R}^n, \ b(gx, gy) = b(x, y)\}.$$

Let B be the matrix of the bilinear form b:

$$b(x, y) = y^T B x.$$

The condition $g \in O(b)$ can be written

$$g^T B g = B.$$

Let us observe that, since the matrix B is invertible, this condition implies that g is invertible. The subgroup $O(b)$ is closed in $GL(n, \mathbb{R})$ and, for $g \in O(b)$,

$$g^{-1} = B^{-1} g^T B.$$

If b is the symmetric bilinear form

$$b(x, y) = \sum_{i=1}^{p} x_i y_i - \sum_{i=1}^{q} x_{p+i} y_{p+i}, \quad p + q = n,$$

then one can write $O(b) = O(p, q)$:

$$O(p, q) = \{g \in GL(n, \mathbb{R}) \mid g^T I_{p,q} g = I_{pq}\},$$

where

$$I_{p,q} = \begin{pmatrix} I_p & 0 \\ 0 & -I_q \end{pmatrix}.$$

The subgroup $O(p, q)$ is called the *pseudo-orthogonal group*.

If b is a symmetric bilinear form with signature (p, q), there exists $g_0 \in GL(n, \mathbb{R})$ such that $B = g_0^T I_{p,q} g_0$. (This is Sylvester's law of inertia.) Therefore, the subgroup $O(b)$ is conjugate to $O(p, q)$:

$$O(b) = g_0^{-1} O(p, q) g_0.$$

The subgroup $O(1, 3)$ plays an important role in relativity theory. This is in fact the group of linear transformations of space-time $\mathbb{R}^4$ which preserve the Lorentz form

$$t^2 - x^2 - y^2 - z^2.$$

(d) Another important example is the case of a non-degenerate skewsymmetric bilinear form. Such a form only exists if n is even, $n = 2m$, and then there exists a basis with respect to which

$$b(x, y) = -\sum_{i=1}^{m} x_i y_{m+i} + \sum_{i=1}^{m} x_{m+i} y_i.$$

The matrix of this form is

$$J = \begin{pmatrix} 0 & I_m \\ -I_m & 0 \end{pmatrix}.$$

In this case the subgroup $O(b)$ is the *symplectic group*

$$Sp(m, \mathbb{R}) = \{g \in GL(2m, \mathbb{R}) \mid g^T J g = J\}.$$

(e) Let us mention the group of upper triangular matrices:

$$T(n, \mathbb{R}) = \{g \in GL(n, \mathbb{R}) \mid g_{ij} = 0 \text{ if } i > j\},$$

which is called the *upper triangular group*. We also have the *strict upper triangular group*:

$$T_0(n, \mathbb{R}) = \{g \in GL(n, \mathbb{R}) \mid g_{ij} = 0 \text{ if } i > j, \text{ and } g_{ii} = 1\}.$$

One can check that $T_0(n, \mathbb{R})$ is a normal subgroup of $T(n, \mathbb{R})$.

(f) Consider on $\mathbb{C}^n$ the Hermitian inner product

$$(x|y) = \sum_{i=1}^{n} x_i \bar{y}_i.$$

The *unitary group* $U(n)$ is the subgroup of $GL(n, \mathbb{C})$ consisting of matrices which preserve this inner product. This can be written

$$U(n) = \{g \in GL(n, \mathbb{C}) \mid g^* g = I\}.$$

The *special unitary group* $SU(n)$ is the group of unitary matrices with determinant one. The *pseudo-unitary group* $U(p, q)$ is defined as

$$U(p, q) = \{g \in GL(n, \mathbb{C}) \mid g^* I_{p,q} g = I_{p,q}\}.$$

1.4 Polar decomposition in $GL(n, \mathbb{R})$

Let us denote by $\mathcal{P}_n$ the set of positive definite real symmetric $n \times n$ matrices. This is an open convex cone in the vector space $Sym(n, \mathbb{R})$ of real symmetric matrices. One can see that $\mathcal{P}_n$ is open as follows. To a matrix $p \in \mathcal{P}_n$ one associates the quadratic form

$$Q(x) = (px|x).$$

The function Q is continuous on the unit sphere S of $\mathbb{R}^n$. It is strictly positive and, since S is compact,

$$\alpha := \inf_{x \in S} Q(x) > 0.$$

One can show that the open ball with centre p and radius α is contained in $\mathcal{P}_n$.

Theorem 1.4.1 (Polar decomposition) *Every $g \in GL(n, \mathbb{R})$ decomposes uniquely as*

$$g = kp,$$

with $k \in O(n)$, $p \in \mathcal{P}_n$. Furthermore the map

$$O(n) \times \mathcal{P}_n \to GL(n, \mathbb{R}), \quad (k, p) \mapsto g = kp,$$

is a homeomorphism

Proof. (a) *Existence.* Let $g \in GL(n, \mathbb{R})$. If $x \neq 0$ then

$$(g^T g x | x) = \|gx\|^2 > 0,$$

therefore $A = g^T g \in \mathcal{P}_n$. It follows that the symmetric matrix A, which is diagonalisable in an orthogonal basis:

$$A = h \begin{pmatrix} \lambda_1 & & \\ & \ddots & \\ & & \lambda_n \end{pmatrix} h^{-1} \quad (h \in O(n)),$$

has positive eigenvalues λ_i. The matrix

$$p = h \begin{pmatrix} \sqrt{\lambda_1} & & \\ & \ddots & \\ & & \sqrt{\lambda_n} \end{pmatrix} h^{-1}$$

belongs to $\mathcal{P}_n$, and $p^2 = A$. Define

$$k = gp^{-1},$$

then

$$k^T k = p^{-1} g^T g p^{-1} = p^{-1} A p^{-1} = I,$$

hence the matrix k is orthogonal, and $g = kp$.

(b) *Unicity.* Let $g \in GL(n, \mathbb{R})$ and assume that

$$g = kp = k_1 p_1,$$

where k and p are the matrices we considered in (a), and $k_1 \in O(n)$, $p_1 \in \mathcal{P}_n$. Let us show that $k_1 = k$, $p_1 = p$. Consider the eigenvalues $\lambda_1, \ldots, \lambda_n$ of $A = g^T g$, and let f be a polynomial in one variable such that

$$f(\lambda_i) = \sqrt{\lambda_i}, \quad i = 1, \ldots, n.$$

Then $p = f(A)$ and, since $p_1^2 = A$,

$$A p_1 = p_1^3 = p_1 A,$$

therefore A and p_1 commute. It follows that $p = f(A)$ and p_1 commute and

$$k_1^{-1} k = p_1 p^{-1}.$$

The matrix $k_1 k^{-1}$, the product of two orthogonal matrices, is orthogonal. In general the product of two symmetric matrices A and B is not symmetric. However, if A and B commute, then the product AB is symmetric. One can diagonalise simultaneously the matrices A and B: there exists $h \in O(n)$ such that

$$A = h \begin{pmatrix} \lambda_1 & & \\ & \ddots & \\ & & \lambda_n \end{pmatrix} h^{-1}, \quad B = h \begin{pmatrix} \mu_1 & & \\ & \ddots & \\ & & \mu_n \end{pmatrix} h^{-1},$$

and

$$AB = h \begin{pmatrix} \lambda_1 \mu_1 & & \\ & \ddots & \\ & & \lambda_n \mu_n \end{pmatrix} h^{-1}.$$

Hence, if A and B are positive definite symmetric matrices, then the product AB is a positive definite symmetric matrix as well. Therefore, since the symmetric matrices p and p_1 commute and are positive definite, the matrix $p_1 p^{-1}$ is symmetric and positive definite. It follows that $k = k_1$, $p = p_1$ since

$$O(n) \cap \mathcal{P}_n = \{I\}.$$

In fact, assume that $g \in O(n) \cap \mathcal{P}_n$. Being orthogonal and symmetric, the matrix g satisfies $g = g^{-1}$. Its eigenvalues are then equal to ± 1. But since g is positive definite, its eigenvalues are all equal to 1, and $g = I$.

(c) *Continuity.* Clearly the map

$$O(n) \times \mathcal{P}_n \to GL(n, \mathbb{R}),$$

$$(k, p) \mapsto g = kp,$$

is continuous. To show that the inverse map is continuous, let us consider a convergent sequence (g_m) in $GL(n, \mathbb{R})$,

$$\lim_{m \to \infty} g_m = g.$$

Decompose each matrix g_m as $g_m = k_m p_m$. Let us show that $k_m \to k$ and $p_m \to p$, with $g = kp$. Since the group $O(n)$ is compact it is possible to extract from the sequence (k_m) a convergent subsequence (k_{m_j}),

$$\lim_{j \to \infty} k_{m_j} = k_0.$$

The sequence $(p_{m_j}) = (k_{m_j}^{-1} g_{m_j})$ also converges, with limit $p_0 = k_0^{-1} g$. Since it is the limit of positive definite symmetric matrices, the matrix p_0 is symmetric and semi-positive definite. Since g is invertible, p_0 is invertible too, hence $p_0 \in \mathcal{P}_n$, and

$$g = k_0 p_0.$$

By the uniqueness of the polar decomposition, $k_0 = k$, and k is the only accumulation point of the sequence (k_m), therefore (k_m) is a convergent sequence with limit k, and (p_m) converges to p. $\square$

By diagonalising the matrix p in an orthogonal basis one obtains the following corollary.

Corollary 1.4.2 *Every element g in $GL(n, \mathbb{R})$ decomposes as*

$$g = k_1 d k_2,$$

with $k_1, k_2 \in O(n)$, and d is a diagonal matrix whose diagonal entries are strictly positive.

Note that the decomposition is not unique.

Let $GL(n, \mathbb{R})_+$ denote the subgroup of $GL(n, \mathbb{R})$ of matrices with positive determinant. Every element g in $GL(n, \mathbb{R})_+$ decomposes as

$$g = kp,$$

with $k \in SO(n)$, $p \in \mathcal{P}_n$, and also

$$g = k_1 d k_2,$$

with $k_1, k_2 \in SO(n)$, and d is a diagonal matrix with positive diagonal entries.

One can establish statements similar to Theorem 1.4.1 and Corollary 1.4.2 by considering $GL(n, \mathbb{C})$ instead of $GL(n, \mathbb{R})$, the unitary group $U(n)$ instead of the orthogonal group $O(n)$, and the set of positive definite Hermitian matrices instead of $\mathcal{P}_n$.

1.5 The orthogonal group

Let S^{n-1} be the unit sphere of $\mathbb{R}^n$:

$$S^{n-1} = \{x \in \mathbb{R}^n \mid \|x\| = 1\}.$$

The group $SO(n)$ acts on S^{n-1}. Let K be the isotropy subgroup of $e_n = (0, \ldots, 0, 1)$:

$$K = \{k \in SO(n) \mid ke_n = e_n\}.$$

This is the group of matrices

$$k = \begin{pmatrix} u & 0 \\ 0 & 1 \end{pmatrix}, \quad u \in SO(n-1).$$

Hence K is isomorphic to $SO(n-1)$.

Proposition 1.5.1 *If $n \geq 2$, the group $SO(n)$ acts transitively on the sphere S^{n-1}.*

Proof. The theorem will be proved recursively with respect to n.

(a) For $n = 2$, $SO(2)$ is the group of rotations in the plane, and S^1 is the unit circle. The statement holds clearly in this case.

(b) Assume that the statement holds for $n - 1$, and let us prove that it holds for n. Let us show that, for x in S^{n-1}, there exists $k \in SO(n)$ such that $x = ke_n$. One can write

$$x = \cos\theta e_n + \sin\theta x',$$

with x' in the subspace generated by $e_1, \ldots, e_{n-1}$, which can be identified with $\mathbb{R}^{n-1}$. The point x' belongs to the sphere S^{n-2}. By the recursion hypothesis there exists $u \in SO(n-1)$ such that $x' = ue_{n-1}$. Put

$$k = \begin{pmatrix} u & 0 \\ 0 & 1 \end{pmatrix}, \quad h_\theta = \begin{pmatrix} I_{n-2} & 0 & 0 \\ 0 & \cos\theta & \sin\theta \\ 0 & -\sin\theta & \cos\theta \end{pmatrix}.$$

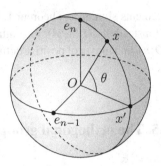

Figure 1

Then

$$kh_\theta e_n = \sin\theta u e_{n-1} + \cos\theta e_n = x. \qquad \square$$

Corollary 1.5.2 (i) *Every element k in $SO(n)$ can be written*

$$k = k_1 h_\theta k_2, \quad k_1, k_2 \in K \simeq SO(n-1), \quad \theta \in \mathbb{R}.$$

(ii) *The group $SO(n)$ is connected.*

Proof. (i) Let $k \in SO(n)$, and put $x = ke_n$. By the proof of Proposition 1.5.1 one can write $x = k_1 h_\theta e_n$, hence $h_\theta^{-1} k_1^{-1} ke_n = e_n$, therefore $k_2 = h_\theta^{-1} k_1^{-1} k \in K$, or $k = k_1 h_\theta k_2$.

(ii) Let us show recursively with respect to n that $SO(n)$ is connected. For $n = 2$, $SO(2)$ is homeomorphic to a circle hence connected. Assume that $SO(n-1)$ is connected. By (i) the map

$$SO(n-1) \times SO(2) \times SO(n-1) \to SO(n), \quad (k_1, h_\theta, k_2) \mapsto k_1 h_\theta k_2,$$

is surjective. Since it is continuous it follows that $SO(n)$ is connected. $\qquad \square$

It follows that $O(n)$ has two connected components:

$$O(n)_+ = \{k \in O(n) \mid \det k = 1\} = SO(n),$$
$$O(n)_- = \{k \in O(n) \mid \det k = -1\}.$$

Note that $SO(n)$ is arcwise connected.

Corollary 1.5.3 *The groups $GL(n, \mathbb{R})_+$ and $SL(n, \mathbb{R})$ are connected.*

Proof. This follows from Corollary 1.4.2 and the polar decomposition in $GL(n, \mathbb{R})_+$ and $SL(n, \mathbb{R})$. $\qquad \square$

1.6 Gram decomposition

Let $G = GL(n, \mathbb{R})$ be the linear group, $K = O(n)$ the orthogonal group, and $T = \mathrm{T}(n, \mathbb{R})_+$ the group of upper triangular matrices with positive diagonal entries.

Theorem 1.6.1 (Gram decomposition) *Every g in G decomposes as*

$$g = kt,$$

with $k \in K$, $t \in T$. The decomposition is unique. The map

$$K \times T \to G, \quad (k, t) \mapsto kt,$$

is a homeomorphism.

This decomposition is called *QR factorisation* in matrix numerical analysis.

Proof. (a) Let us show that the decomposition is unique. Assume that

$$g = k_1 t_1 = k_2 t_2, \quad k_1, k_2 \in K, \ t_1, t_2 \in T,$$

then

$$k_2^{-1} k_1 = t_2 t_1^{-1}.$$

It follows that $k_1 = k_2$, $t_1 = t_2$ since

$$K \cap T = \{I\}.$$

In fact assume that $g \in K \cap T$. Then, since $K \cap T$ is a group, $g^{-1} \in K \cap T$. But, since g is orthogonal, $g = (g^{-1})^T$. Hence, since g is both upper triangular and lower triangular, g is diagonal. Since g is orthogonal, its diagonal entries are equal to ± 1, and since $g \in T$, they are positive, hence equal to 1, and $g = I$.

(b) Recall first the Gram–Schmidt orthogonalisation. Let us consider n independent vectors $v_1, \ldots, v_n$ in $\mathbb{R}^n$. One constructs a sequence $f_1, \ldots, f_n$ of orthogonal vectors as follows :

$$f_1' = v_1, \qquad\qquad f_1 = \frac{f_1'}{\|f_1'\|},$$

$$f_2' = -(v_2|f_1)f_1 + v_2, \qquad f_2 = \frac{f_2'}{\|f_2'\|},$$

$$\ldots$$

$$f_j' = -\sum_{i=1}^{j-1}(v_j|f_j)f_i + v_j, \quad f_j = \frac{f_j'}{\|f_j'\|},$$

$$\ldots$$

$$f_n' = -\sum_{i=1}^{n-1}(v_n|f_i)f_i + v_n, \quad f_n = \frac{f_n'}{\|f_n'\|}.$$

The vectors $f_1, \ldots, f_n$ can be written

$$f_1 = \alpha_{11} v_1,$$
$$f_2 = \alpha_{12} v_1 + \alpha_{22} v_2,$$
$$\cdots$$
$$f_n = \alpha_{1n} v_1 + \cdots + \alpha_{nn} v_n,$$

with $\alpha_{ii} > 0$ $(\alpha_{ii} = 1/\| f_i' \|)$. The matrix $\alpha = (\alpha_{ij})$ belongs to T. Let t be its inverse. There exists an orthogonal matrix k such that

$$f_j = \sum_{i=1}^{n} k_{ij} e_i,$$

where $e_1, \ldots, e_n$ denote the vectors of the canonical basis of $\mathbb{R}^n$. Then

$$v_i = \sum_{j=1}^{i} t_{ji} f_j = \sum_{\ell=1}^{n} \left(\sum_{j=1}^{i} t_{ji} k_{\ell j} \right) e_\ell.$$

By performing the orthogonalisation of the rows of a matrix g in G one obtains

$$g = kt,$$

with $k \in K, t \in T$.

(c) The map

$$K \times T \to G, \quad (k, t) \mapsto kt$$

is continuous. Its inverse is continuous too. In fact this results from the sequence of operations which constitute the Gram–Schmidt orthogonalisation. $\square$

If $g \in GL(n, \mathbb{R})_+$ (i.e. if $\det g > 0$), then $k \in SO(n)$.

One can establish a similar result for $G = GL(n, \mathbb{C})$, $K = U(n)$ and T being the group of upper triangular matrices with compex entries, and positive diagonal entries.

1.7 Exercises

1. Show that a topological group is Hausdorff if and only if $\{e\}$ is closed.
2. Show that a discrete subgroup of a Hausdorff topological group is closed.
3. Show that, if H is a subgroup of a topological group G, the canonical map $G \to G/H$ is open. Furthermore show that, if G is Hausdorff and H closed, then G/H is Hausdorff.
4. Show that, if G/H and H are connected, then G is connected.

5. Let G be a closed subgroup of $\mathbb{R}$. Show that, if $G \neq \mathbb{R}$, then $G = \alpha \mathbb{Z}$ ($\alpha \in \mathbb{R}$).

6. Let V be a vector space over $\mathbb{R}$ of finite dimension n. It is a group for addition. Let $\Gamma \neq \{0\}$ be a discrete subgroup of V. Show that there exist p independent vectors $e_1, \dots, e_p$ ($1 \leq p \leq n$) such that

$$\Gamma = \left\{ \sum_{j=1}^{p} m_j e_j \mid m_j \in \mathbb{Z} \right\}.$$

Hint. The proof uses recursion with respect to the dimension n of V. For $n = 1$ this is Exercise 5. Assume $n > 1$. Let $b \in \Gamma, b \neq 0$, and put $V_1 = \mathbb{R}b$. Show that

$$\Gamma \cap V_1 = \mathbb{Z}a \quad (a \in \Gamma).$$

Assume $\Gamma \neq \Gamma_1$. Let Γ' be the image of Γ in $V' = V/V_1$. We show that Γ' is discrete. If Γ' were not discrete, there would be a sequence (γ_k) in $\Gamma \backslash V_1$ and a sequence (λ_k) of real numbers such that $\lim_{\to \infty}(\gamma_k - \lambda_k a) = 0$. Put $\lambda_k = [\lambda_k] + r_k$, where $[\lambda_k]$ is the integer part of λ_k and $0 \leq r_k < 1$. Then there exists a subsequence (γ_{k_j}) such that $\lim_{j \to \infty} r_{k_j} = r$, hence

$$\lim_{j \to \infty} (\gamma_{k_j} - [\lambda_{k_j}]a) = ra.$$

Show that this leads to a contradiction. (For that one observes that a convergent sequence (u_j) in a discrete set is constant for j large enough.)

7. Let G be a topological group, and $E \subset G$. Let f be a real or complex valued function defined on E. The function f is said to be left (respectively right) uniformly continuous if, for every $\varepsilon > 0$, there exists a neighbourhood V of the identity element e such that, if $x \in E$, $y \in Vx$ (respectively $y \in xV$), then

$$|f(y) - f(x)| \leq \varepsilon.$$

Show that, for E compact, every continuous function defined on E is left and right uniformly continuous.

8. Show that the centre of $GL(n, \mathbb{K})$ ($\mathbb{K} = \mathbb{R}$ or $\mathbb{C}$) is equal to $\{\lambda I \mid \lambda \in \mathbb{K}^*\}$.

9. For $\mathbb{K} = \mathbb{R}$ or $\mathbb{C}$, show that $GL(n, \mathbb{K})$ is dense in $M(n, \mathbb{K})$.

10. Show that, in $GL(n, \mathbb{C})$, the set of diagonalisable matrices is dense. Is this true in $GL(n, \mathbb{R})$?

11. Show that every continuous group morphism $h : GL(n, \mathbb{R}) \to \mathbb{R}_+^*$ is of the form

$$h(g) = |\det g|^\alpha \quad (\alpha \in \mathbb{R}).$$

And show that every continuous morphism $h : GL(n, \mathbb{R}) \to \mathbb{R}^*$ is of the form

$$h(g) = |\det(g)|^\alpha \operatorname{sign}(\det g)^\varepsilon \quad (\alpha \in \mathbb{R}, \ \varepsilon = 0 \text{ or } 1).$$

12. The aim of this exercise is to show that $O(n)$ is a maximal compact subgroup of $GL(n, \mathbb{R})$.
    (a) Let p be a positive definite symmetric matrix for which there exists a constant C such that

    $$\forall k \in \mathbb{Z}, \quad \|p^k\| \le C.$$

    Show that $p = I$.
    (b) Let H be a compact subgroup of $GL(n, \mathbb{R})$ containing $O(n)$, and let $g = kp$ be the polar decomposition of an element g in H. By using (a) show that $p = I$. And then show that $H = O(n)$.

13. *Gaussian decomposition.* Let $T = \mathrm{T}(n, \mathbb{R})$ denote the upper triangular group, and N the strict lower triangular group. Show that a matrix $x \in M(n, \mathbb{R})$ can be written

    $$x = nt,$$

    with $n \in N, t \in T$, if and only if

    $$\Delta_i(x) \ne 0, \quad i = 1, \ldots, n,$$

    where $\Delta_i(x)$ denotes the ith principal minor determinant of x. Show that, if it holds, then the decomposition $x = nt$ is unique. Show that the set NT is open and dense in $GL(n, \mathbb{R})$.

    This decomposition is called *LU factorisation* in matrix numerical analysis.

14. (a) Let $\mathcal{Q}_n$ be the set of positive definite Hermitian $n \times n$ matrices. Show that every matrix $g \in GL(n, \mathbb{C})$ decomposes as

    $$g = uq,$$

    with $u \in U(n)$ and $q \in \mathcal{Q}_n$, and that the map

    $$U(n) \times \mathcal{Q}_n \to GL(n, \mathbb{C}), \quad (u, q) \mapsto uq,$$

    is a homeomorphism.
    (b) Show that $U(n)$ is connected. Then, by using (a), show that $GL(n, \mathbb{C})$ is connected. (One can follow the proofs in Section 1.5)

15. Let U be an open set in $\mathbb{C}$, and $\mathcal{E}$ be the set of matrices $X \in M(n, \mathbb{C})$ whose eigenvalues belong to U. Show that $\mathcal{E}$ is open in $M(n, \mathbb{C})$.

Hint. One can use the following property. Let $K \subset U$ be a compact set with regular boundary, and γ its oriented boundary. Let f be a holomorphic function defined in U which does not vanish on γ. Let $Z(f)$ denote the number of zeros of f belonging to the interior of K, each being counted a number of times equal to its order. Then

$$Z(f) = \frac{1}{2i\pi} \int_{\gamma} \frac{f'(z)}{f(z)} dz.$$

2

The exponential map

Using functional calculus it is possible to extend exponential and logarithm functions as matrix valued functions of a matrix variable. In this setting the exponential of the sum of two matrices is no longer equal to the product of the exponentials. However, this is true up to first order, and the second order involves the commutator of these matrices. This fact is at the origin of the notion of a Lie algebra.

2.1 Exponential of a matrix

The *exponential of a matrix* $X \in M(n, \mathbb{K})$ ($\mathbb{K} = \mathbb{R}$ or $\mathbb{C}$) is defined as the sum of the series

$$\exp(X) = \sum_{k=0}^{\infty} \frac{X^k}{k!}.$$

Since $\|X^k\| \leq \|X\|^k$, the series converges normally for every matrix X, and uniformly on any compact set in $M(n, \mathbb{K})$. If X and Y commute, $XY = YX$, then $\exp(X + Y) = \exp X \exp Y$. It follows that $\exp(X)$ is invertible, and $(\exp X)^{-1} = \exp(-X)$. For $g \in GL(n, \mathbb{K})$,

$$g \exp X g^{-1} = \exp(g X g^{-1}).$$

If X is diagonalisable:

$$X = g \begin{pmatrix} \lambda_1 & & \\ & \ddots & \\ & & \lambda_n \end{pmatrix} g^{-1},$$

then

$$\exp X = g \begin{pmatrix} e^{\lambda_1} & & \\ & \ddots & \\ & & e^{\lambda_n} \end{pmatrix} g^{-1}.$$

If $\mathbb{K} = \mathbb{C}$, it is possible to use Jordan reduction to compute the exponential. In fact let us consider the case of a Jordan block of order k:

$$X = \lambda I + N,$$

where

$$N = \begin{pmatrix} 0 & 1 & & 0 \\ & \ddots & \ddots & \\ & & \ddots & 1 \\ & & & 0 \end{pmatrix}.$$

The matrix N is nilpotent: $N^k = 0$, hence

$$\exp(tX) = e^{\lambda t} \exp(tN) = e^{\lambda t} \sum_{j=0}^{k-1} \frac{t^j}{j!} N^j$$

$$= e^{\lambda t} \begin{pmatrix} 1 & t & \frac{t^2}{2} & \cdots & \frac{t^{k-1}}{(k-1)!} \\ & 1 & t & & \vdots \\ & & \ddots & & \vdots \\ & & & 1 & t \\ & & & & 1 \end{pmatrix}.$$

The following equality is an important property of the exponential map:

$$\det(\exp X) = e^{\operatorname{tr}(X)}.$$

To establish this relation observe that the function f defined by

$$f(t) = \det(\exp tX),$$

satisfies

$$f(t+s) = f(t)f(s), \quad f(0) = 1, \quad f'(0) = \operatorname{tr} X,$$

hence $f(t) = e^{t \operatorname{tr} X}$. To see that $f'(0) = \operatorname{tr} X$, one can use the following formula giving the derivative of a determinant: let $X(t) = (x_{ij}(t))$ be a matrix whose entries are C^1-functions of the real variable t. Let $X_1(t), \ldots, X_n(t)$ be the rows of $X(t)$. By multilinearity of the determinant it follows that

$$\frac{d}{dt} \det X(t)$$
$$= \det\big(X_1'(t), X_2(t), \ldots, X_n(t)\big) + \det\big(X_1(t), X_2'(t), X_3(t), \ldots, X_n(t)\big)$$
$$+ \cdots + \det\big(X_1(t), \ldots, X_{n-1}(t), X_n'(t)\big).$$

One can also use the fact that every matrix $X \in M(n, \mathbb{C})$ is triangularisable: there exists an invertible matrix g, and an upper triangular matrix Y such that

$$X = gYg^{-1}, \quad \exp X = g \exp Y g^{-1}.$$

If

$$Y = \begin{pmatrix} y_{11} & * & * \\ 0 & \ddots & * \\ 0 & 0 & y_{nn} \end{pmatrix},$$

then

$$\exp Y = \begin{pmatrix} e^{y_{11}} & * & * \\ 0 & \ddots & * \\ 0 & 0 & e^{y_{nn}} \end{pmatrix},$$

hence

$$\det(\exp X) = \det(\exp Y) = e^{\operatorname{tr} Y} = e^{\operatorname{tr} X}.$$

For $\mathbb{K} = \mathbb{R}$, the exponential map is a map from $M(n, \mathbb{R})$ into $GL(n, \mathbb{R})_+$. For $n \geq 2$ it is not injective. In fact,

$$\exp \begin{pmatrix} 0 & \theta \\ -\theta & 0 \end{pmatrix} = \begin{pmatrix} \cos\theta & \sin\theta \\ -\sin\theta & \cos\theta \end{pmatrix},$$

and, for every $k \in \mathbb{Z}$,

$$\exp \begin{pmatrix} 0 & 2k\pi \\ -2k\pi & 0 \end{pmatrix} = I.$$

It is not surjective either. Let λ and μ be the eigenvalues of $X \in M(2, \mathbb{R})$. The eigenvalues of $\exp X$ are e^λ and e^μ. If λ and μ are real then e^λ and e^μ are positive. If λ and μ are complex conjugate, the numbers e^λ and e^μ are complex conjugate, and if real these numbers are equal. Therefore, if a and b are negative real numbers, $a \neq b$, there does not exist a matrix $X \in M(2, \mathbb{R})$ such that

$$\exp X = \begin{pmatrix} a & 0 \\ 0 & b \end{pmatrix}.$$

Let us recall that $\mathcal{P}_n$ denotes the set of positive definite real symmetric $n \times n$ matrices.

Theorem 2.1.1 *The exponential map is a homeomorphism from $Sym(n, \mathbb{R})$ onto $\mathcal{P}_n$.*

Proof. (a) *Surjectivity.* Let $p \in \mathcal{P}_n$, and $\lambda_1 > 0, \ldots, \lambda_n > 0$ be its eigenvalues. There exists $k \in O(n)$ such that

$$p = k \begin{pmatrix} \lambda_1 & & \\ & \ddots & \\ & & \lambda_n \end{pmatrix} k^{-1}.$$

Put

$$X = k \begin{pmatrix} \log \lambda_1 & & \\ & \ddots & \\ & & \log \lambda_n \end{pmatrix} k^{-1}.$$

Then $\exp X = p$.

(b) *Injectivity.* Let X and $Y \in Sym(n, \mathbb{R})$ be such that $\exp X = \exp Y$. Let us diagonalise X and Y:

$$X = k \begin{pmatrix} \lambda_1 & & \\ & \ddots & \\ & & \lambda_n \end{pmatrix} k^{-1}, \quad k \in O(n),$$

$$\exp X = k \begin{pmatrix} e^{\lambda_1} & & \\ & \ddots & \\ & & e^{\lambda_n} \end{pmatrix} k^{-1},$$

$$Y = h \begin{pmatrix} \mu_1 & & \\ & \ddots & \\ & & \mu_n \end{pmatrix} h^{-1}, \quad h \in O(n),$$

$$\exp Y = h \begin{pmatrix} e^{\mu_1} & & \\ & \ddots & \\ & & e^{\mu_n} \end{pmatrix} h^{-1}.$$

Let us show that X and Y commute. If f is a polynomial in one variable such that

$$f(e^{\mu_i}) = \mu_i, \quad i = 1, \ldots, n,$$

then $f(\exp Y) = Y$, hence

$$YX = f(\exp Y)X = f(\exp X)X = Xf(\exp X) = Xf(\exp Y) = XY.$$

It follows that X and Y are diagonalisable with respect to the same basis: one can take $h = k$, and then $e^{\lambda_i} = e^{\mu_i}$, hence $\lambda_i = \mu_i$.

(c) *Continuity.* The exponential map is continuous. For $\alpha > 0$ let E be the closed ball

$$E = \{X \in Sym(n, \mathbb{R}) \mid \|X\| \leq \alpha\}.$$

The exponential maps E onto the set

$$F = \{p \in \mathcal{P}_n \mid \|p\| \leq e^{\alpha}, \|p^{-1}\| \leq e^{\alpha}\}$$

which is compact (Proposition 1.2.3). The exponential maps continuously and injectively the compact set E onto the compact set F, and hence is a homeomorphism from E onto F. It follows that it is a homeomorphism from $Sym(n, \mathbb{R})$ onto $\mathcal{P}_n$. □

Corollary 2.1.2 *Every matrix $g \in GL(n, \mathbb{R})$ can be written $g = k \exp X$, with $k \in O(n)$, $X \in Sym(n, \mathbb{R})$, and the map*

$$(k, X) \mapsto k \exp X, \quad O(n) \times Sym(n, \mathbb{R}) \to GL(n, \mathbb{R}),$$

is a homeomorphism

The exponential map is real analytic, hence $\mathcal{C}^{\infty}$.

Theorem 2.1.3 (i) *The differential of the exponential map at 0 is the identity map:*

$$(D \exp)_0 = I.$$

(ii) *There exists a neighbourhood $\mathcal{U}$ of 0 in $M(n, \mathbb{R})$ such that the restriction to $\mathcal{U}$ of the exponential map is a diffeomorphism from $\mathcal{U}$ onto $\exp \mathcal{U}$.*

Proof. (i) One can write

$$\exp X = I + X + R(X),$$

with

$$R(X) = \sum_{k=2}^{\infty} \frac{X^k}{k!},$$

and

$$\|R(X)\| = \|X\| \varepsilon(X), \quad \lim_{X \to 0} \varepsilon(X) = 0.$$

(ii) This follows from the local inversion theorem. □

We will compute the differential of the exponential map. Let us introduce the following notation: for $A, X \in M(n, \mathbb{R})$,

$$L_A X = AX, \quad R_A X = XA, \quad \text{ad } A \, X = [A, X] = AX - XA.$$

The maps L_A, R_A and ad A are endomorphisms of the vector space $M(n, \mathbb{R})$. Observe that

$$L_A R_A = R_A L_A, \quad \text{ad } A = L_A - R_A.$$

Theorem 2.1.4 *The differential of the exponential map at A is given by*

$$(D \exp)_A X = \exp A \sum_{k=0}^{\infty} \frac{(-1)^k}{(k+1)!} (\text{ad } A)^k X.$$

By putting, for $z \in \mathbb{C}$,

$$\Phi(z) = \sum_{k=0}^{\infty} \frac{(-1)^k}{(k+1)!} z^k = \frac{1 - e^{-z}}{z},$$

the statement can be written

$$(D \exp)_A = L_{\exp A} \circ \Phi(\text{ad } A) = L_{\exp A} \circ \frac{I - \text{Exp}(-\text{ad } A)}{\text{ad } A},$$

where Exp T denotes the exponential of an endomorphism T of the vector space $M(n, \mathbb{R})$.

Proof. (a) Let us consider the maps

$$F_k : M(n, \mathbb{R}) \to M(n, \mathbb{R}), \quad X \mapsto X^k,$$

and compute the differential of F_k at A:

$$(DF_k)_A X = \frac{d}{dt} (A + tX)^k \big|_{t=0}$$

$$= \sum_{j=0}^{k-1} A^{k-j-1} X A^j$$

$$= \sum_{j=0}^{k-1} L_A^{k-j-1} R_A^j X.$$

One can write

$$R_A^j = (L_A - \text{ad } A)^j = \sum_{i=0}^{j} (-1)^i \binom{j}{i} L_A^{j-i} (\text{ad } A)^i,$$

since L_A and ad A commute, hence

$$(DF_k)_A = \sum_{j=0}^{k-1} L_A^{k-j-1} \left(\sum_{i=0}^{j} (-1)^i \binom{j}{i} L_A^{j-i} (\text{ad } A)^i \right)$$

$$= \sum_{i=0}^{k-1} (-1)^i \left(\sum_{j=i}^{k-1} \binom{j}{i} \right) L_A^{k-i-1} (\text{ad } A)^i.$$

We will establish below (see (c)) the identity:

$$\sum_{j=i}^{k-1} \binom{j}{i} = \binom{k}{i+1}.$$

Then

$$(DF_k)_A = \sum_{i=0}^{k-1} (-1)^i \binom{k}{i+1} L_A^{k-i-1} (\text{ad } A)^i.$$

(b) By (a)

$$\|(DF_k)_A\| \le k\|A\|^{k-1},$$

($\|(DF_k)_A\|$ denotes the norm of the endomorphism $(DF_k)_A$ of the normed vector space $M(n, \mathbb{R})$) and the series of the differentials

$$\sum_{k=1}^{\infty} \frac{1}{k!} (DF_k)_A$$

converges uniformly on every ball of $M(n, \mathbb{R})$. It follows that the differential of the exponential map is given by

$$(D \exp)_A = \sum_{k=1}^{\infty} \frac{1}{k!} (DF_k)_A$$

$$= \sum_{k=1}^{\infty} \frac{1}{k!} \left(\sum_{i=0}^{k-1} (-1)^i \binom{k}{i+1} L_A^{k-i-1} (\text{ad } A)^i \right)$$

$$= \left(\sum_{j=0}^{\infty} \frac{1}{j!} L_A^j \right) \left(\sum_{i=0}^{\infty} \frac{(-1)^i}{(i+1)!} (\text{ad } A)^i \right)$$

$$= \exp A \sum_{i=0}^{\infty} \frac{(-1)^i}{(i+1)!} (\text{ad } A)^i.$$

(c) Let us now establish the identity

$$\sum_{j=i}^{k-1} \binom{j}{i} = \binom{k}{i+1}.$$

For k fixed put

$$a_i = \sum_{j=i}^{k-1} \binom{j}{i}.$$

Then

$$\sum_{i=0}^{k-1} a_i z^i = \sum_{i=0}^{k-1}\sum_{j=i}^{k-1} \binom{j}{i} z^i = \sum_{j=0}^{k-1}\sum_{i=0}^{j} \binom{j}{i} z^i = \sum_{j=0}^{k-1}(z+1)^j$$

$$= \frac{(z+1)^k - 1}{z} = \frac{1}{z}\sum_{i=1}^{k} \binom{k}{i} z^i = \sum_{i=0}^{k-1} \binom{k}{i+1} z^i.$$

Hence

$$a_i = \binom{k}{i+1}.$$

$\square$

2.2 Logarithm of a matrix

We will define an inverse map for the exponential map in a neighbourhood of the identity I. We know that the ball

$$B(I, 1) = \{X \in M(n, \mathbb{R}) \mid \|X - I\| < 1\}$$

is contained in $GL(n, \mathbb{R})$ (see the proof of Proposition 1.2.1). If $\|g - I\| < 1$, we define the logarithm of the matrix g by

$$\log(g) = \sum_{k=1}^{\infty} \frac{(-1)^{k+1}}{k}(g - I)^k.$$

Theorem 2.2.1 (i) *For $g \in B(I, 1)$,*

$$\exp(\log g) = g.$$

(ii) *For $X \in B(0, \log 2)$,*

$$\log(\exp X) = X.$$

Before giving the proof of this theorem we make some remarks about the functional calculus. Let

$$f(z) = \sum_{k=0}^{\infty} a_k z^k$$

be a power series with convergence radius $R > 0$. For $X \in M(n, \mathbb{R})$ with $\|X\| < R$ the functional calculus associates to the function f the matrix

$$f(X) = \sum_{k=0}^{\infty} a_k X^k.$$

The map $f \mapsto f(X) \in M(n, \mathbb{R})$ is an algebra morphism: it is linear and

$$(f_1 f_2)(X) = f_1(X) f_2(X).$$

Let

$$g(z) = \sum_{m=1}^{\infty} b_m z^m$$

be another power series, with $g(0) = b_0 = 0$. The function $f \circ g$ is the sum of a power series in a neighbourhood of 0,

$$f \circ g(z) = \sum_{p=0}^{\infty} c_p z^p.$$

Lemma 2.2.2 *If*

(∗) $$\sum_{m=1}^{\infty} |b_m| \|X\|^m < R,$$

then $g(X)$, $f(g(X))$ and $(f \circ g)(X)$ are well defined, and

$$(f \circ g)(X) = f\big(g(X)\big).$$

This means that

$$\sum_{p=0}^{\infty} c_p X^p = \sum_{k=0}^{\infty} a_k \left(\sum_{m=1}^{\infty} b_m X^m \right)^k.$$

Proof. We can write

$$g(z)^k = \sum_{m=k}^{\infty} b_{m,k} z^m,$$

where

$$b_{m,k} = \sum_{m_1 + \cdots + m_k = m} b_{m_1} b_{m_2} \ldots b_{m_k},$$

and

$$\sum_{m=k}^{\infty} |b_{m,k}| r^m \leq \left(\sum_{m=1}^{\infty} |b_m| r^m \right)^k.$$

Then

$$f \circ g(z) = \sum_{m=0}^{\infty} c_m z^m,$$

with

$$c_m = \sum_{k=0}^{m} a_k b_{m,k},$$

and

$$\sum_{m=0}^{\infty} |c_m| r^m \le \sum_{m=0}^{\infty} \left(\sum_{k=0}^{m} |a_k||b_{m,k}| \right) r^m = \sum_{k=0}^{\infty} |a_k| \left(\sum_{m=k}^{\infty} |b_{m,k}| r^m \right)$$

$$\le \sum_{k=0}^{\infty} |a_k| \left(\sum_{m=1}^{\infty} |b_m| r^m \right)^k.$$

Assume that

$$\sum_{m=1}^{\infty} |b_m| \|X\|^m < R,$$

then the series

$$\sum_{m=0}^{\infty} |c_m| \|X\|^m$$

converges and

$$(f \circ g)(X) = \sum_{m=0}^{\infty} c_m X^m = \sum_{m=0}^{\infty} \left(\sum_{k=0}^{m} a_k b_{m,k} \right) X^m.$$

Since

$$\sum_{m=0}^{\infty} \sum_{k=0}^{m} |a_k||b_{m,k}| \|X\|^m < \infty,$$

one can invert the summations

$$(f \circ g)(X) = \sum_{k=0}^{\infty} a_k \left(\sum_{m=k}^{\infty} b_{m,k} X^m \right)$$

$$= \sum_{k=0}^{\infty} a_k \left(\sum_{m=1}^{\infty} b_m X^m \right)^k.$$

$\square$

Proof of Theorem 2.2.1 To prove (i) one puts

$$f(z) = \exp(z), \quad g(z) = \log(1 + z).$$

The condition (∗) is then equivalent to $\|X\| < 1$.
    To prove (ii) one puts

$$f(z) = \log(1 + z), \quad g(z) = \exp(z) - 1.$$

Since $R = 1$, the condition $(*)$ can be written

$$\sum_{m=1}^{\infty} \frac{\|X\|^m}{m!} < 1,$$

or $\exp(\|X\|) - 1 < 1$, or $\|X\| < \log 2$. $\square$

Proposition 2.2.3 *For $X, Y \in M(n, \mathbb{R})$,*

(1) $\exp(tX)\exp(tY) = \exp\left(t(X + Y) + \frac{t^2}{2}[X, Y] + O(t^3)\right),$

(2) $\exp(tX)\exp(tY)\exp(-tX)\exp(-tY) = \exp\left(t^2[X, Y] + O(t^3)\right).$

Proof. Put $F(t) = \exp(tX)\exp(tY)$,

$$F(t) = \left(I + tX + \frac{t^2}{2}X^2 + O(t^3)\right)\left(I + tY + \frac{t^2}{2}Y^2 + O(t^3)\right)$$

$$= I + t(X + Y) + \frac{t^2}{2}(X^2 + 2XY + Y^2) + O(t^3).$$

For t small enough $\|F(t) - I\| < 1$, and

$$\log F(t) = t(X + Y) + \frac{t^2}{2}(X^2 + 2XY + Y^2) - \frac{t^2}{2}(X + Y)^2 + O(t^3)$$

$$= t(X + Y) + \frac{t^2}{2}[X, Y] + O(t^3).$$

This proves (i). To prove (ii) put

$$G(t) = \exp(tX)\exp(tY)\exp(-tX)\exp(-tY)$$

$$= \left(I + t(X + Y) + \frac{t^2}{2}(X^2 + 2XY + Y^2) + O(t^3)\right)$$

$$\cdot \left(I - t(X + Y) + \frac{t^2}{2}(X^2 + 2XY + Y^2) + O(t^3)\right)$$

$$= \left(I + t^2[X, Y] + O(t^3)\right),$$

and (ii) follows by considering $\log G(t)$. $\square$

Corollary 2.2.4 *For $X, Y \in M(n, \mathbb{R})$,*

(i) $\lim_{k \to \infty}\left(\exp\frac{X}{k}\exp\frac{Y}{k}\right)^k = \exp(X + Y),$

(ii) $\lim_{k \to \infty}\left(\exp\frac{X}{k}\exp\frac{Y}{k}\exp-\frac{X}{k}\exp-\frac{Y}{k}\right)^{k^2} = \exp([X, Y]).$

Proof. From the preceding proposition

$$\exp \frac{X}{k} \exp \frac{Y}{k} = \exp \left(\frac{1}{k}(X+Y) + O\left(\frac{1}{k^2}\right) \right),$$

$$\left(\exp \frac{X}{k} \exp \frac{Y}{k} \right)^k = \exp \left((X+Y) + O\left(\frac{1}{k}\right) \right),$$

and this proves (i). Similarly

$$\left(\exp \frac{X}{k} \exp \frac{Y}{k} \exp -\frac{X}{k} \exp -\frac{Y}{k} \right)^{k^2} = \exp \left([X,Y] + O\left(\frac{1}{k}\right) \right),$$

and this proves (ii). □

2.3 Exercises

1. Show that every matrix in the group $SU(2)$ is conjugate to one of the following matrices

$$\begin{pmatrix} e^{i\theta} & 0 \\ 0 & e^{-i\theta} \end{pmatrix} \quad (\theta \in \mathbb{R}).$$

Then show that the exponential map

$$\exp : \left\{ \begin{pmatrix} ix & y+iz \\ -y+iz & -ix \end{pmatrix} \middle| x, y, z \in \mathbb{R} \right\} \rightarrow SU(2)$$

is surjective.

2. (a) Show that every matrix in $SL(2, \mathbb{R})$ is conjugate to one of the following matrices

$$\begin{pmatrix} a & 0 \\ 0 & \frac{1}{a} \end{pmatrix}, \quad \begin{pmatrix} 1 & t \\ 0 & 1 \end{pmatrix}, \quad \begin{pmatrix} -1 & t \\ 0 & -1 \end{pmatrix}, \quad \begin{pmatrix} \cos\theta & \sin\theta \\ -\sin\theta & \cos\theta \end{pmatrix}$$

   $(a \in \mathbb{R}, a \neq 0, t \in \mathbb{R}, \theta \in \mathbb{R})$.

   (b) Show that the range of the exponential map

$$\exp : \left\{ \begin{pmatrix} x & y \\ z & -x \end{pmatrix} \middle| x, y, z \in \mathbb{R} \right\} \rightarrow SL(2, \mathbb{R}),$$

   is equal to

$$\{g \in SL(2, \mathbb{R}) \mid \text{tr}(g) > -2\} \cup \{-I\}.$$

3. *Polar decomposition of complex matrices.* Show that every matrix in $g \in GL(n, \mathbb{C})$ can be written $g = k \exp X$ with $k \in U(n)$ and $X \in Herm(n, \mathbb{C})$.

Show that the decomposition is unique, and that the map

$$(k, X) \mapsto g = k \exp X, \quad U(n) \times Herm(n, \mathbb{C}) \to GL(n, \mathbb{C}),$$

is a homeomorphism.

4. *Polar decomposition of unitary matrices.*

(a) Let $u \in M(n, \mathbb{C})$ be a complex matrix which is symmetric and unitary: $u \in Sym(n, \mathbb{C}) \cap U(n)$. Show that there is a real symmetric matrix $X \in Sym(n, \mathbb{R})$ such that

$$u = \exp(i X).$$

(b) Let the matrix u be unitary: $u \in U(n)$. Show that there is a real orthogonal matrix $k \in O(n)$, and a real symmetric matrix $X \in Sym(n, \mathbb{R})$, such that

$$u = k \exp(i X).$$

Are the matrices k and X unique ?

5. *Polar decomposition of complex orthogonal matrices.* The complex orthogonal group $O(n, \mathbb{C})$ is defined by

$$O(n, \mathbb{C}) = \{g \in M(n, \mathbb{C}) \mid g^{-1} = g^{T}\}.$$

Show that every matrix $g \in O(n, \mathbb{C})$ can be written

$$g = k \exp(i X),$$

with $k \in O(n)$ and $X \in Asym(n, \mathbb{R})$, the space of real skewsymmetric matrices. Is the decomposition unique ?

Using that show that the identity component in $O(n, \mathbb{C})$ is equal to

$$SO(n, \mathbb{C}) = \{g \in O(n, \mathbb{C}) \mid \det g = 1\}.$$

6. Show that, for $X \in M(n, \mathbb{R})$,

$$\det(I + X) = 1 + \operatorname{tr} X + O(\|X\|^{2}).$$

7. *Integral formula for the differential of the exponential map.*

(a) Let $A, X \in M(n, \mathbb{R})$. Define

$$F(t) = \exp(t(A + X)).$$

Show that the function F is a solution of the following integral equation

$$F(t) - \int_{0}^{t} \exp((t - s)A)X F(s)ds = \exp(t A).$$

(b) Define

$$W_0(t) = \exp(tA),$$

$$W_k(t) = \int_0^t \exp((t-s)A)X W_{k-1}(s)ds.$$

Show that the series

$$\sum_{k=0}^{\infty} W_k(t)$$

converges for every $t \in \mathbb{R}$, and that its sum is equal to $F(t)$.

(c) Prove the formula

$$(D\exp)_A X = W_1(1) = \int_0^1 \exp((1-s)A)X \exp(sA)ds.$$

(d) For $X \in M(n, \mathbb{R})$ and $g \in GL(n, \mathbb{R})$ one puts

$$\mathrm{Ad}(g)X = gXg^{-1}.$$

We will see (Proposition 3.2.2) that

$$\mathrm{Exp}(\mathrm{ad}\,A) = \mathrm{Ad}(\exp A).$$

Show that the above formula can be written

$$(D\exp)_A X = \exp A \int_0^1 \mathrm{Exp}(-s\,\mathrm{ad}\,A)Xds,$$

and deduce that

$$(D\exp)_A = L_{\exp A} \circ \frac{I - \mathrm{Exp}(-\,\mathrm{ad}\,A)}{\mathrm{ad}\,A}.$$

8. Let $A \in M(n, \mathbb{C})$ with eigenvalues $\lambda_1, \ldots, \lambda_n$.
   (a) Show that the eigenvalues of L_A are the numbers $\lambda_1, \ldots, \lambda_n$, each of them being repeated n times. (Consider a basis with respect to which A is triangular.) Show that

   $$\mathrm{Det}(L_A) = \det(A)^n.$$

   (b) Show that the eigenvalues of $\mathrm{ad}\,A$ are the numbers $\lambda_j - \lambda_k$.
   (c) Show that $(D\exp)_A$ is invertible if and only if $\lambda_j - \lambda_k \notin 2i\pi\mathbb{Z}^*$.
9. Show that, for $X \in M(n, \mathbb{R})$ with $\|X\| < 1$,

$$\log(I+X) = X \int_0^1 (I+tX)^{-1}dt.$$

This integral formula makes it possible to extend the definition domain of the matricial logarithm map. In fact this integral is well defined if $]-\infty, -1]$ does not contain any eigenvalue of X. With this new definition, does the following identity still hold:

$$\exp\bigl(\log(I + X)\bigr) = I + X ?$$

Hint. Show that the matrix valued function f of the complex variable z,

$$f(z) = zX \int_0^1 (I + tzX)^{-1} dt,$$

is defined and holomorphic in a complex neighbourhood of $[0, 1]$, and that, for z in a neighbourhood of 0, $\exp f(z) = I + zX$.

10. Show that the exponential map

$$\exp : M(n, \mathbb{C}) \to GL(n, \mathbb{C})$$

is surjective.

Hint. Use Jordan reduction.

11. Let $\mathcal{N}_p$ denote the set of nilpotent matrices of order p,

$$\mathcal{N}_p = \{X \in M(n, \mathbb{C}) \mid X^p = 0\},$$

and $\mathcal{U}_p$ the set of unipotent matrices of order p,

$$\mathcal{U}_p = \{g \in GL(n, \mathbb{C}) \mid (g - I)^p = 0\}.$$

Show that the exponential map is a bijection from $\mathcal{N}_p$ onto $\mathcal{U}_p$, whose inverse is the logarithm map.

Hint. For $X \in \mathcal{N}_p$, $\log(\exp tX) - tX$ is a polynomial in t, vanishing on a neighbourhood of 0, hence identically zero.

12. Let $A \in M(n, \mathbb{C})$ be a complex matrix for which there exists a constant C such that

$$\forall t \in \mathbb{R}, \quad \| \exp(t A)\| \le C.$$

(a) Show that the eigenvalues of A are pure imaginary, and that A is diagonalisable.

Hint. Consider first the case of a Jordan block of order k:

$$A = \lambda I + N, \quad N = \begin{pmatrix} 0 & 1 & & 0 \\ & \ddots & \ddots & \\ & & \ddots & 1 \\ & & & 0 \end{pmatrix},$$

and then show that there exists $\alpha > 0$ such that

$$\| \exp(t A)\| \sim \alpha e^{\Re(\lambda) t} |t|^{k-1} \quad (t \to \pm\infty).$$

(b) Show that $\operatorname{tr}(A^2) \le 0$, and, if $\operatorname{tr}(A^2) = 0$, then $A = 0$.

13. Let $\mathcal{E}$ denote the set of real matrices X in $M(n, \mathbb{R})$ whose eigenvalues λ_j satisfy $|\,\mathrm{Im}\,\lambda_j| < \pi$. The aim of this exercise is to show that the exponential map is a diffeomorphism from $\mathcal{E}$ onto its image $\exp(\mathcal{E})$.

   (a) Show that the set $\mathcal{E}$ is open and connected, and invariant under the maps

$$X \mapsto gXg^{-1} \quad (g \in GL(n, \mathbb{R})).$$

   Hint. To show that $\mathcal{E}$ is open, use Exercise 15 of Chapter 1, and to show that $\mathcal{E}$ is connected, use the fact that $\mathcal{E}$ is starlike with respect to 0.

   (b) Show that $\exp(\mathcal{E})$ contains the ball $B(I, 1)$.

   (c) Let $X \in M(n, \mathbb{R})$ with eigenvalues λ_j. Show that the eigenvalues of ad X are the numbers $\lambda_j - \lambda_k$. Show that the differential of the exponential map is invertible at every point in $\mathcal{E}$.

   (d) Let X and Y be two diagonalisable matrices in $\mathcal{E}$. Show that, if $\exp X = \exp Y$, then $X = Y$.

   Hint. Apply an argument used for the polar decomposition in $GL(n, \mathbb{R})$.

   (e) Show that every matrix $X \in M(n, \mathbb{C})$ can be written uniquely $X = S + N$, where $S \in M(n, \mathbb{C})$ is diagonalisable, $N \in M(n, \mathbb{C})$ is nilpotent and $SN = NS$, and that every matrix $g \in GL(n, \mathbb{C})$ can be written uniquely $g = su$, where $s \in GL(n, \mathbb{C})$ is diagonalisable, $u \in GL(n, \mathbb{C})$ is unipotent and $su = us$. Show that the exponential map is injective on $\mathcal{E}$. Draw a conclusion.

   (f) For $n = 2$ define

$$\mathcal{E}_0 = \{X \in M(2, \mathbb{R}) \mid \mathrm{tr}\, X = 0\} \cap \mathcal{E}.$$

   Show that

$$\mathcal{E}_0 = \left\{ \begin{pmatrix} a & b \\ c & -a \end{pmatrix} \;\middle|\; a^2 + bc + \pi^2 > 0 \right\}.$$

14. (a) For a C^1 function on $\mathbb{R}$ define

$$m_f(\lambda, \mu) = \begin{cases} \dfrac{f(\lambda) - f(\mu)}{\lambda - \mu} & \text{if } \lambda \neq \mu, \\ f'(\lambda) & \text{if } \lambda = \mu. \end{cases}$$

   (i) Show that, if $f(\lambda) = \lambda^m$, then

$$m_f(\lambda, \mu) = \sum_{k=0}^{m-1} \lambda^{m-k-1} \mu^k.$$

(ii) Let $f \in C^1(\mathbb{R})$, and $\alpha < \beta$ be two real numbers. Show that there exists a sequence $\{p_N\}$ of polynomials in one variable such that

$$\lim_{N \to \infty} p_N(\lambda) = f(\lambda), \quad \lim_{N \to \infty} p'_N(\lambda) = f'(\lambda),$$

uniformly on $[\alpha, \beta]$.

Hint. Consider a sequence $\{q_N\}$ of polynomials which converges uniformly to f' and put

$$p_N(\lambda) = f(\alpha) + \int_\alpha^\lambda q_N(\mu) d\mu.$$

(b) Let $V = Sym(n, \mathbb{R})$ denote the space of $n \times n$ real symmetric matrices, endowed with the norm defined by

$$\|X\| = \left(\mathrm{Tr}(X^2)\right)^{1/2}.$$

To every polynomial p with real coefficients,

$$p(\lambda) = \sum_{k=0}^m a_k \lambda^k,$$

one associates the map $\tilde{p}$ from V into V defined by

$$X \mapsto Y = \tilde{p}(X) = \sum_{k=0}^m a_k X^k.$$

(i) Show that, if

$$X = k \begin{pmatrix} \lambda_1 & & \\ & \ddots & \\ & & \lambda_n \end{pmatrix} k^T,$$

where k is an orthogonal matrix and $\lambda_1, \ldots, \lambda_n \in \mathbb{R}$, then

$$\tilde{p}(X) = k \begin{pmatrix} p(\lambda_1) & & \\ & \ddots & \\ & & p(\lambda_n) \end{pmatrix} k^T.$$

(ii) Show that, if the eigenvalues $\lambda_1, \ldots, \lambda_n$ of X belong to $[\alpha, \beta]$, then

$$\|\tilde{p}(X)\| \leq \sqrt{n} \sup_{\alpha \leq \lambda \leq \beta} |p(\lambda)|.$$

(iii) Let f be a continuous function on $\mathbb{R}$, and $\{p_N\}$ a sequence of polynomials which converges uniformly to f on $[\alpha, \beta]$. Show that, if the eigenvalues of X belong to $[\alpha, \beta]$, then the sequence of the matrices $Y_N = \tilde{p}_N(X)$ converges. Define

$$\tilde{f}(X) = \lim_{N \to \infty} \tilde{p}_N(X).$$

(c) Let $f \in C^1(\mathbb{R})$, and Λ a real diagonal matrix:

$$\Lambda = \begin{pmatrix} \lambda_1 & & \\ & \ddots & \\ & & \lambda_n \end{pmatrix}.$$

Let $M_{f,\Lambda}$ denote the linear map from V into V which, to a matrix $X = (x_{ij})$, associates the matrix $Y = (y_{ij})$ given by

$$y_{ij} = m_f(\lambda_i, \lambda_j)x_{ij}.$$

(i) Show that, if the numbers $\lambda_1, \ldots, \lambda_n$ belong to $[\alpha, \beta]$, then

$$\|Y\| \leq \sup_{\alpha \leq \lambda \leq \beta} |f'(\lambda)|\, \|X\|.$$

(ii) Let p be a polynomial. The differential of $\tilde{p}$ at A is defined by

$$(D\tilde{p})_A(X) = \frac{d}{dt}\tilde{p}(A + tX)\big|_{t=0}.$$

One assumes here that the matrix $A = \Lambda$ is diagonal. Show that

$$(D\tilde{p})_\Lambda(X) = M_{p,\Lambda}(X).$$

Hint. Consider first the case of $p(\lambda) = \lambda^m$. Recall that

$$\frac{d}{dt}(A + tX)^m\big|_{t=0} = \sum_{k=0}^{m-1} A^{m-k-1} X A^k.$$

(iii) Show that, if $A = k\Lambda k^T$, where k is an orthogonal matrix and Λ diagonal, then

$$(D\tilde{p})_A(X) = k M_{p,\Lambda}(k^T X k)k^T.$$

(iv) Show that, if $f \in C^1(\mathbb{R})$, then the map $\tilde{f}$ is differentiable, and that, if Λ is diagonal,

$$(D\tilde{f})_\Lambda(X) = M_{f,\Lambda}(X).$$

For more information on this topic:

Ju. Daleckii, S. G. Krein (1965). Integration and differentiation of functions of Hermitian operators and applications to the theory of perturbations. *American Mathematical Society Translations, Series 2*, **47**, 1–30.

3

Linear Lie groups

A *linear Lie group* is a closed subgroup of $GL(n, \mathbb{R})$. To a linear Lie group one associates its Lie algebra. In this way the properties of the group are translated in terms of the linear algebra properties of its Lie algebra. We saw several examples in Section 1.3. Let us observe that $GL(n, \mathbb{C})$ is a linear Lie group since it can be seen as a closed subgroup of $GL(2n, \mathbb{R})$. In fact, to a matrix $Z = X + iY$ in $M(n, \mathbb{C})$ one associates the matrix

$$\tilde{Z} = \begin{pmatrix} X & -Y \\ Y & X \end{pmatrix}$$

in $M(2n, \mathbb{R})$, and the map $Z \mapsto \tilde{Z}$ is an algebra morphism which maps $GL(n, \mathbb{C})$ onto a closed subgroup of $GL(2n, \mathbb{R})$.

3.1 One parameter subgroups

Let G be a topological group. A *one parameter subgroup* of G is a continuous group morphism

$$\gamma : \mathbb{R} \to G,$$

$\mathbb{R}$ being equipped with the additive group structure.

Theorem 3.1.1 *Let* $\gamma : \mathbb{R} \to GL(n, \mathbb{R})$ *be a one parameter subgroup of* $GL(n, \mathbb{R})$. *Then* γ *is* C^∞ *and*

$$\gamma(t) = \exp(tA),$$

with $A = \gamma'(0)$. *In fact* γ *is even real analytic, as can be proved.*

Proof. Assume that γ is C^1. Then

$$\gamma'(t) = \lim_{s \to 0} \frac{\gamma(t+s) - \gamma(t)}{s}$$

$$= \gamma(t) \lim_{s \to 0} \frac{\gamma(s) - \gamma(0)}{s}$$

$$= \gamma(t)\gamma'(0) = \gamma'(0)\gamma(t).$$

Put $A = \gamma'(0)$. Then

$$\gamma'(t) = A\gamma(t).$$

This differential equation has a unique solution γ such that $\gamma(0) = I$, which is given by

$$\gamma(t) = \exp(tA).$$

In fact, if γ is such a solution

$$\frac{d}{dt}\big(\exp(-tA)\gamma(t)\big) = \exp(-tA)\big(\gamma'(t) - A\gamma(t)\big) = 0.$$

We will now show that γ is C^1. Let α be a C^∞ function on $\mathbb{R}$ with compact support, and consider the regularised function f of γ:

$$f(t) = \int_{-\infty}^{\infty} \alpha(t - s)\gamma(s)ds.$$

Then $f : \mathbb{R} \to M(n, \mathbb{R})$ is C^∞, and

$$f(t) = \int_{-\infty}^{\infty} \alpha(s)\gamma(t - s)ds$$

$$= \left(\int_{-\infty}^{\infty} \alpha(s)\gamma(-s)ds \right) \cdot \gamma(t).$$

We will choose the function α in such a way that the matrix

$$B = \int_{-\infty}^{\infty} \alpha(s)\gamma(-s)ds$$

is invertible. It will follow that γ is C^∞. If $\|B - I\| < 1$ then it holds. Let $\alpha \geq 0$, with integral equal to one. Then

$$\|B - I\| \leq \int_{-\infty}^{\infty} \alpha(s)\|\gamma(-s) - I\|ds.$$

Since γ is continuous at 0, for every $\epsilon > 0$ there exists $\eta > 0$ such that, if $|s| \leq \eta$, then $\|\gamma(s) - I\| \leq \epsilon$. If the support of α is contained in $[-\eta, \eta]$, then $\|B - I\| \leq \epsilon$. $\square$

3.2 Lie algebra of a linear Lie group

Let G be a linear Lie group, that is a closed subgroup of $GL(n, \mathbb{R})$. We associate to the group G the set

$$\mathfrak{g} = Lie(G) = \{X \in M(n, \mathbb{R}) \mid \forall t \in \mathbb{R},\ \exp(tX) \in G\}.$$

Theorem 3.2.1 (i) *The set $\mathfrak{g}$ is a vector subspace of $M(n, \mathbb{R})$.*
  (ii) *If $X, Y \in \mathfrak{g}$, then $[X, Y] := XY - YX \in \mathfrak{g}$.*

Proof. (a) If $X, Y \in \mathfrak{g}$, then

$$\left(\exp \frac{t}{k} X \exp \frac{t}{k} Y\right)^k \in G,$$

and, since G is closed, as $k \to \infty$,

$$\exp\bigl(t(X + Y)\bigr) \in G$$

by Corollary 2.2.4, hence $X + Y \in \mathfrak{g}$.
  (b) Similarly, for $t > 0$,

$$\lim_{k \to \infty} \left(\exp \frac{\sqrt{t}}{k} X \exp \frac{\sqrt{t}}{k} Y \exp -\frac{\sqrt{t}}{k} X \exp -\frac{\sqrt{t}}{k} Y\right)^{k^2} = \exp(t[X, Y]) \in G,$$

hence $[X, Y] \in \mathfrak{g}$. $\square$

A real (respectively complex) *Lie algebra* is a vector space $\mathfrak{g}$ over $\mathbb{R}$ (respectively $\mathbb{C}$) equipped with a linear map

$$\mathfrak{g} \times \mathfrak{g} \to \mathfrak{g},$$
$$(X, Y) \mapsto [X, Y],$$

called the bracket or commutator of X and Y, such that

(1) $$[X, Y] = -[Y, X],$$
(2) $$\bigl[X, [Y, Z]\bigr] = \bigl[[X, Y], Z\bigr] + \bigl[Y, [X, Z]\bigr].$$

Relation (2) is called the *Jacobi identity*.
  The space $M(n, \mathbb{R})$ equipped with the product

$$[X, Y] = XY - YX$$

is a Lie algebra. If $G \subset GL(n, \mathbb{R})$ is a linear Lie group, then $\mathfrak{g} = Lie(G)$ is a subalgebra of $M(n, \mathbb{R})$, it is the Lie algebra of G.

Examples.

$$Lie\bigl(GL(n, \mathbb{R})\bigr) = M(n, \mathbb{R}),$$
$$Lie\bigl(SL(n, \mathbb{R})\bigr) = \{X \in M(n, \mathbb{R}) \mid \mathrm{tr}\, X = 0\},$$
$$Lie\bigl(SO(n)\bigr) = \{X \in M(n, \mathbb{R}) \mid X^T = -X\},$$
$$Lie\bigl(Sp(n, \mathbb{R})\bigr) = \left\{ \begin{pmatrix} A & B \\ C & -A^T \end{pmatrix} \middle| A \in M(n, \mathbb{R}),\ B, C \in Sym(n, \mathbb{R}) \right\},$$
$$Lie\bigl(U(n)\bigr) = \{X \in M(n, \mathbb{C}) \mid X^* = -X\}.$$

Consider $G = SL(2, \mathbb{R})$ and let $\mathfrak{g} = \mathfrak{sl}(2, \mathbb{R})$ be its Lie algebra. The following matrices constitute a basis of $\mathfrak{g}$:

$$H = \begin{pmatrix} 1 & 0 \\ 0 & -1 \end{pmatrix}, \quad E = \begin{pmatrix} 0 & 1 \\ 0 & 0 \end{pmatrix}, \quad F = \begin{pmatrix} 0 & 0 \\ 1 & 0 \end{pmatrix},$$

and

$$[H, E] = 2E, \quad [H, F] = -2F, \quad [E, F] = H.$$

Let G be the group '$ax + b$', that is the group of affine linear transformations of $\mathbb{R}$. It is the set $\mathbb{R}^* \times \mathbb{R}$ equipped with the product

$$(a_1, b_1)(a_2, b_2) = (a_1 a_2, a_1 b_2 + b_1).$$

This is not a group of matrices, but it can be identified with the closed subgroup of $GL(2, \mathbb{R})$ whose elements are the matrices

$$\begin{pmatrix} a & b \\ 0 & 1 \end{pmatrix}.$$

The matrices

$$X_1 = \begin{pmatrix} 1 & 0 \\ 0 & 0 \end{pmatrix}, \quad X_2 = \begin{pmatrix} 0 & 1 \\ 0 & 0 \end{pmatrix},$$

constitute a basis of its Lie algebra and $[X_1, X_2] = X_2$.

Let G be the motion group of $\mathbb{R}^2$, that is the group of affine linear transformations of the form

$$(x, y) \mapsto (x \cos \theta - y \sin \theta + a, x \sin \theta + y \cos \theta + b).$$

The group G can be identified with the subgroup of $GL(3, \mathbb{R})$ whose elements are the matrices

$$\begin{pmatrix} \cos \theta & -\sin \theta & a \\ \sin \theta & \cos \theta & b \\ 0 & 0 & 1 \end{pmatrix}.$$

Its Lie algebra $\mathfrak{g}$ has dimension 3. The following matrices constitute a basis for $\mathfrak{g}$:

$$X_1 = \begin{pmatrix} 0 & -1 & 0 \\ 1 & 0 & 0 \\ 0 & 0 & 0 \end{pmatrix}, \quad X_2 = \begin{pmatrix} 0 & 0 & 1 \\ 0 & 0 & 0 \\ 0 & 0 & 0 \end{pmatrix}, \quad X_3 = \begin{pmatrix} 0 & 0 & 0 \\ 0 & 0 & 1 \\ 0 & 0 & 0 \end{pmatrix},$$

and

$$[X_1, X_2] = X_3, \quad [X_1, X_3] = -X_2, \quad [X_2, X_3] = 0.$$

Let $\mathfrak{g}$ and $\mathfrak{h}$ be two Lie algebras over $\mathbb{R}$ (or $\mathbb{C}$). A *Lie algebra morphism* of $\mathfrak{g}$ into $\mathfrak{h}$ is a linear map $A : \mathfrak{g} \to \mathfrak{h}$ satisfying

$$[AX, AY] = A[X, Y].$$

The group of automorphisms of the Lie algebra $\mathfrak{g}$ is denoted by $Aut(\mathfrak{g})$.

Let G be a linear Lie group, and $\mathfrak{g} = Lie(G)$ its Lie algebra. By the definition of the Lie algebra of G, the exponential map maps $\mathfrak{g}$ into G:

$$\exp : \mathfrak{g} \to G.$$

For $g \in G, X \in \mathfrak{g}, t \in \mathbb{R}$,

$$g \exp(tX)g^{-1} = \exp(tgXg^{-1}).$$

Hence $gXg^{-1} \in \mathfrak{g}$. The map $\mathrm{Ad}(g) : X \mapsto \mathrm{Ad}(g)X = gXg^{-1}$ is an automorphism of the Lie algebra $\mathfrak{g}$,

$$\mathrm{Ad}(g)[X, Y] = [\mathrm{Ad}(g)X, \mathrm{Ad}(g)Y] \quad (X, Y \in \mathfrak{g}).$$

Furthermore

$$\mathrm{Ad}(g_1 g_2) = \mathrm{Ad}(g_1) \circ \mathrm{Ad}(g_2),$$

and this means that the map

$$\mathrm{Ad} : G \to Aut(\mathfrak{g})$$

is a group morphism.

Proposition 3.2.2 (i) *For $X \in \mathfrak{g}$,*

$$\frac{d}{dt} \mathrm{Ad}(\exp tX)\Big|_{t=0} = \mathrm{ad}\, X.$$

(ii) *Let us denote by* Exp *the exponential map from* $End(\mathfrak{g})$ *into* $GL(\mathfrak{g})$. *Then*

$$\mathrm{Exp}(\mathrm{ad}\, X) = \mathrm{Ad}(\exp X) \quad (X \in \mathfrak{g}).$$

Proof. (a)

$$\frac{d}{dt} \operatorname{Ad}(\exp t X) Y \bigg|_{t=0} = \frac{d}{dt} \exp(t X) Y \exp(-t X) \bigg|_{t=0} = [X, Y].$$

(b) Put

$$\gamma_1(t) = \operatorname{Exp}(t \operatorname{ad} X),$$
$$\gamma_2(t) = \operatorname{Ad}(\exp t X).$$

They are two one parameter subgroups of $GL(\mathfrak{g})$, and

$$\gamma_1'(0) = \operatorname{ad} X,$$
$$\gamma_2'(0) = \operatorname{ad} X.$$

Therefore $\gamma_1(t) = \gamma_2(t)$ $(t \in \mathbb{R})$ by Theorem 3.1.1. $\qquad\square$

3.3 Linear Lie groups are submanifolds

Let us recall first the definition of a submanifold in a finite dimensional real vector space. A *submanifold* of dimension m in $\mathbb{R}^N$ is a subset M with the following property: for every $x \in M$ there exists a neighbourhood U of 0 in $\mathbb{R}^N$, a neighbourhood W of x in $\mathbb{R}^N$ and a diffeomorphism Φ from U onto W such that

$$\Phi(U \cap \mathbb{R}^m) = W \cap M.$$

Theorem 3.3.1 *Let G be a linear Lie group and $\mathfrak{g} = Lie(G)$ be its Lie algebra. There exists a neighbourhood U of 0 in $\mathfrak{g}$ and a neighbourhood V of I in G such that*

$$\exp : U \to V$$

is a homeomorphism.

Proof. Let $G \subset GL(n, \mathbb{R})$ be a linear Lie group, and $\mathfrak{g} \subset M(n, \mathbb{R})$ be its Lie algebra. Let U_0 be a neighbourhood of 0 in $M(n, \mathbb{R})$ and V_0 a neighbourhood of I in $GL(n, \mathbb{R})$ for which $\exp : U_0 \to V_0$ is a diffeomorphism. Then $U_0 \cap \mathfrak{g}$ is a neighbourhood of 0 in $\mathfrak{g}$, the restriction of the exponential map to $U_0 \cap \mathfrak{g}$ is injective and maps $U_0 \cap \mathfrak{g}$ into $V_0 \cap G$, but one does not know yet whether $\exp(U_0 \cap \mathfrak{g}) = V_0 \cap G$, even if one assumes that G is connected. (See Exercise 5.)

Lemma 3.3.2 *Let (g_k) be a sequence of elements in G which converges to I. One assumes that, for all k, $g_k \neq I$. Then the accumulation points of the* *sequence*

$$X_k = \frac{\log g_k}{\|\log g_k\|}$$

belong to $\mathfrak{g}$.

Proof. We may assume that

$$\lim_{k \to \infty} X_k = X \in M(n, \mathbb{R}).$$

Put $Y_k = \log g_k$ and, for $t \in \mathbb{R}$,

$$\lambda_k = \frac{t}{\|\log g_k\|},$$

then

$$\exp(tX) = \lim_{k \to \infty} \exp(\lambda_k Y_k).$$

Let us denote by $[\lambda_k]$ the integer part of λ_k. We can write

$$\exp(\lambda_k Y_k) = (\exp Y_k)^{[\lambda_k]} \exp\big((\lambda_k - [\lambda_k])Y_k\big),$$

and

$$\|(\lambda_k - [\lambda_k])Y_k\| \leq \|Y_k\| \to 0,$$

hence, since $\exp Y_k = g_k$,

$$\exp(tX) = \lim_{k \to \infty} (g_k)^{[\lambda_k]} \in G,$$

and this proves that X belongs to $\mathfrak{g}$. $\qquad\qquad\qquad\qquad\qquad\square$

Lemma 3.3.3 *Let* $\mathfrak{m}$ *be a subspace of* $M(n, \mathbb{R})$, *complementary to* $\mathfrak{g}$. *Then there exists a neighbourhood* U *of* 0 *in* $\mathfrak{m}$ *such that* $\exp U \cap G = \{I\}$.

Proof. Let us assume the opposite. In this case there exists a sequence $X_k \in \mathfrak{m}$ with limit 0 such that

$$g_k = \exp X_k, \quad g_k \neq I, \ g_k \in G.$$

Let Y be an accumulation point of the sequence $X_k / \|X_k\|$. By Lemma 3.3.2, $Y \in \mathfrak{g} \cap \mathfrak{m} = \{0\}$, and this is impossible since $\|Y\| = 1$. $\qquad\qquad\square$

Lemma 3.3.4 *Let* E *and* F *be two complementary subspaces in* $M(n, \mathbb{R})$. *Then the map*

$$\Phi : E \times F \to GL(n, \mathbb{R}),$$
$$(X, Y) \mapsto \exp X \exp Y$$

is differentiable, and

$$D\Phi_{(0,0)}(X, Y) = X + Y.$$

The proof is left to the reader.

We can now finish the proof of Theorem 3.3.1. Let $\mathfrak{m}$ be a subspace of $M(n, \mathbb{R})$ complementary to $\mathfrak{g}$, and consider the map

$$\Phi : \mathfrak{g} \times \mathfrak{m} \to GL(n, \mathbb{R}),$$

$$(X, Y) \mapsto \exp X \exp Y.$$

There exists a neighbourhood U of 0 in $\mathfrak{g}$, a neighbourhood V of 0 in $\mathfrak{m}$, and a neighbourhood W of I in $GL(n, \mathbb{R})$ such that the restriction of Φ to $U \times V$ is a diffeomorphism onto W. Observe that

$$\exp U = \Phi(U \times \{0\}) \subset W \cap G.$$

By Lemma 3.3.3 the neighbourhood V can be chosen such that

$$\exp V \cap G = \{I\}.$$

Let us show that $\exp U = W \cap G$. Let $g \in W \cap G$. One can write $g = \exp X \exp Y$ ($X \in U, Y \in V$), and then

$$\exp Y = \exp(-X)g \in \exp V \cap G = \{I\},$$

hence $g = \exp X$. □

Corollary 3.3.5 *A linear Lie group $G \subset GL(n, \mathbb{R})$ is a submanifold of $M(n, \mathbb{R})$ of dimension $m = \dim \mathfrak{g}$.*

Proof. Let $g \in G$ and let $L(g)$ be the map

$$L(g) : GL(n, \mathbb{R}) \to GL(n, \mathbb{R}),$$

$$h \mapsto gh.$$

Let U be a neighbourhood of 0 in $M(n, \mathbb{R})$ and W_0 a neighbourhood of I in $GL(n, \mathbb{R})$ such that the exponential map is a diffeomorphism from U onto W_0 which maps $U \cap \mathfrak{g}$ onto $W_0 \cap G$. The composed map $\Phi = L(g) \circ \exp$ maps U onto $W = gW_0$, and $U \cap \mathfrak{g}$ onto $W \cap G$. □

An important consequence of Theorem 3.3.1 is that the set $\exp \mathfrak{g}$ is a neighbourhood of I in G, hence generates the identity component G_0 of G by Proposition 1.1.1:

$$\bigcup_{k=1}^{\infty}(\exp \mathfrak{g})^k = G_0.$$

Corollary 3.3.6 *If two closed subgroups G_1 and G_2 of $GL(n, \mathbb{R})$ have the same Lie algebra then the identity components of G_1 and G_2 are the same.*

It also follows from Theorem 3.3.1 that the group G is discrete if and only if its Lie algebra reduces to $\{0\}$: $Lie(G) = \{0\}$.

To every closed subgroup G of $GL(n, \mathbb{R})$ one associates its Lie algebra $\mathfrak{g} = Lie(G) \subset M(n, \mathbb{R})$. However, not every Lie subalgebra of $M(n, \mathbb{R})$ corresponds to a closed subgroup of $GL(n, \mathbb{R})$. (See Exercise 1.)

3.4 Campbell–Hausdorff formula

Let G be a linear Lie group and $\mathfrak{g} = Lie(G)$ its Lie algebra. The Campbell–Hausdorff formula expresses $\log(\exp X \exp Y)$ $(X, Y \in \mathfrak{g})$ in terms of a series, each term of which is a homogeneous polynomial in X and Y involving iterated brackets.

Let us introduce the functions

$$\Phi(z) = \frac{1 - e^{-z}}{z} = \sum_{k=0}^{\infty} (-1)^k \frac{z^k}{(k+1)!} \quad (z \in \mathbb{C}),$$

$$\Psi(z) = \frac{z \log z}{z - 1} = z \sum_{k=0}^{\infty} \frac{(-1)^k}{k+1} (z-1)^k \quad (|z-1| < 1).$$

If $|z| < \log 2$, then $|e^z - 1| \le e^{|z|} - 1 < 1$, and

$$\Psi(e^z)\Phi(z) = \frac{e^z z}{e^z - 1} \frac{1 - e^{-z}}{z} = 1.$$

Therefore, if L is an endomorphism such that $\|L\| < \log 2$, then

$$\Psi(\operatorname{Exp} L)\Phi(L) = Id.$$

With this notation the differential of the exponential map (Theorem 2.1.4) can be written

$$(D \exp)_A = \exp A \, \Phi(\operatorname{ad} A).$$

Theorem 3.4.1 *If $\|X\|$, $\|Y\| < r = \frac{1}{2}\log(2 - \frac{1}{2}\sqrt{2})$, then*

$$\log(\exp X \exp Y) = X + \int_0^1 \Psi\big(\operatorname{Exp}(\operatorname{ad} X)\operatorname{Exp}(t \operatorname{ad} Y)\big) Y \, dt.$$

Lemma 3.4.2 *If $\|X\|$, $\|Y\| \le \alpha$, then*

$$\| \exp X \exp Y - I \| \le e^{2\alpha} - 1.$$

Proof.

$$\exp X \exp Y - I = (\exp X - I)(\exp Y - I) + (\exp X - I) + (\exp Y - I),$$

and, since $\| \exp X - I \| \le e^{\|X\|} - 1 \le e^{\alpha} - 1$,

$$\| \exp X \exp Y - I \| \le (e^{\alpha} - 1)^2 + 2(e^{\alpha} - 1) = e^{2\alpha} - 1. \qquad \square$$

Lemma 3.4.3 *If $\|g - I\| \le \beta < 1$, then*

$$\| \log g \| \le \log \frac{1}{1 - \beta}.$$

Proof.

$$\| \log g \| \le \sum_{k=1}^{\infty} \frac{\|(g - I)^k\|}{k} \le \sum_{k=1}^{\infty} \frac{\beta^k}{k} = \log \frac{1}{1 - \beta}. \qquad \square$$

Let us now prove Theorem 3.4.1. For $\|X\|, \|Y\| < \frac{1}{2} \log 2$, put

$$F(t) = \log(\exp X \exp tY).$$

By Lemma 3.4.2, the function F is defined for $|t| \le 1$. If furthermore $\|X\|$, $\|Y\| < r$ (observe that $r < \frac{1}{2} \log 2$), then, by Lemmas 3.4.2 and 3.4.3,

$$\|F(t)\| < \frac{1}{2} \log 2.$$

From the inequality

$$\|XY - YX\| \le 2\|X\|\|Y\|$$

it follows that $\| \operatorname{ad} X \| \le 2\|X\|$, hence

$$\| \operatorname{ad} F(t) \| < \log 2.$$

Let us prove that the function F satisfies the differential equation

$$F'(t) = \Psi\big(\operatorname{Exp}(\operatorname{ad} F(t))\big)Y.$$

One can write

$$\exp F(t) = \exp X \exp tY.$$

Taking the derivative at t:

$$(D \exp)_{F(t)}\big(F'(t)\big) = (\exp X \exp tY)Y.$$

By Theorem 2.1.4, we obtain

$$\Phi\big(\operatorname{ad} F(t)\big)F'(t) = Y.$$

Since $\| \operatorname{ad} F(t) \| < \log 2$ this can be written

$$F'(t) = \Psi\big(\operatorname{Exp}(\operatorname{ad} F(t))\big)Y.$$

We can also write

$$F'(t) = \Psi\big(\text{Ad}(\exp F(t))Y\big)$$
$$= \Psi\big(\text{Ad}(\exp X)\,\text{Ad}(\exp tY)\big)Y$$
$$= \Psi\big(\text{Exp}(\text{ad}\,X)\,\text{Exp}(\text{ad}\,tY)\big)Y.$$

Furthermore $F(0) = \log(\exp X) = X$, and

$$F(1) = F(0) + \int_0^1 F'(t)dt,$$

hence

$$\log(\exp X \exp Y) = X + \int_0^1 \Psi\big(\text{Exp}(\text{ad}\,X)\,\text{Exp}(t\,\text{ad}\,Y)\big)Y\,dt. \qquad \square$$

Theorem 3.4.4 (Campbell–Hausdorff formula) *If* $\|X\|$, $\|Y\| < \frac{1}{2}\log(2 - \frac{1}{2}\sqrt{2})$, *then*

$$\log(\exp X \exp Y) = X + \sum_{k=0}^{\infty} \frac{(-1)^k}{k+1} \sum_{\mathcal{E}(k)} \frac{1}{(q_1 + \cdots + q_k + 1)}$$
$$\cdot \frac{(\text{ad}\,X)^{p_1}(\text{ad}\,Y)^{q_1}\ldots(\text{ad}\,X)^{p_k}(\text{ad}\,Y)^{q_k}(\text{ad}\,X)^m}{p_1!q_1!\ldots p_k!q_k!m!}Y,$$

where, for $k \geq 1$,

$$\mathcal{E}(k) = \{p_1, q_1, \ldots, p_k, q_k, m \in \mathbb{N} \mid p_i + q_i > 0,\ i = 1, \ldots, k\},$$

and

$$\mathcal{E}(0) = \{m \in \mathbb{N}\}.$$

Proof. If A and B are two endomorphisms

$$(\exp A \exp B - I)^k \exp A = \sum_{\mathcal{E}(k)} \frac{A^{p_1}B^{q_1}\ldots A^{p_k}B^{q_k}A^m}{p_1!q_1!\ldots p_k!q_k!m!}.$$

Since

$$\Psi(z) = \sum_{k=0}^{\infty} \frac{(-1)^k}{k+1}(z-1)^k z,$$

we have

$$\Psi\big(\text{Exp}(\text{ad}\,X)\,\text{Exp}(t\,\text{ad}\,Y)\big)Y$$
$$= \sum_{k=0}^{\infty} \frac{(-1)^k}{k+1}\big(\text{Exp}(\text{ad}\,X)\,\text{Exp}(t\,\text{ad}\,Y) - I\big)^k \text{Exp}(\text{ad}\,X)\,\text{Exp}(t\,\text{ad}\,Y)Y.$$

Observing that

$$\text{Exp}(t \text{ ad } Y)Y = Y,$$

we obtain

$$\Psi\big(\text{Exp}(\text{ad } X)\,\text{Exp}(t \text{ ad } Y)\big)Y$$

$$= \sum_{k=0}^{\infty} \frac{(-1)^k}{k+1} \cdot \sum_{\mathcal{E}(k)} t^{q_1+\cdots+q_k} \frac{(\text{ad } X)^{p_1}(\text{ad } Y)^{q_1} \dots (\text{ad } X)^{p_k}(\text{ad } Y)^{q_k}(\text{ad } X)^m}{p_1!q_1!\dots p_k!q_k!m!} Y.$$

The convergence of the series is uniform for t in $[0, 1]$. The statement is obtained by termwise integration since

$$\int_0^1 t^{q_1+\cdots+q_k}\,dt = \frac{1}{q_1+\cdots+q_k+1}. \qquad \square$$

Corollary 3.4.5

$$\log(\exp X \exp Y) = X + Y + \frac{1}{2}[X, Y] + \frac{1}{12}[X, [X, Y]] + \frac{1}{12}[Y, [Y, X]]$$

$$+ \text{ terms of degree} \geq 4.$$

Proof. The terms of degree 2 and 3 are written in the following table.

k	p_1	q_1	p_2	q_2	m	
0					1	$[X, Y]$
0					2	$\frac{1}{2}[X, [X, Y]]$
1	1	0			0	$-\frac{1}{2}[X, Y]$
1	1	0			1	$-\frac{1}{2}[X, [X, Y]]$
1	0	1			1	$-\frac{1}{4}[Y, [X, Y]]$
1	2	0			0	$-\frac{1}{4}[X, [X, Y]]$
2	1	0	1	0	0	$\frac{1}{3}[X, [X, Y]]$
2	0	1	1	0	0	$\frac{1}{6}[Y, [X, Y]]$

$\square$

3.5 Exercises

1. Let α be an irrational real number.
   (a) Show that $\mathbb{Z} + \alpha\mathbb{Z}$ is dense in $\mathbb{R}$.

(b) Let G be the subgroup of $GL(2, \mathbb{C})$ defined by

$$G = \left\{ \begin{pmatrix} e^{2i\pi t} & 0 \\ 0 & e^{2i\pi\alpha t} \end{pmatrix} \middle| \ t \in \mathbb{R} \right\}.$$

Determine the closure $\bar{G}$ of G in $GL(2, \mathbb{C})$.

(c) Show that there does not exist any closed subgroup of $GL(2, \mathbb{C})$ with Lie algebra

$$\mathfrak{g} = \left\{ \begin{pmatrix} it & 0 \\ 0 & i\alpha t \end{pmatrix} \middle| \ t \in \mathbb{R} \right\}.$$

2. Let G be a linear Lie group and $\mathfrak{g}$ its Lie algebra. One assumes that G is Abelian.

(a) Show that $\mathfrak{g}$ is Abelian, that is

$$\forall X, Y \in \mathfrak{g}, \ [X, Y] = 0.$$

(b) Show that $\exp \mathfrak{g} = G_0$, where G_0 is the identity component in G.

(c) Show that G_0 is isomorphic to a group of the form $\mathbb{R}^p \times \mathbb{T}^q$, where $\mathbb{T} = \mathbb{R}/\mathbb{Z}$.

Hint. Use Exercise 6 of Chapter 1.

3. Let G be a linear Lie group, and φ a differentiable morphism from $GL(n, \mathbb{R})$ into G. Define $\Phi = (D\varphi)_I$.

(a) Show that, for every $X \in M(n, \mathbb{R})$,

$$\varphi(\exp X) = \exp(\Phi X).$$

(b) Deduce

$$\det(\exp X) = e^{\operatorname{tr} X}.$$

4. Show that in $GL(n, \mathbb{R})$ there is no arbitrary small subgroup. More precisely, show that there is a neighbourhood V of I in $GL(n, \mathbb{R})$ such that, if H is a subgroup contained in V, then $H = \{I\}$.

5. The aim of this exercise is to illustrate the difficulty we pointed out at the beginning of the proof of Theorem 3.3.1.

Let

$$U = \{X \in M(2, \mathbb{C}) \mid \|X\| < r\};$$

the number r is chosen such that the exponential map is a diffeomorphism from U onto its image $V = \exp U$.

There exists $R > 0$ such that V contains the ball

$$B(I, R) = \{g \in GL(2, \mathbb{C}) \mid \|g - I\| < R\}.$$

For a positive integer m put

$$X = \begin{pmatrix} i & 0 \\ 0 & \frac{i}{m} \end{pmatrix},$$

and, for $t \in \mathbb{R}$,

$$F(t) = \exp tX = \begin{pmatrix} e^{it} & 0 \\ 0 & e^{i\frac{t}{m}} \end{pmatrix}.$$

Show that $G = F(\mathbb{R})$ is a closed subgroup in $GL(2, \mathbb{C})$, and that $\mathfrak{g} = Lie(G) = \mathbb{R}X$. Show that, for $g = F(2\pi)$,

$$\|g - I\| = 2 \sin \frac{\pi}{m}.$$

Show that, for m large enough, $g \in V$ and $g \notin \exp(U \cap \mathfrak{g})$, hence

$$\exp(U \cap \mathfrak{g}) \nsubseteq V \cap G.$$

4

Lie algebras

In this chapter we consider Lie algebras from an algebraic point of view. We will see how some properties of linear Lie groups can be deduced from the corresponding properties of their Lie algebras. Then we present the basic properties of nilpotent, solvable, and semi-simple Lie algebras.

4.1 Definitions and examples

A *Lie algebra* over $\mathbb{K} = \mathbb{R}$ or $\mathbb{C}$ is a vector space $\mathfrak{g}$ equipped with a bilinear map

$$\mathfrak{g} \times \mathfrak{g} \to \mathfrak{g}, \quad (X, Y) \mapsto [X, Y],$$

satisfying

(1) $$[Y, X] = -[X, Y],$$

(2) $$\big[[X, Y], Z\big] + \big[[Y, Z], X\big] + \big[[Z, X], Y\big] = 0.$$

The equality (2) is called the *Jacobi identity*.

Assume $\mathfrak{g}$ is finite dimensional, and let $(X_1, \ldots, X_n)$ be a basis of $\mathfrak{g}$. One can write

$$[X_i, X_j] = \sum_{k=1}^{n} c_{ij}^k X_k.$$

The numbers c_{ij}^k are called the *structure constants* of the Lie algebra $\mathfrak{g}$. Property (1) can be written $c_{ij}^k = -c_{ji}^k$, and property (2) says that, for any m,

$$\sum_{\ell=1}^{n} \left(c_{ij}^\ell c_{\ell k}^m + c_{jk}^\ell c_{\ell i}^m + c_{ki}^\ell c_{\ell j}^m \right) = 0.$$

An *automorphism* of a Lie algebra is a linear automorphism $g \in GL(\mathfrak{g})$ such that

$$[gX, gY] = g[X, Y].$$

The group of all automorphisms of the Lie algebra $\mathfrak{g}$ is denoted by $Aut(\mathfrak{g})$. If $\mathfrak{g}$ is finite dimensional, it is a closed subgroup of $GL(\mathfrak{g})$. A *derivation* of $\mathfrak{g}$ is a linear endomorphism $D \in End(\mathfrak{g})$ such that

$$D([X, Y]) = [DX, Y] + [X, DY].$$

For $X \in \mathfrak{g}$ let ad X denote the endomorphism of $\mathfrak{g}$ defined by

$$\text{ad } X \cdot Y = [X, Y].$$

The Jacobi identity (2) says that ad X is a derivation. The space $Der(\mathfrak{g})$ of the derivations of $\mathfrak{g}$ is a Lie algebra for the bracket defined by

$$[D_1, D_2] = D_1 D_2 - D_2 D_1,$$

and the map ad : $\mathfrak{g} \to Der(\mathfrak{g})$ is a Lie algebra morphism:

$$\text{ad}[X, Y] = [\text{ad } X, \text{ad } Y].$$

Proposition 4.1.1 *Let $\mathfrak{g}$ be a finite dimensional Lie algebra. The Lie algebra of $Aut(\mathfrak{g})$ is equal to $Der(\mathfrak{g})$.*

Proof. Let $D \in Lie\big(Aut(\mathfrak{g})\big)$. For every $t \in \mathbb{R}$, $\text{Exp}(tD)$ is an automorphism of $\mathfrak{g}$: for $X, Y \in \mathfrak{g}$,

$$\text{Exp}(tD)[X, Y] = \big[\text{Exp}(tD)X, \text{Exp}(tD)Y\big].$$

Taking derivatives of both sides at $t = 0$ we obtain

$$D[X, Y] = [DX, Y] + [X, DY],$$

which means that D is a derivation: $D \in Der(\mathfrak{g})$.

Conversely, let $D \in Der(\mathfrak{g})$ and put, for $X, Y \in \mathfrak{g}$,

$$F_1(t) = \text{Exp}(tD)[X, Y],$$
$$F_2(t) = \big[\text{Exp}(tD)X, \text{Exp}(tD)Y\big].$$

We have

$$F_1'(t) = D \text{ Exp}(tD)[X, Y] = D F_1(t),$$
$$F_2'(t) = \big[D \text{ Exp}(tD)X, \text{Exp}(tD)Y\big] + \big[\text{Exp}(tD)X, D \text{ Exp}(tD)Y\big],$$

and, since D is a derivation of $\mathfrak{g}$,

$$F_2'(t) = D\big[\text{Exp}(tD)X, \text{Exp}(tD)Y\big] = D F_2(t).$$

Thus F_1 and F_2 are solutions of the same differential equation with the same initial data: $F_1(0) = F_2(0) = [X, Y]$. Hence, for every $t \in \mathbb{R}$, $F_1(t) = F_2(t)$. This means that, for every t, $\mathrm{Exp}(tD)$ is an automorphism of $\mathfrak{g}$, and that $D \in Lie(Aut(\mathfrak{g}))$. $\qquad \square$

An *ideal* $\mathfrak{J}$ of a Lie algebra $\mathfrak{g}$ is a subalgebra which furthermore satisfies

$$\forall X \in \mathfrak{J}, \quad \forall Y \in \mathfrak{g}, \quad [X, Y] \in \mathfrak{J}.$$

Let G be a linear Lie group, and H a closed subgroup. Then $\mathfrak{h} = Lie(H)$ is a subalgebra of $\mathfrak{g} = Lie(G)$ and, if H is a normal subgroup of G, then $\mathfrak{h}$ is an ideal of $\mathfrak{g}$. The converse holds if G and H are connected.

Let G be a topological group and $\mathcal{V}$ a finite dimensional vector space over $\mathbb{R}$ or $\mathbb{C}$. A *representation* of G on $\mathcal{V}$ is a continuous map

$$\pi : G \to GL(\mathcal{V}),$$

which is a group morphism:

$$\pi(g_1 g_2) = \pi(g_1)\pi(g_2) \quad (g_1, g_2 \in G), \quad \pi(e) = Id,$$

A vector subspace $\mathcal{W} \subset \mathcal{V}$ is said to be *invariant* if, for every $g \in G$, $\pi(g)\mathcal{W} = \mathcal{W}$. Let us denote by $\pi_0(g)$ the restriction of $\pi(g)$ to $\mathcal{W}$:

$$\pi_0(g) = \pi(g)\big|_{\mathcal{W}}.$$

Then π_0 is a representation of G on $\mathcal{W}$, one says that π_0 is a *subrepresentation* of π. The representation π_1 of G on the quotient space $\mathcal{V}/\mathcal{W}$ is called a *quotient representation*. The representation π is said to be *irreducible* if the only invariant subspaces are $\{0\}$ and $\mathcal{V}$.

Two representations $(\pi_1, \mathcal{V}_1)$ and $(\pi_2, \mathcal{V}_2)$ are said to be *equivalent* if there exists an isomorphism $A : \mathcal{V}_1 \to \mathcal{V}_2$ (A is an invertible linear map) such that

$$A\pi_1(g) = \pi_2(g)A,$$

for every $g \in G$. One says that A is an *intertwinning operator* or that A *intertwins* the representations π_1 and π_2.

A *representation* of a Lie algebra $\mathfrak{g}$ on a vector space $\mathcal{V}$ is a linear map

$$\rho : \mathfrak{g} \to End(\mathcal{V})$$

which is a Lie algebra morphism:

$$\rho([X, Y]) = [\rho(X), \rho(Y)] = \rho(X)\rho(Y) - \rho(Y)\rho(X).$$

One also says that $\mathcal{V}$ is a module over $\mathfrak{g}$, or that $\mathcal{V}$ is a $\mathfrak{g}$-module.

The map ad : $\mathfrak{g} \to Der(\mathfrak{g}) \subset End(\mathfrak{g})$ is a representation of $\mathfrak{g}$, which is called the *adjoint representation*.

Let G be a linear Lie group with Lie algebra $\mathfrak{g}$ and let π be a representation of G on a finite dimensional vector space $\mathcal{V}$. Then, for $X \in \mathfrak{g}$, $t \mapsto \gamma(t) = \pi(\exp tX)$ is a one parameter subgroup of $GL(\mathcal{V})$, hence differentiable by Theorem 3.1.1. Put

$$d\pi(X) = \frac{d}{dt}\pi(\exp tX)\Big|_{t=0} \quad (X \in \mathfrak{g}),$$

then $d\pi$ is a representation of the Lie algebra of $\mathfrak{g}$ on $\mathcal{V}$, which is called the *derived representation* of π. Let us prove this fact. By Theorem 3.1.1,

$$\pi(\exp X) = \operatorname{Exp} d\pi(X) \quad (X \in \mathfrak{g}).$$

From the definition of $d\pi$ it follows at once that, for $t \in \mathbb{R}$, $d\pi(tX) = td\pi(X)$. By Corollary 2.2.4

$$
\begin{aligned}
\pi\left(\exp t(X+Y)\right) &= \lim_{k\to\infty} \left(\pi\left(\exp \frac{tX}{k}\right)\pi\left(\exp \frac{tY}{k}\right)\right)^k \\
&= \lim_{k\to\infty} \left(\operatorname{Exp} \frac{d\pi(tX)}{k}\operatorname{Exp} \frac{d\pi(tY)}{k}\right)^k \\
&= \operatorname{Exp}\left(d\pi(tX)+d\pi(tY)\right) = \operatorname{Exp}\left(td\pi(X)+td\pi(Y)\right),
\end{aligned}
$$

and, by taking the derivatives at $t = 0$, we get

$$d\pi(X+Y) = d\pi(X)+d\pi(Y).$$

Furthermore

$$\pi\left(\exp\left(t\operatorname{Ad}(g)Y\right)\right) = \pi(g)\pi(\exp tY)\pi(g^{-1}).$$

By taking the derivatives at $t = 0$, we get

$$d\pi\left(\operatorname{Ad}(g)Y\right) = \pi(g)d\pi(Y)\pi(g^{-1}).$$

Then put $g = \exp sX$ and take the derivatives at $s = 0$,

$$d\pi([X,Y]) = d\pi(X)d\pi(Y) - d\pi(Y)d\pi(X).$$

The adjoint representation $\pi = \operatorname{Ad}$ of G on $\mathfrak{g}$ is a special case for which the derived representation is the adjoint representation ad of $\mathfrak{g}$ on $\mathfrak{g}$.

If π_1 and π_2 are two equivalent representations, then the derived representations $d\pi_1$ and $d\pi_2$ are also equivalent. The converse holds if G is connected.

The kernel of a representation of a Lie algebra is an ideal. The *centre* of a Lie algebra $\mathfrak{g}$, denoted by $Z(\mathfrak{g})$, is defined as

$$Z(\mathfrak{g}) = \{X \in \mathfrak{g} \mid \forall Y \in \mathfrak{g}, \ [X, Y] = 0\}.$$

It is an Abelian ideal. It is the kernel of the adjoint representation.

Remark. One can show that every finite dimensional Lie algebra admits a faithful (i.e. injective) finite dimensional representation. This is the theorem of Ado. Hence every finite dimensional Lie algebra can be seen as a subalgebra of $\mathfrak{gl}(N, \mathbb{K}) = M(N, \mathbb{K})$, for some N.

Let G and H be two linear Lie groups and ϕ a continuous morphism of G into H. One puts, for $X \in \mathfrak{g} = Lie(G)$,

$$d\phi(X) = \frac{d}{dt}\phi(\exp tX)\Big|_{t=0}.$$

From what we have seen, $d\phi$ is a Lie algebra morphism from $\mathfrak{g}$ into $\mathfrak{h} = Lie(H)$. Observe that $d\phi$ is the differential of ϕ at the identity element I of G:

$$d\phi = (D\phi)_I.$$

Proposition 4.1.2 (i) *The Lie algebra of the kernel of the morphism ϕ is equal to the kernel of $d\phi$:*

$$Lie(\ker(\phi)) = \ker(d\phi).$$

Therefore the kernel of ϕ is discrete if and only if $d\phi$ is injective.

(ii) *If $d\phi$ is surjective, then the image of ϕ contains the identity component H_0 of H.*

(iii) *If G and H are connected and if $d\phi$ is an isomorphism, then (G, ϕ) is a covering of H.*

Let us recall the definition of a covering. Let X and Y be two connected topological spaces and $\phi : X \to Y$ a continuous map. The pair (X, ϕ) is called a *covering* of Y if ϕ is surjective and if, for every $x \in X$, there exist neighbourhoods V of x and W of $y = \phi(x)$ such that the restriction of φ to V is a homeomorphism from V onto W.

Let (X, ϕ) be a covering of Y; if, for $y_0 \in Y$, the pullback $\phi^{-1}(y_0) \subset X$ is a finite set, then the same holds for every $y \in Y$, and the pullbacks $\phi^{-1}(y)$ all have the same number of elements. Let k be that number. Then one says that (X, ϕ) is a covering of order k of Y (or a covering with k sheets).

Proof. (a) From Theorem 3.1.1 it follows that, for $X \in \mathfrak{g}, t \in \mathbb{R}$,

$$\phi(\exp tX) = \exp(t d\phi(X)).$$

Hence

$$Lie(\ker \phi) = \ker (d\phi).$$

In particular, $d\phi$ is injective if and only if the Lie algebra of $\ker \phi$ reduces to $\{0\}$, that is if $\ker \phi$ is discrete.

(b) Recall that G_0 denotes the identity component of G. Assume that $d\phi$ is surjective. This means that the differential of ϕ at the identity element e of G is surjective. Then $V = \phi(G_0)$ is a neighbourhood of the identity element e' of H. We saw that

$$H_0 = \bigcup_{k=1}^{\infty} V^k$$

(Proposition 1.1.1). Since ϕ is a group morphism, $V = \phi(G_0)$ is a subgroup of H, and $V^k = V$, hence $H_0 = \phi(G_0)$.

(c) Assume that G and H are connected and that $d\phi$ is an isomorphism. Let us show that (G, ϕ) is a covering of H. From (ii) it follows that ϕ is surjective. By using Theorem 3.3.1, and the relation

$$\phi(\exp X) = \exp(d\phi(X)) \quad (X \in \mathfrak{g}),$$

one can show that there is a neighbourhood $V \subset G$ of the identity element of G, and a neighbourhood $W \subset H$ of the identity element of H such that ϕ is an isomorphism from V onto W. It follows that, for every $g \in G$, ϕ is a homeomorphism of the neighbourhood gV of g onto the neighbourhood hW of $h = \phi(g)$ since

$$\phi(gv) = h\phi(v) \quad (v \in V).$$

If $\ker \phi$ is a finite group, then (G, ϕ) is a covering of order k of H, where k is the number of elements in $\ker \phi$. □

Examples. Let $\mathcal{V}$ be the vector space of 2×2 Hermitian matrices with zero trace. Such a matrix can be written

$$x = \begin{pmatrix} x_1 & x_2 + ix_3 \\ x_2 - ix_3 & -x_1 \end{pmatrix} \quad (x_1, x_2, x_3 \in \mathbb{R}).$$

Then $\mathcal{V} \simeq \mathbb{R}^3$. For $g \in G = SU(2)$ the transformation

$$x \mapsto gxg^{-1} = gxg^*,$$

is a linear map $\pi(g)$ from $\mathcal{V}$ onto $\mathcal{V}$. From the relation

$$\det x = -x_1^2 - x_2^2 - x_3^2,$$

it follows that the transformation $\pi(g)$ is orthogonal. Then one gets a morphism

ϕ from $SU(2)$ into $O(3)$. For $T \in \mathfrak{su}(2)$,

$$d\pi(T)x = Tx - xT.$$

If

$$T = \begin{pmatrix} iu & v+iw \\ -v+iw & -iv \end{pmatrix},$$

one can establish easily that the matrix of $d\pi(T)$ is

$$\begin{pmatrix} 0 & 2v & 2w \\ -2v & 0 & -2u \\ -2w & 2u & 0 \end{pmatrix}.$$

Therefore $d\phi$ is a bijection from $\mathfrak{su}(2)$ onto $Lie\big(O(3)\big) = Asym(3, \mathbb{R})$. The group $SU(2)$ is connected. It follows that the group $\phi(G)$ is the identity component of $O(3)$, that is $SO(3)$. The kernel of ϕ is discrete. In fact one can check that $\ker\phi = \{I, -I\}$. This establishes that

$$SO(3) \simeq SU(2)/\{\pm I\},$$

and that $(SU(2), \phi)$ is a covering of order two of $SO(3)$.

4.2 Nilpotent and solvable Lie algebras

Let us recall some definitions and notation in group theory. Let G be a group. If $\{e\}$ and G are the only normal subgroups, G is said to be *simple*. If G is commutative, every subgroup is normal. The *commutator* of two elements x and y of G is defined as

$$[x, y] = x^{-1}y^{-1}xy.$$

The *derived group* $D(G)$ is the subgroup of G which is generated by the commutators. If H is a normal subgroup, then G/H is a group. It is commutative if and only if H contains the derived group $D(G)$.

One defines the successive derived groups: $D_0(G) = G$ and $D_{i+1}(G) = D(D_i(G))$. The group G is said to be *solvable* if there exists an integer $n \geq 0$ such that $D_n(G) = \{e\}$. (The terminology comes from the fact that, in Galois theory, such groups make it possible to characterise polynomial equations which are solvable by radicals.)

Let $\mathfrak{g}$ be a finite dimensional Lie algebra over $\mathbb{K} = \mathbb{R}$ or $\mathbb{C}$. If A and B are two vector subspaces of $\mathfrak{g}$, then $[A, B]$ denotes the vector subspace of $\mathfrak{g}$ generated by the brackets $[X, Y]$ with $X \in A$ and $Y \in B$. One puts

$$\mathcal{D}(\mathfrak{g}) = [\mathfrak{g}, \mathfrak{g}].$$

This is an ideal of $\mathfrak{g}$ which is called the *derived ideal*. The *descending central series* $C^k(\mathfrak{g})$ is defined recursively by:

$$C^1(\mathfrak{g}) = \mathfrak{g}, \quad C^k(\mathfrak{g}) = [C^{k-1}(\mathfrak{g}), \mathfrak{g}].$$

It is also denoted by $C^k(\mathfrak{g}) = \mathfrak{g}^k$. Observe that $C^2(\mathfrak{g}) = \mathcal{D}(\mathfrak{g})$. The *derived series* is defined by

$$\mathcal{D}^1(\mathfrak{g}) = \mathcal{D}(\mathfrak{g}), \quad \mathcal{D}^k(\mathfrak{g}) = \mathcal{D}\big(\mathcal{D}^{k-1}(\mathfrak{g})\big) = \big[\mathcal{D}^{k-1}(\mathfrak{g}), \mathcal{D}^{k-1}(\mathfrak{g})\big].$$

It is also denoted by $\mathcal{D}^k(\mathfrak{g}) = \mathfrak{g}^{(k)}$.

The subspaces $C^k(\mathfrak{g})$ and $\mathcal{D}^k(\mathfrak{g})$ $(k = 1, 2, \ldots)$ are ideals. The sequence $C^k(\mathfrak{g})$ is decreasing, hence constant for k large enough. The Lie algebra $\mathfrak{g}$ is said to be *nilpotent* if there exists an integer $n \geq 1$ such that $C^n(\mathfrak{g}) = \{0\}$. Similarly the sequence $\mathcal{D}^k(\mathfrak{g})$ is decreasing, hence constant for k large enough. The Lie algebra $\mathfrak{g}$ is said to be *solvable* if there exists $n \geq 1$ such that $\mathcal{D}^n(\mathfrak{g}) = \{0\}$. Observe that a nilpotent Lie algebra is solvable. A subalgebra of a nilpotent Lie algebra is nilpotent. A subalgebra of a solvable Lie algebra is solvable. Let X be an element in a nilpotent Lie algebra $\mathfrak{g}$, then ad X is a nilpotent endomorphism. (Recall that an endomorphism T is said to be nilpotent if there exists an integer $k \geq 1$ such that $T^k = 0$.) In fact ad X maps $C^k(\mathfrak{g})$ into $C^{k+1}(\mathfrak{g})$ and, if $C^n(\mathfrak{g}) = \{0\}$, then $(\text{ad } X)^{n-1} = 0$.

Examples. (1) Let G be the group '$ax + b$', that is the group of affine transformations of $\mathbb{R}$. The Lie algebra $\mathfrak{g} = Lie(G)$ has dimension 2. It has a basis $\{X_1, X_2\}$ satisfying

$$[X_1, X_2] = X_1.$$

Hence $\mathcal{D}(\mathfrak{g}) = \mathbb{R}X_1, C^3(\mathfrak{g}) = C^2(\mathfrak{g}) = \mathbb{R}X_1, \mathcal{D}^2(\mathfrak{g}) = \{0\}$. Therefore $\mathfrak{g}$ is solvable, but not nilpotent.

(2) The *Heisenberg Lie algebra* $\mathfrak{g}$ of dimension 3 has a basis $\{X_1, X_2, X_3\}$ satisfying

$$[X_1, X_2] = X_3, \quad [X_1, X_3] = 0, \quad [X_2, X_3] = 0.$$

Hence $C^2(\mathfrak{g}) = \mathbb{K}X_3$, which is the centre of $\mathfrak{g}$, and $C^3(\mathfrak{g}) = \{0\}$. Therefore $\mathfrak{g}$ is nilpotent.

(3) Let $\mathfrak{g} = \mathfrak{sl}(2, \mathbb{K})$ be the Lie algebra of the group $SL(2, \mathbb{K})$. It has a basis $\{X_1, X_2, X_3\}$ satisfying

$$[X_1, X_2] = 2X_2, \quad [X_1, X_3] = -2X_3, \quad [X_2, X_3] = X_1.$$

Hence $\mathcal{D}(\mathfrak{g}) = \mathfrak{g}$. Therefore $\mathfrak{g}$ is neither nilpotent, nor solvable.

(4) Let $T_0(n, \mathbb{K})$ be the group of upper triangular matrices with diagonal entries equal to one. Its Lie algebra $\mathfrak{g} = \mathfrak{t}_0(n, \mathbb{K})$ consists of the upper triangular matrices with zero diagonal entries

$$\mathfrak{t}_0(n, \mathbb{K}) = \{x \in M(n, \mathbb{K}) \mid x_{ij} = 0 \text{ if } i \geq j\}.$$

For $1 \leq k \leq n - 1$,

$$C^k(\mathfrak{g}) = \{x \in \mathfrak{g} \mid x_{ij} = 0 \text{ if } i \geq j - k + 1\}.$$

In particular $C^n(\mathfrak{g}) = \{0\}$, and $\mathfrak{g}$ is nilpotent. This is the basic example of a nilpotent Lie algebra.

(5) Let $T(n, \mathbb{K})$ be the group of upper triangular matrices with non-zero diagonal entries. Its Lie algebra $\mathfrak{g} = \mathfrak{t}(n, \mathbb{K})$ consists of the upper triangular matrices,

$$\mathfrak{t}(n, \mathbb{K}) = \{x \in M(n, \mathbb{K}) \mid x_{ij} = 0 \text{ if } i > j\}.$$

We have

$$C^2(\mathfrak{g}) = C^3(\mathfrak{g}) = \cdots = \mathfrak{t}_0(n, \mathbb{K}),$$
$$\mathcal{D}^k(\mathfrak{g}) = \{x \in M(n, \mathbb{K}) \mid x_{ij} = 0 \text{ if } i > j - 2^{k-1}\}.$$

Hence $\mathcal{D}^k(\mathfrak{g}) = \{0\}$ if $2^{k-1} \geq n - 1$. Therefore $\mathfrak{g}$ is solvable, but is not nilpotent. This is the basic example of a solvable Lie algebra.

Let $\mathfrak{g}$ be a Lie algebra and ρ a representation of $\mathfrak{g}$ on a finite dimensional vector space V. The representation ρ is said to be *nilpotent* if, for every X of $\mathfrak{g}$, the endomorphism $\rho(X)$ is nilpotent.

Lemma 4.2.1 *If X is a nilpotent endomorphism acting on a vector space V, then* $\mathrm{ad}\, X$ *is nilpotent.*

Proof. Let $k \geq 1$ be such that $X^k = 0$. We have

$$(\mathrm{ad}\, X)^N = (L_X - R_X)^N = \sum_{j=0}^N (-1)^{N-j} \binom{N}{j} L_{X^j} R_{X^{N-j}}.$$

Hence, if $N \geq 2k - 1$, then $(\mathrm{ad}\, X)^N = 0$. $\qquad\square$

Theorem 4.2.2 *Let ρ be a nilpotent representation of a Lie algebra $\mathfrak{g}$ on a vector space V. There exists a vector $v_0 \neq 0$ in V such that, for every $X \in \mathfrak{g}$,*

$$\rho(X)v_0 = 0.$$

Proof. Let $\ker(\rho)$ be the kernel of ρ. It is an ideal of $\mathfrak{g}$. It is enough to prove the statement for the representation $\dot{\rho}$ of the quotient algebra $\mathfrak{g}/\ker(\rho)$. This representation is faithful (i.e. injective). Hence we may assume that $\mathfrak{g}$ is a subalgebra of $\mathfrak{gl}(V)$. We have to show the following statement: if $\mathfrak{g}$ is a Lie subalgebra of $\mathfrak{gl}(V)$ made of nilpotent endomorphisms, then there exists $v_0 \neq 0$ in V such that, for every $X \in \mathfrak{g}$, $X v_0 = 0$.

The statement will be proved recursively with respect to the dimension of $\mathfrak{g}$. If $\dim \mathfrak{g} = 1$, then $\mathfrak{g} = \mathbb{K}X$, and X is nilpotent. Hence there exists $v_0 \neq 0$ in V such that $X v_0 = 0$. Assume that the property holds for every Lie algebra with dimension $\leq n - 1$.

(a) Let $\mathfrak{g}$ be a subalgebra of dimension n of $\mathfrak{gl}(V)$ made of nilpotent endomorphisms, and let $\mathfrak{h}$ be a proper subalgebra of $\mathfrak{g}$ with maximal dimension. We will show that $\mathfrak{h}$ is an ideal of dimension $n - 1$. Let us consider the representation α of $\mathfrak{h}$ on $W = \mathfrak{g}/\mathfrak{h}$ defined by

$$\alpha(X) : Y + \mathfrak{h} \mapsto [X, Y] + \mathfrak{h}.$$

By Lemma 4.2.1 it follows that the representation α is nilpotent. By the recursion assumption it follows that there exists $w_0 \neq 0$ in W such that, for every X in $\mathfrak{h}$,

$$\alpha(X) w_0 = 0.$$

Let $X_0 \in \mathfrak{g}$ be a representative of w_0. Then X_0 does not belong to $\mathfrak{h}$ and $[X_0, \mathfrak{h}] \subset \mathfrak{h}$. Hence $\mathbb{K}X_0 + \mathfrak{h}$ is a subalgebra of $\mathfrak{g}$ whose dimension is greater than that of $\mathfrak{h}$, therefore $\mathfrak{g} = \mathbb{K}X_0 + \mathfrak{h}$, and $\dim \mathfrak{h} = n - 1$. Furthermore $[\mathfrak{g}, \mathfrak{h}] \subset \mathfrak{h}$, and this means that $\mathfrak{h}$ is an ideal.

(b) Let us use for a second time the recursion assumption: there exists $v_1 \neq 0$ in V such that, for every X in $\mathfrak{h}$,

$$X v_1 = 0.$$

Put

$$V_0 = \{v \in V \mid \forall X \in \mathfrak{h}, \ X v = 0\}.$$

Since $v_1 \in V_0$, $V_0 \neq \{0\}$. The subspace V_0 is invariant under $\mathfrak{g}$. In fact let $X \in \mathfrak{g}$, $v \in V_0$, and show that $X v \in V_0$. For $Y \in \mathfrak{h}$,

$$Y X v = X Y v - [X, Y] v = 0.$$

In particular, $X_0 V_0 \subset V_0$. Since X_0 is nilpotent there exists in V_0 a vector $v_0 \neq 0$ such that $X_0 v_0 = 0$, and then, for every X in $\mathfrak{g}$,

$$X v_0 = 0. \qquad \square$$

Theorem 4.2.3 *Let ρ be a nilpotent representation of a Lie algebra $\mathfrak{g}$ on a vector space V. There exists a basis of V such that, for every X in $\mathfrak{g}$, the matrix of $\rho(X)$ is upper triangular with zero diagonal entries.*

Proof. Let us prove the statement recursively with respect to the dimension of V. By Theorem 4.2.2 there exists a vector v_1 such that, for every $X \in \mathfrak{g}$,

$$\rho(X)v_1 = 0.$$

From the recursion assumption applied to the quotient $W = V/\mathbb{K}v_1$ we get the result. □

Corollary 4.2.4 (Engel's Theorem) *A Lie algebra is nilpotent if and only if, for every $X \in \mathfrak{g}$, ad X is a nilpotent endomorphism*

Proof. (a) Assume that the Lie algebra $\mathfrak{g}$ is nilpotent: there exists an integer n such that $C^n(\mathfrak{g}) = \{0\}$. For every X in $\mathfrak{g}$, ad X maps $C^k(\mathfrak{g})$ into $C^{k+1}(\mathfrak{g})$, hence $(\text{ad } X)^{n-1} = 0$.

(b) Assume that, for every X in $\mathfrak{g}$, ad X is nilpotent. By Theorem 4.2.3 the Lie algebra ad $\mathfrak{g}$ is isomorphic to a subalgebra of $t_0(N, \mathbb{K})$, hence ad $\mathfrak{g}$ is nilpotent. There exists an integer n such that $C^n(\text{ad } \mathfrak{g}) = \{0\}$, hence $C^n(\mathfrak{g}) \subset Z(\mathfrak{g})$, the centre of $\mathfrak{g}$. Therefore $C^{n+1}(\mathfrak{g}) = \{0\}$. □

Let $\mathfrak{J}$ be an ideal of $\mathfrak{g}$. If $\mathfrak{g}$ is solvable, then $\mathfrak{g}/\mathfrak{J}$ is solvable too. In fact,

$$\mathcal{D}^k(\mathfrak{g}/\mathfrak{J}) \simeq \mathcal{D}^k(\mathfrak{g})/\big(\mathfrak{J} \cap \mathcal{D}^k(\mathfrak{g})\big).$$

Proposition 4.2.5 *If $\mathfrak{J}$ and $\mathfrak{g}/\mathfrak{J}$ are solvable, then $\mathfrak{g}$ is solvable.*

Proof. There is an integer m such that $\mathcal{D}^m(\mathfrak{g}/\mathfrak{J}) = \{0\}$, hence $\mathcal{D}^m(\mathfrak{g}) \subset \mathfrak{J}$. There exists n such that $\mathcal{D}^n(\mathfrak{J}) = \{0\}$. Therefore $\mathcal{D}^{m+n}(\mathfrak{g}) = \{0\}$. □

Proposition 4.2.6 *If $\mathfrak{J}_1$ and $\mathfrak{J}_2$ are two solvable ideals then the ideal $\mathfrak{J}_1 + \mathfrak{J}_2$ is also solvable.*

Proof. The Lie algebra $(\mathfrak{J}_1 + \mathfrak{J}_2)/\mathfrak{J}_2$ is isomorphic to $\mathfrak{J}_1/\mathfrak{J}_1 \cap \mathfrak{J}_2$. This follows from the preceding proposition. □

Hence, if $\mathfrak{g}$ is finite dimensional, there exists a largest solvable ideal: the sum of all solvable ideals. It is called the *radical* of $\mathfrak{g}$, and is denoted by $rad(\mathfrak{g})$.

Theorem 4.2.7 (Lie's Theorem) *Let $\mathfrak{g}$ be a solvable Lie algebra over $\mathbb{C}$, and let ρ be a representation of $\mathfrak{g}$ on a finite dimensional complex vector space V. There exists a vector $v_0 \neq 0$ in V, and a linear form λ on $\mathfrak{g}$ such that, for every X in $\mathfrak{g}$,*

$$\rho(X)v_0 = \lambda(X)v_0.$$

Proof. We will prove the statement recursively with respect to the dimension of $\mathfrak{g}$. If $\dim \mathfrak{g} = 1$, then $\mathfrak{g} = \mathbb{C}X_0$, and $\rho(X_0)$ has an eigenvector.

Assume that the property holds for every solvable Lie algebra of dimension $\leq n - 1$. Let $\mathfrak{g}$ be a solvable Lie algebra of dimension n, and let $\mathfrak{h}$ be a subspace of $\mathfrak{g}$ of dimension $n - 1$ containing $\mathcal{D}(\mathfrak{g})$. Such a subspace exists since, because $\mathfrak{g}$ is solvable, $\mathcal{D}(\mathfrak{g}) \neq \mathfrak{g}$. The subspace $\mathfrak{h}$ is an ideal because

$$[\mathfrak{g}, \mathfrak{h}] \subset [\mathfrak{g}, \mathfrak{g}] = \mathcal{D}(\mathfrak{g}) \subset \mathfrak{h}.$$

By the recursion assumption there is a vector $w_0 \neq 0$ in V and a linear form λ on $\mathfrak{h}$ such that, for every Y in $\mathfrak{h}$,

$$\rho(Y)w_0 = \lambda(Y)w_0.$$

Let $X_0 \in \mathfrak{g} \setminus \mathfrak{h}$, and put

$$w_j = \rho(X_0)^j w_0, \quad j \geq 1.$$

Let k be the largest integer for which the vectors $w_0, \ldots, w_k$ are linearly independent, and let W_j be the subspace which is generated by $w_0, \ldots, w_j$ $(0 \leq j \leq k)$. Observe that $w_j \in W_k$ for $j \geq k$. Hence $\rho(X_0)$ maps W_k into W_k and, for $0 \leq j < k$, W_j into W_{j+1}. We will show that, for $Y \in \mathfrak{h}$, the restriction of $\rho(Y)$ to W_k is equal to $\lambda(Y)I$. In a first step we will show that the matrix of $\rho(Y)$ with respect to the basis $\{w_0, \ldots, w_k\}$ is upper triangular with diagonal entries equal to $\lambda(Y)$. Let us show recursively with respect to j $(0 \leq j \leq k)$ that

$$\rho(Y)w_j = \lambda(Y)w_j \mod W_{j-1}.$$

(One puts $W_{-1} = \{0\}$.) This holds clearly for $j = 0$. Assume that it holds for $j < k$. Then, for $Y \in \mathfrak{h}$,

$$\rho(Y)w_{j+1} = \rho(Y)\rho(X_0)w_j = \rho(X_0)\rho(Y)w_j + \rho([Y, X_0])w_j,$$

and, since $[Y, X_0] \in \mathfrak{h}$,

$$\rho([Y, X_0])w_j = \lambda([Y, X_0])w_j \mod W_{j-1}$$

by the recursion assumption. Hence

$$\rho(Y)w_{j+1} = \lambda(Y)w_{j+1} \mod W_j.$$

This shows that the subspace W_k is invariant under the representation ρ. For $Y \in \mathfrak{h}$,

$$\mathrm{Tr}\left(\rho([Y, X_0])\big|_{W_k}\right) = 0.$$

On the other hand, for $Z \in \mathfrak{h}$,

$$\mathrm{Tr}\left(\rho(Z)\big|_{W_k}\right) = (k+1)\lambda(Z).$$

Hence, if $Z = [Y, X_0]$, then $\lambda(Z) = 0$.

In a second step we will show that, for $Y \in \mathfrak{h}$, and $w \in W_k$, $\rho(Y)w = \lambda(Y)w$. Let us show recursively with respect to j $(0 \le j \le k)$ that, for $Y \in \mathfrak{h}$,

$$\rho(Y)w_j = \lambda(Y)w_j.$$

This holds for $j = 0$. Assume that $\rho(Y)w_j = \lambda(Y)w_j$. Then

$$\rho(Y)w_{j+1} = \rho(X_0)\rho(Y)w_j + \rho([Y, X_0])w_j$$
$$= \lambda(Y)\rho(X_0)w_j + \lambda([Y, X_0])w_j = \lambda(Y)w_{j+1}.$$

Let $v_0 \in W_k$ be an eigenvector of $\rho(X_0)$,

$$\rho(X_0)v_0 = \mu v_0,$$

and extend the linear form λ to $\mathfrak{g}$ by putting

$$\lambda(X_0) = \mu.$$

Then, for every X in $\mathfrak{g}$,

$$\rho(X)v_0 = \lambda(X)v_0. \qquad \square$$

Corollary 4.2.8 *Let $\mathfrak{g}$ be a solvable Lie algebra over $\mathbb{C}$, and ρ be a representation of $\mathfrak{g}$ on a finite dimensional complex vector space V. There exists a basis of V such that, for every X in $\mathfrak{g}$, the matrix of $\rho(X)$ is upper triangular. The diagonal entries can be written $\lambda_1(X), \dots, \lambda_m(X)$, where $\lambda_1, \dots, \lambda_m$ are linear forms on $\mathfrak{g}$.*

The statements of Theorem 4.2.7 and Corollary 4.2.8 do not hold if $\mathfrak{g}$ is a solvable Lie algebra over $\mathbb{R}$. (In fact one knows that, if A is an endomorphism of a finite dimensional real vector space, in general there is no basis with respect to which the matrix of A is upper triangular.) See Exercise 4.

4.3 Semi-simple Lie algebras

A Lie algebra is said to be *simple* if it has no non-trivial ideal, and if it is not commutative. In other words a Lie algebra is simple if its dimension is greater than 1 and if the adjoint representation ad is irreducible. If $\mathfrak{g}$ is simple then $[\mathfrak{g}, \mathfrak{g}] = \mathfrak{g}$, because $[\mathfrak{g}, \mathfrak{g}]$ is an ideal. The Lie algebra $\mathfrak{sl}(n, \mathbb{K})$ is simple

($n \geq 2$). Let us show that $\mathfrak{sl}(2, \mathbb{C})$ is simple. (For $n \geq 3$, see Exercise 5.) For that consider the following basis of $\mathfrak{sl}(2, \mathbb{C})$:

$$H = \begin{pmatrix} 1 & 0 \\ 0 & -1 \end{pmatrix}, \quad E = \begin{pmatrix} 0 & 1 \\ 0 & 0 \end{pmatrix}, \quad F = \begin{pmatrix} 0 & 0 \\ 1 & 0 \end{pmatrix}.$$

The commutation relations are:

$$[H, E] = 2E, \quad [H, F] = -2F, \quad [E, F] = H.$$

Let $\mathcal{I}$ be an ideal of $\mathfrak{sl}(2, \mathbb{C})$ which does not reduce to $\{0\}$. If one of the elements H, E, or F belongs to $\mathcal{I}$, then $\mathcal{I} = \mathfrak{sl}(2, \mathbb{C})$. The basis elements H, E, F are eigenvectors of ad H for the eigenvalues $0, 2, -2$, and $\mathcal{I}$ is invariant under ad H, hence one of the eigenvectors belongs to $\mathcal{I}$.

A Lie algebra $\mathfrak{g}$ is said to be *semi-simple* if the only commutative ideal is $\{0\}$. A simple Lie algebra is semi-simple. There is no semi-simple Lie agebra of dimension 1 or 2. But there exist semi-simple Lie algebras of dimension 3; in fact $\mathfrak{sl}(2, \mathbb{C})$, $\mathfrak{sl}(2, \mathbb{R})$ and $\mathfrak{su}(2)$ are semi-simple Lie algebras.

The centre of a semi-simple Lie algebra reduces to $\{0\}$. Hence, if $\mathfrak{g}$ is semi-simple, then the adjoint representation is faithful, ad$(\mathfrak{g}) \simeq \mathfrak{g}$.

A direct sum of semi-simple Lie algebras is semi-simple

Let ρ be a representation of a Lie algebra $\mathfrak{g}$ on a finite dimensional vector space V. For $X, Y \in \mathfrak{g}$ one puts

$$B_\rho(X, Y) = \mathrm{Tr}\big(\rho(X)\rho(Y)\big).$$

This is a symmetric bilinear form on $\mathfrak{g}$ which is associative:

$$B_\rho([X, Y], Z) = B_\rho(X, [Y, Z]).$$

This means that the transformations ad X are skewsymmetric with respect to the form B_ρ. The orthogonal of an ideal with respect to the form B_ρ is an ideal.

The *Killing form* is the symmetric bilinear form associated to the adjoint representation ($\rho = \mathrm{ad}$):

$$B(X, Y) = \mathrm{Tr}(\mathrm{ad}\, X\, \mathrm{ad}\, Y).$$

Examples. (1) Let $\mathfrak{g} = M(n, \mathbb{K})$.

$$B(X, Y) = 2n\, \mathrm{Tr}(XY) - 2\, \mathrm{Tr}\, X\, \mathrm{Tr}\, Y.$$

In order to establish this formula let us consider the canonical basis $\{E_{ij}\}$ of $\mathfrak{g} = M(n, \mathbb{K})$. If

$$X = \sum_{i,j=1}^n x_{ij} E_{ij} \in \mathfrak{g},$$

then

$$\text{ad}\, X \cdot E_{k\ell} = [X, E_{k,\ell}] = \sum_{i=1}^{n} (x_{ik} E_{i\ell} - x_{\ell i} E_{ki}) \quad (k, \ell = 1, \ldots, n).$$

Hence, if

$$Y = \sum_{i,j} y_{ij} E_{ij},$$

then

$$(\text{ad}\, X \circ \text{ad}\, Y) E_{k\ell} = \sum_{i,j}^{n} (x_{ik} y_{ji} E_{j\ell} + x_{\ell i} y_{ij} E_{kj})$$

$$- \sum_{i,j=1}^{n} (x_{ik} y_{\ell j} + x_{\ell j} y_{ik}) E_{ij}.$$

Therefore

$$\text{Tr}(\text{ad}\, X\, \text{ad}\, Y) = n \sum_{i,j}^{n} x_{ij} y_{ji} + x_{ji} y_{ij} - 2 \sum_{i,j}^{n} x_{ii} y_{jj}$$

$$= 2n\, \text{Tr}(XY) - 2\, \text{Tr}\, X\, \text{Tr}\, Y.$$

(2) Let $\mathfrak{g} = \mathfrak{sl}(n, \mathbb{K})$. If $n \geq 2$,

$$B(X, Y) = 2n\, \text{Tr}(XY).$$

(This follows from Proposition 4.3.1 below.)

(3) Let $\mathfrak{g} = \mathfrak{so}(n, \mathbb{K})$. If $n \geq 2$,

$$B(X, Y) = (n - 2)\, \text{Tr}(XY).$$

The proof is left as an exercise.

Proposition 4.3.1 *Let $\mathfrak{I}$ be an ideal in a Lie algebra $\mathfrak{g}$. The Killing form of the Lie algebra $\mathfrak{I}$ is the restriction to $\mathfrak{I}$ of the Killing form of $\mathfrak{g}$.*

Proof. Let $X, Y \in \mathfrak{I}$. The endomorphisms $S = \text{ad}\, X$, $T = \text{ad}\, Y$ map $\mathfrak{g}$ into $\mathfrak{I}$. Let us consider a basis of $\mathfrak{g}$ obtained by completing a basis of $\mathfrak{I}$. With respect to this basis the matrices of S and T have the following shape

$$Mat(S) = \begin{pmatrix} S_1 & * \\ 0 & 0 \end{pmatrix}, \quad Mat(T) = \begin{pmatrix} T_1 & * \\ 0 & 0 \end{pmatrix},$$

and

$$Mat(ST) = \begin{pmatrix} S_1 T_1 & * \\ 0 & 0 \end{pmatrix}.$$

Therefore $\text{Tr}(ST) = \text{Tr}(S_1 T_1)$. $\qquad \square$

We will see that a Lie algebra is semi-simple if and only if the Killing form is non-degenerate. To prove this we will use Cartan's criterion for solvable Lie algebras.

We will need the properties of the decomposition of an endomorphism into semi-simple and nilpotent parts. Let V be a finite dimensional vector space over $\mathbb{C}$. Recall that an endomorphism T of V decomposes as

$$T = T_s + T_n,$$

where T_s is semi-simple (i.e. diagonalisable), and T_n is nilpotent, in such a way that T_s and T_n are polynomials in T. The endomorphisms T_s and T_n commute. This decomposition is unique in the following sense: if

$$T = D + N,$$

with D semi-simple, N nilpotent, and $DN = ND$, then $D = T_s$, $N = T_n$. T_s is called the *semi-simple* part of T, and T_n the *nilpotent* part

We have

$$\operatorname{ad} T = \operatorname{ad} T_s + \operatorname{ad} T_n,$$

$\operatorname{ad} T_s$ is semi-simple, $\operatorname{ad} T_n$ is nilpotent (Lemma 4.2.1). In order to show that $\operatorname{ad} T_s$ and $\operatorname{ad} T_n$ are the semi-simple and nilpotent parts of $\operatorname{ad} T$ it is enough to show that $\operatorname{ad} T_s$ and $\operatorname{ad} T_n$ commute. But

$$[\operatorname{ad}(T_s), \operatorname{ad}(T_n)] = \operatorname{ad}[T_s, T_n] = 0.$$

It follows that $\operatorname{ad} T_s$ and $\operatorname{ad} T_n$ are polynomials in $\operatorname{ad} T$.

Theorem 4.3.2 (Cartan's criterion) *Let $\mathfrak{g}$ be a Lie subalgebra of $M(m, \mathbb{C})$. Assume that $\operatorname{Tr}(XY) = 0$ for every $X, Y \in \mathfrak{g}$. Then $\mathfrak{g}$ is solvable.*

Proof. We will show that every $X \in [\mathfrak{g}, \mathfrak{g}]$ is a nilpotent endomorphism.

(a) Let $X = X_s + X_n$ be the decomposition of $X \in \mathfrak{g}$ into semi-simple and nilpotent parts. We may assume that

$$X_s = \begin{pmatrix} \lambda_1 & & \\ & \ddots & \\ & & \lambda_m \end{pmatrix},$$

the numbers λ_j being the eigenvalues of X. Let p be a polynomial in one variable, and put

$$U = p(X_s) = \begin{pmatrix} p(\lambda_1) & & \\ & \ddots & \\ & & p(\lambda_m) \end{pmatrix}.$$

Since X_s is a polynomial in X, one can write $U = p_0(X)$, and since X_n is also a polynomial in X, U and X_n commute. Then

$$(UX_n)^k = U^k X_n^k,$$

hence UX_n is nilpotent, therefore $\mathrm{Tr}(UX_n) = 0$, or

$$\mathrm{Tr}(UX_s) = \mathrm{Tr}(UX).$$

(b) The eigenvalues of $\mathrm{ad}\, X_s$ are the numbers $\lambda_i - \lambda_j$, and the corresponding eigenvectors are the matrices E_{ij},

$$\mathrm{ad}\, X_s E_{ij} = (\lambda_i - \lambda_j)E_{ij}.$$

Let us now choose a polynomial p in one variable with complex coefficients such that

$$p(\lambda_i) = \overline{\lambda_i} \quad (i = 1, \ldots, m).$$

Hence, if $\lambda_i - \lambda_j = \lambda_k - \lambda_\ell$, then

$$p(\lambda_i) - p(\lambda_j) = p(\lambda_k) - p(\lambda_\ell).$$

Therefore there exists a polynomial P such that, if $U = p(X_s)$, then $\mathrm{ad}\, U = P(\mathrm{ad}\, X_s)$ and, since $\mathrm{ad}\, X_s$ is a polynomial in $\mathrm{ad}\, X$, there exists a polynomial P_0 such that $\mathrm{ad}\, U = P_0(\mathrm{ad}\, X)$. Therefore $\mathrm{ad}\, U(\mathfrak{g}) \subset \mathfrak{g}$.

(c) Let us now take $X \in [\mathfrak{g}, \mathfrak{g}]$, and show that

$$\mathrm{Tr}(UX) = 0.$$

We can write

$$X = \sum_{j=1}^{N} [Y_j, Z_j], \quad Y_j, Z_j \in \mathfrak{g},$$

and then

$$\mathrm{Tr}(UX) = \sum_{j=1}^{N} \mathrm{Tr}(U[Y_j, Z_j]) = \sum_{j=1}^{N} \mathrm{Tr}([U, Y_j]Z_j) = 0,$$

by assumption, since $[U, Y_j] = \mathrm{ad}\, U\, Y_j \in \mathfrak{g}$. But, by (a)

$$\mathrm{Tr}(UX) = \mathrm{Tr}(UX_s),$$

hence

$$\mathrm{Tr}(UX) = \sum_{j=1}^{m} p(\lambda_j)\lambda_j = \sum_{j=1}^{m} |\lambda_j|^2.$$

Therefore the eigenvalues λ_j of X vanish, and X is nilpotent. By Engel's Theorem (Corollary 4.2.4) it follows that $[\mathfrak{g}, \mathfrak{g}]$ is nilpotent, hence $\mathfrak{g}$ is solvable $\qquad\qquad\qquad\square$

Corollary 4.3.3 *If the Killing form of $\mathfrak{g}$ vanishes identically, then $\mathfrak{g}$ is solvable.*

Theorem 4.3.4 *Let $\mathfrak{g}$ be a Lie algebra. The following properties are equivalent:*

(i) $\mathfrak{g}$ *is semi-simple,*
(ii) *the radical of $\mathfrak{g}$ reduces to $\{0\}$,*
(iii) *the Killing form of $\mathfrak{g}$ is non-degenerate.*

Proof. (i) $\Rightarrow$ (ii). Assume that there exists a solvable ideal $\mathfrak{I} \neq \{0\}$ in $\mathfrak{g}$. Let $\mathcal{D}^{k-1}(\mathfrak{I})$ be the last non-zero derived ideal of $\mathfrak{I}$. Then $\mathcal{D}^{k-1}(\mathfrak{I})$ is a non-zero commutative ideal in $\mathfrak{g}$, and this contradicts (i).

(ii) $\Rightarrow$ (iii). Put $\mathfrak{I} = \mathfrak{g}^\perp$,

$$\mathfrak{I} = \{X \in \mathfrak{g} \mid \forall Y \in \mathfrak{g}, \ B(X, Y) = 0\}.$$

This is an ideal and the restriction of B to $\mathfrak{I}$ vanishes identically. By Corollary 4.3.3, $\mathfrak{I}$ is a solvable ideal, and $\mathfrak{I} = \{0\}$ by (ii).

(iii) $\Rightarrow$ (i). Let $\mathfrak{I}$ be a commutative ideal in $\mathfrak{g}$. For $X \in \mathfrak{I}$, $Y \in \mathfrak{g}$, the endomorphism $\mathrm{ad}\, X\, \mathrm{ad}\, Y$ maps $\mathfrak{g}$ into $\mathfrak{I}$, and $(\mathrm{ad}\, X\, \mathrm{ad}\, Y)^2$ maps $\mathfrak{g}$ into $[\mathfrak{I}, \mathfrak{I}] = \{0\}$, hence $\mathrm{ad}\, X\, \mathrm{ad}\, Y$ is nilpotent. Therefore

$$B(X, Y) = \mathrm{Tr}(\mathrm{ad}\, X\, \mathrm{ad}\, Y) = 0.$$

Since B is non-degenerate it follows that $\mathfrak{I} = \{0\}$. $\qquad\qquad\square$

Proposition 4.3.5 *A semi-simple Lie algebra $\mathfrak{g}$ is a direct sum of simple subalgebras. Furthermore,*

$$[\mathfrak{g}, \mathfrak{g}] = \mathfrak{g}.$$

Proof. Let $\mathfrak{I}$ be an ideal of $\mathfrak{g}$, and let $\mathfrak{I}^\perp$ be its orthogonal complement with respect to the Killing form,

$$\mathfrak{I}^\perp = \{X \in \mathfrak{g} \mid \forall Y \in \mathfrak{I}, \ B(X, Y) = 0\}.$$

Since the Killing form is associative it follows that $\mathfrak{I}^\perp$ is an ideal and, by Corollary 4.3.3, that the ideal $\mathfrak{I} \cap \mathfrak{I}^\perp$ is solvable, hence reduces to $\{0\}$ since the radical of $\mathfrak{g}$ reduces to $\{0\}$ (Theorem 4.3.4). Therefore

$$\mathfrak{g} = \mathfrak{I} \oplus \mathfrak{I}^\perp.$$

To get the stated decomposition one starts from a minimal non-zero ideal $\mathfrak{I}_1$ in $\mathfrak{g}$, which is necessarily simple, then one obtains recursively a decomposition

$$\mathfrak{g} = \mathfrak{I}_1 \oplus \mathfrak{I}_2 \oplus \cdots \oplus \mathfrak{I}_m,$$

where $\mathfrak{I}_1, \ldots, \mathfrak{I}_m$ are simple ideals. It follows furthermore that

$$[\mathfrak{g}, \mathfrak{g}] = \bigoplus_{i=1}^{m} [\mathfrak{I}_i, \mathfrak{I}_i] = \bigoplus_{i=1}^{m} \mathfrak{I}_i = \mathfrak{g}. \qquad \square$$

From this theorem it follows that, if $\mathfrak{I}$ is a solvable ideal in $\mathfrak{g}$, then $\mathfrak{I} = rad(\mathfrak{g})$ if and only if $\mathfrak{g}/\mathfrak{I}$ is semi-simple.

Finally let us state without proof the *theorem of Levi–Malcev*. Let $\mathfrak{g}$ be a Lie algebra. A *Levi subalgebra* of $\mathfrak{g}$ is a Lie subalgebra which is a complement to $rad(\mathfrak{g})$. It is a semi-simple algebra since it is isomorphic to $\mathfrak{g}/rad(\mathfrak{g})$. The theorem of Levi–Malcev says that, in every Lie algebra $\mathfrak{g}$, there is a Levi subalgebra $\mathfrak{s}$. Therefore every Lie algebra decomposes as

$$\mathfrak{g} = \mathfrak{s} + rad(\mathfrak{g}),$$

the sum of a semi-simple Lie algebra, and a solvable Lie algebra. This is the so-called *Levi decomposition*.

Examples. Let $\mathfrak{g}$ be the Lie subalgebra of $M(n+1, \mathbb{R})$ consisting of the matrices

$$\begin{pmatrix} x & y \\ 0 & 0 \end{pmatrix} \quad (x \in \mathfrak{so}(n), y \in \mathbb{R}^n)$$

($\mathfrak{g}$ is isomorphic to the Lie algebra of the motion group of $\mathbb{R}^n$). Let $\mathfrak{I}$ be the ideal of $\mathfrak{g}$ consisting of the matrices

$$\begin{pmatrix} 0 & y \\ 0 & 0 \end{pmatrix} \quad (y \in \mathbb{R}^n).$$

It is Abelian, hence solvable. Let $\mathfrak{s}$ be the subalgebra of $\mathfrak{g}$ consisting of the matrices

$$\begin{pmatrix} x & 0 \\ 0 & 0 \end{pmatrix} \quad (x \in \mathfrak{so}(n)).$$

If $n \geq 3$, the subalgebra $\mathfrak{s}$, which is isomorphic to $\mathfrak{so}(n)$, is semi-simple (even simple for $n \neq 4$). Therefore, since $\mathfrak{g}/\mathfrak{I} \simeq \mathfrak{s}$, $\mathfrak{I}$ is the radical of $\mathfrak{g}$, and

$$\mathfrak{g} = \mathfrak{s} + \mathfrak{I}$$

is a Levi decomposition of $\mathfrak{g}$.

4.4 Exercises

1. Let $G \subset GL(n, \mathbb{R})$ be a connected linear Lie group, and $\mathfrak{g} = Lie(G) \subset M(n, \mathbb{R})$. Assume that $[\mathfrak{g}, \mathfrak{g}] = \mathfrak{g}$. Show that $G \subset SL(n, \mathbb{R})$.
2. Let G be a linear Lie group and H a closed subgroup. Assume that G and H are connected, and that $\mathfrak{h} = Lie(H)$ is an ideal in $\mathfrak{g} = Lie(G)$. Show that H is a normal subgroup of G.
3. Show that the vector fields on $\mathbb{R}$ of the form

$$Xf(t) = p(t)\frac{df}{dt}(t),$$

where p is polynomial of degree ≤ 2 with real coefficients,

$$p(t) = at^2 + bt + c,$$

form a Lie algebra isomorphic to $\mathfrak{sl}(2, \mathbb{R})$.
4. Let $\mathfrak{g}$ be a solvable Lie algebra over $\mathbb{R}$, and ρ a representation of $\mathfrak{g}$ on a finite dimensional real vector space V. Show that there is a sequence of vector subspaces V_j in V such that:
   (a) $\{0\} = V_0 \subset V_1 \subset \cdots \subset V_k = V$;
   (b) for $1 \leq j \leq k$ the representation ρ_j which is induced by ρ on the quotient space V_j/V_{j-1} is irreducible;
   (c) $\dim V_j/V_{j-1} \leq 2$.
   Deduce that there is a sequence $\mathfrak{g}_j$ of ideals in $\mathfrak{g}$ such that

$$\{0\} = \mathfrak{g}_0 \subset \mathfrak{g}_1 \subset \cdots \subset \mathfrak{g}_k = \mathfrak{g},$$
$$\dim(\mathfrak{g}_j/\mathfrak{g}_{j-1}) \leq 2.$$

5. Show that $\mathfrak{sl}(n, \mathbb{C})$ is a simple Lie algebra.
   Hint. Let $\mathcal{I}$ be an ideal in $\mathfrak{g} = \mathfrak{sl}(n, \mathbb{C})$ which does not reduce to $\{0\}$. One can show the following.
   (a) If one of the matrices E_{ij} ($i \neq j$) belongs to $\mathcal{I}$, then $\mathcal{I} = \mathfrak{g}$.
   (b) Assume that $\mathcal{I}$ contains a diagonal matrix H,

$$H = \sum_{i=1}^{n} a_i E_{ii}.$$

   There are $i \neq j$ such that $a_i - a_j \neq 0$. Show that E_{ij} belongs to $\mathcal{I}$.
   (c) Let H be a diagonal matrix,

$$H = \sum_{i=1}^{n} a_i E_{ii},$$

such that, if $(i, j) \neq (k, \ell)$, then $a_i - a_j \neq a_k - a_\ell$. Show that one of the eigenvectors of ad H belongs to $\mathcal{I}$, and then that the situation is as in (a) or as in (b).

6. Show that the Killing form B of $\mathfrak{g} = \mathfrak{so}(n, \mathbb{R})$ is equal to

$$B(X, Y) = (n - 2)Tr(XY),$$

and that $\mathfrak{g}$ is semi-simple if $n \geq 3$.

7. One says that a Lie algebra $\mathfrak{g}$ is *reductive* if every Abelian ideal is included in the centre $\mathfrak{z}$ of $\mathfrak{g}$, and if $\mathfrak{z} \cap \mathcal{D}(\mathfrak{g}) = \{0\}$.
   (a) Show that a Lie algebra $\mathfrak{g}$ is reductive if and only if it is the direct sum of its centre and a semi-simple ideal $\mathfrak{m}$:

$$\mathfrak{g} = \mathfrak{z} \oplus \mathfrak{m}.$$

   (b) Show that in fact $\mathfrak{m} = \mathcal{D}(\mathfrak{g})$.
   (c) Deduce that the radical $\mathfrak{r}$ of a reductive Lie algebra is equal to its centre $\mathfrak{z}$.

8. Show that a reductive Lie algebra is solvable if and only if it is Abelian.

9. Let $\mathfrak{g} = \mathfrak{gl}(n, \mathbb{R})$.
   (a) Show that $\mathcal{D}(\mathfrak{g}) = \mathfrak{sl}(n, \mathbb{R})$.
   (b) Show that $\mathfrak{g}$ is not semi-simple but reductive.

10. The aim of this exercise is to determine the Lie algebras with dimension ≤ 3 over $\mathbb{K}$ ($\mathbb{K} = \mathbb{R}$ or $\mathbb{C}$).
    (a) dim $\mathfrak{g} = 2$. Show that dim $\mathcal{D}(\mathfrak{g}) \leq 1$. If dim $\mathcal{D}(\mathfrak{g}) = 1$, show that there are two elements X and Y of $\mathfrak{g}$ such that

$$[X, Y] = Y.$$

    Show that $\mathfrak{g}$ is isomorphic to the Lie algebra consisting of the matrices

$$\begin{pmatrix} x & y \\ 0 & 0 \end{pmatrix} \quad (x, y \in \mathbb{K}).$$

    (b) dim $\mathfrak{g} = 3$.
       (i) Assume that dim $\mathcal{D}(\mathfrak{g}) = 1$. Let Z be a non-zero element in $\mathcal{D}(\mathfrak{g})$. One can write

$$[X, Y] = b(X, Y)Z \quad (X, Y \in \mathfrak{g}),$$

       where b is a skewsymmetric bilinear form on $\mathfrak{g}$.
          If $Z \in rad(b)$, show that there is a basis $\{X, Y, Z\}$ of $\mathfrak{g}$ such that

$$[X, Y] = Z, \quad [X, Z] = 0, \quad [Y, Z] = 0.$$

Show that $\mathfrak{g}$ is isomorphic to the Lie algebra which consists of the matrices

$$\begin{pmatrix} 0 & x & z \\ 0 & 0 & y \\ 0 & 0 & 0 \end{pmatrix} \quad (x, y, z \in \mathbb{K}).$$

(This Lie algebra is called the *Heisenberg Lie algebra*.)

Otherwise show that there is a basis $\{X, Y, Z\}$ of $\mathfrak{g}$ such that

$$[X, Y] = 0, \quad [Y, Z] = 0, \quad [X, Z] = Z.$$

Show then that $\mathfrak{g}$ is isomorphic to the Lie algebra consisting of the matrices

$$\begin{pmatrix} x & z & 0 \\ 0 & 0 & 0 \\ 0 & 0 & y \end{pmatrix} \quad (x, y, z \in \mathbb{K}).$$

(ii) Assume that $\dim \mathcal{D}(\mathfrak{g}) = 2$. Let $\{X, Y, Z\}$ be a basis of $\mathfrak{g}$ such that $X, Y \in \mathcal{D}(\mathfrak{g})$. Note that $\mathrm{ad}\, Z$ is a derivation of $\mathcal{D}(\mathfrak{g})$. Using this fact and (a) show that $[X, Y] = 0$. Show that there are $\alpha, \beta, \gamma, \delta$ such that

$$[Z, X] = \alpha X + \gamma Y, \quad [Z, Y] = \beta X + \delta Y.$$

Show that $\mathfrak{g}$ is isomorphic to the Lie subalgebra in $M(3, \mathbb{K})$ generated by the matrices

$$Z = \begin{pmatrix} \alpha & \beta & 0 \\ \gamma & \delta & 0 \\ 0 & 0 & 0 \end{pmatrix}, \quad X = \begin{pmatrix} 0 & 0 & 1 \\ 0 & 0 & 0 \\ 0 & 0 & 0 \end{pmatrix}, \quad Y = \begin{pmatrix} 0 & 0 & 0 \\ 0 & 0 & 1 \\ 0 & 0 & 0 \end{pmatrix}.$$

(iii) Assume that $\dim \mathcal{D}(\mathfrak{g}) = 3$. We will see that, if $\mathbb{K} = \mathbb{C}$, then $\mathfrak{g}$ is isomorphic to $\mathfrak{sl}(2, \mathbb{C})$, and that, if $\mathbb{K} = \mathbb{R}$, then $\mathfrak{g}$ is isomorphic either to $\mathfrak{sl}(2, \mathbb{R})$ or to $\mathfrak{su}(2)$. If $\mathbb{K} = \mathbb{C}$, show that there is an element $X \in \mathfrak{g}$ such that $\mathrm{ad}\, X$ has eigenvalues $0, \lambda, -\lambda$ ($\lambda \in \mathbb{C}$, $\lambda \neq 0$). Let Y and Z be eigenvectors of $\mathrm{ad}\, X$ for the eigenvalues λ and $-\lambda$:

$$[X, Y] = \lambda Y, \quad [X, Z] = -\lambda Z.$$

Show that

$$[Y, Z] = \mu X \quad (\mu \neq 0),$$

and conclude that $\mathfrak{g}$ is isomorphic to $\mathfrak{sl}(2, \mathbb{C})$.

Assume that $\mathbb{K} = \mathbb{R}$. If there is $X \in \mathfrak{g}$ such that ad X has eigenvalues $0, \lambda, -\lambda$ ($\lambda \in \mathbb{R}$, $\lambda \neq 0$), show that $\mathfrak{g}$ is isomorphic to $\mathfrak{sl}(2, \mathbb{R})$.

Assume that there is $X \in \mathfrak{g}$ such that ad X has eigenvalues $0, i\lambda, -i\lambda$ ($\lambda \in \mathbb{R}$, $\lambda \neq 0$). Show that there are $Y, Z \in \mathfrak{g}$ such that

$$[X, Y] = \lambda Z, \quad [X, Z] = -\lambda Y,$$

and that

$$[Y, Z] = \nu X \quad (\nu \neq 0).$$

Show then that $\mathfrak{g}$ is isomorphic either to $\mathfrak{sl}(2, \mathbb{R})$ or to $\mathfrak{su}(2)$.

For each case with dim $\mathcal{D}(\mathfrak{g}) \leq 2$ determine a linear Lie group G with Lie algebra $\mathfrak{g}$.

11. Let $\mathfrak{h}$ be the Heisenberg Lie algebra with dimension 3: there is a basis $\{X_1, X_2, X_3\}$ of $\mathfrak{h}$ such that

$$[X_1, X_2] = X_3, \quad [X_1, X_3] = 0, \quad [X_2, X_3] = 0.$$

Let $G = Aut(\mathfrak{h})$ and $\mathfrak{g} = Lie(G)$. Show that $\mathfrak{g}$ is isomorphic to the Lie subalgebra in $M(3, \mathbb{R})$ consisting of the matrices

$$\begin{pmatrix} a_{11} & a_{12} & 0 \\ a_{21} & a_{22} & 0 \\ a_{31} & a_{32} & a_{11} + a_{22} \end{pmatrix}.$$

Is the Lie algebra $\mathfrak{g}$ nilpotent? Is it solvable?

12. Let $\{X_1, X_2, X_3\}$ be the canonical basis in $\mathbb{R}^3$.

(a) Consider the Lie algebra $\mathfrak{m}$ with dimension 3 defined by

$$[X_1, X_2] = 0, \quad [X_1, X_3] = X_2, \quad [X_2, X_3] = -X_1.$$

  (i) Is the Lie algebra $\mathfrak{m}$ nilpotent? Is it solvable?

  (ii) Let G the group of automorphisms of $\mathfrak{m}$, and $\mathfrak{g} = Lie(G)$. Show that $\mathfrak{g}$ consists of the matrices

$$\begin{pmatrix} \alpha & \beta & \gamma \\ -\beta & \alpha & \delta \\ 0 & 0 & 0 \end{pmatrix} \quad (\alpha, \beta, \gamma, \delta \in \mathbb{R}).$$

(b) Consider the Lie algebra $\mathfrak{s}$ with dimension 3 defined by

$$[X_1, X_2] = X_3, \quad [X_1, X_3] = X_2, \quad [X_2, X_3] = -X_1.$$

  (i) Is the Lie algebra $\mathfrak{s}$ nilpotent? Is it solvable?

(ii) For $g \in GL(3, \mathbb{R})$ define, for $X, Y \in \mathbb{R}^3$,

$$[X, Y]' = g^{-1}[gX, gY].$$

Show that $(\mathbb{R}^3, [\cdot, \cdot]')$ is a Lie algebra. It will be denoted by $\mathfrak{s}'$.

(iii) If $g = \text{diag}(\lambda, \lambda, 1)$ $(\lambda > 0)$, one writes $[\cdot, \cdot]' = [\cdot, \cdot]_\lambda$, $\mathfrak{s}' = \mathfrak{s}_\lambda$. Show that, for $X, Y \in \mathbb{R}^3$, $\lim_{\lambda \to 0}[X, Y]_\lambda$ exists. Denote this limit by $[X, Y]_0$. Show that $(\mathbb{R}^3, [\cdot, \cdot]_0)$ is a Lie algebra.

13. Let $\mathfrak{g}$ be a semi-simple Lie algebra.

(a) Show that the map

$$X \mapsto \text{ad } X, \quad \mathfrak{g} \to Der(\mathfrak{g}),$$

is an isomorphism.

Hint. To prove surjectivity, proceed as follows. For $D \in Der(\mathfrak{g})$ show that there is $X \in \mathfrak{g}$ such that for every $Y \in \mathfrak{g}$, then $B(X, Y) = \text{tr}(D \text{ ad } Y)$. Then prove that, for all $Y, Z \in \mathfrak{g}$, $B(D_0 Y, Z) = 0$, where $D_0 = D - \text{ad } X$.

(b) Let G be a connected linear Lie group whose Lie algebra $\mathfrak{g} = Lie(G)$ is semi-simple. Show that $Ad(G)$ is a closed subgroup in $GL(\mathfrak{g})$.

Hint. Show that $Ad(G)$ is equal to the identity component of the group $Aut(\mathfrak{g})$.

14. Let $\mathfrak{g}$ be a finite dimensional real Lie algebra. Assume that the Killing form B of $\mathfrak{g}$ is negative definite.

(a) Show that the group $Aut(\mathfrak{g})$ is compact.

Hint. Show that $Aut(\mathfrak{g})$ is a closed subgroup of the orthogonal group $O(B)$.

(b) Show that $\mathfrak{g}$ is isomorphic to the Lie algebra of a compact linear Lie group.

Hint. Use the preceding exercise.

Remark. Let G be a connected linear Lie group. One can show that, if the Killing form B of the Lie algebra $\mathfrak{g}$ of G is negative definite, then G is compact.

5
Haar measure

On a locally compact group there is a left invariant measure which is called the Haar measure. We show its existence in the case of linear Lie groups by using differential calculus, and determine it explicitly for some groups. We will also see how it can be expressed using the Gram decomposition in the case of the linear group $GL(n, \mathbb{R})$

5.1 Haar measure

Let G be a locally compact group. A Radon measure $\mu \geq 0$ on G is said to be left invariant if

$$\int_G f(gx)\mu(dx) = \int_G f(x)\mu(dx),$$

for every $g \in G$, and for every $f \in \mathcal{C}_c(G)$, the space of continuous functions on G with compact support. This amounts to saying that, for every Borel set $E \subset G$, and for every $g \in G$,

$$\mu(gE) = \mu(E).$$

Theorem 5.1.1 *There exists a (non-zero) left invariant measure on G. It is unique up to a positive factor.*

We will admit this theorem without proof. Such a measure is called a left *Haar measure*. We will establish the existence of such a measure for a linear Lie group G.

If G is compact then the Haar measure μ is said to be normalised if

$$\int_G \mu(dx) = 1.$$

In the following we will denote by μ such a left Haar measure. For g fixed in G the linear form

$$f \mapsto \int_G f(gxg^{-1})\mu(dx)$$

defines a left invariant measure. Hence there is a positive number $\Delta(g)$ such that

$$\int_G f(gxg^{-1})\mu(dx) = \Delta(g) \int_G f(x)\mu(dx).$$

Observe also that

$$\int_G f(xg^{-1})\mu(dx) = \Delta(g) \int_G f(x)d\mu(dx).$$

The function Δ is clearly multiplicative. In fact we have the following.

Proposition 5.1.2 *The function Δ is a continuous group morphism,*

$$\Delta : G \to \mathbb{R}_+^*.$$

Proof. In order to show that Δ is continuous, let us consider a function $f \in \mathcal{C}_c(G)$ such that

$$\int_G f(x)\mu(dx) = 1.$$

Then

$$\Delta(g) = \int_G f(xg^{-1})\mu(dx).$$

Since f is left uniformly continuous, it follows that the function Δ is continuous. (See Exercise 5 of Chapter 1.) $\qquad\square$

The function Δ is called the *module* of the group G. Observe that, for a Borel set $E \subset G$,

$$\mu(Eg) = \Delta(g)\mu(E).$$

If $\Delta \equiv 1$, the group G is said to be *unimodular*. A commutative group is unimodular.

Proposition 5.1.3 *A compact group is unimodular. A discrete group is unimodular.*

Proof. (a) If G is compact, then $\Delta(G)$ is a compact subgroup of the group $\mathbb{R}_+^*$, and $\{1\}$ is the only compact subgroup of $\mathbb{R}_+^*$.

(b) If G is discrete, then a continuous function with compact support is a function with finite support. The measure μ defined on G by

$$\int_G f(x)\mu(dx) = \sum_{x \in G} f(x)$$

is left and right invariant.   □

Proposition 5.1.4 *The measure $\Delta(x^{-1})\mu(dx)$ is a right Haar measure. Furthermore, for $f \in \mathcal{C}_c(G)$,*

$$\int_G f(x^{-1})\mu(dx) = \int_G f(x)\Delta(x^{-1})\mu(dx).$$

Proof. Let us consider the linear form

$$f \mapsto \int_G f(x^{-1})\Delta(x^{-1})\mu(dx).$$

For $g \in G$,

$$\int_G f(gx^{-1})\Delta(x^{-1})\mu(dx) = \int_G f\big((xg^{-1})^{-1}\big)\Delta(x^{-1})\mu(dx).$$

By putting $y = xg^{-1}$ we get

$$\int_G f(gx^{-1})\Delta(x^{-1})\mu(dx) = \Delta(g)\int_G f(y^{-1})\Delta(g^{-1}y^{-1})\mu(dy)$$

$$= \int_G f(y^{-1})\Delta(y^{-1})\mu(dy).$$

Hence this linear form defines a left Haar measure. Therefore there is $C > 0$ such that

$$\int_G f(x^{-1})\Delta(x^{-1})\mu(dx) = C\int_G f(x)\mu(dx).$$

By applying this relation to the function $f_1(x) = f(x^{-1})\Delta(x^{-1})$ one gets $C^2 = 1$, hence $C = 1$.   □

5.2 Case of a group which is an open set in $\mathbb{R}^n$

Assume that the group G is realised as an open set in $\mathbb{R}^m$, and that the maps

$$L(g) : x \mapsto gx, \quad R(g) : x \mapsto xg \quad (g \in G),$$

are restrictions of linear, or affine linear, transformations. In this case it is natural to look for a left Haar measure of the form

$$\mu(dx) = h(x)\lambda(dx),$$

where λ is the Lebesgue measure on $\mathbb{R}^m$. By using the invariance property of μ it is possible to determine the density h. In fact

$$h(gx)|J\big(L(g)\big)| = h(x),$$

where $J(L(g))$ is the Jacobian determinant of the transformation $L(g)$. Therefore

$$h(x) = h(e)|J\big(L(x)\big)|^{-1}.$$

Determination of the measure μ amounts to computing the determinant of the linear part of the affine linear transformation $L(g)$.

Examples. (1) A Lebesgue measure on $\mathbb{R}^n$ is a left and right Haar measure on the group $G = \mathbb{R}^n$.

(2) Let G be the group '$ax + b$', that is the group of affine linear transformations of the real line. It can be identified with the subgroup of $GL(2, \mathbb{R})$ consisting of the matrices

$$g = \begin{pmatrix} a & b \\ 0 & 1 \end{pmatrix}, \quad a, b \in \mathbb{R},\ a \neq 0,$$

and is homeomorphic to the open set in $\mathbb{R}^2$:

$$\{(a, b) \in \mathbb{R}^2 \mid a \neq 0\} = \mathbb{R}^* \times \mathbb{R}.$$

The transformations $L(g)$ and $R(g)$ are given by, if $g = (a, b)$ and $x = (u, v)$,

$$L(g)x = (au, av + b), \quad R(g)x = (au, bu + v).$$

The measure μ_ℓ given by

$$\int_G f(g)\mu_\ell(dg) = \int_{\mathbb{R}^* \times \mathbb{R}} f(a, b)\frac{da\,db}{a^2}$$

is a left Haar measure. The measure μ_r defined by

$$\int_G f(g)\mu_r(dg) = \int_{\mathbb{R}^* \times \mathbb{R}} f(a, b)\frac{da\,db}{|a|}$$

is a right Haar measure. The module function is given by

$$\Delta(g) = \frac{1}{|a|}.$$

(3) Let $G = GL(n, \mathbb{R})$. By definition this is the open set in $M(n, \mathbb{R})$ of invertible matrices. If $x^1, \ldots, x^n$ are the columns of $x \in M(n, \mathbb{R})$ then, for $g \in G$, $L(g)x = gx = (gx^1, \ldots, gx^n)$ and therefore

$$\text{Det} \, L(g) = (\det g)^n.$$

Similarly, by considering rows instead of columns, one gets

$$\text{Det} \, R(g) = (\det g)^n.$$

It follows that the measure

$$|\det x|^{-n} \prod_{i,j=1}^{n} dx_{ij}$$

is left and right invariant. The group $GL(n, \mathbb{R})$ is unimodular.

(4) Let $G = T_0(n, \mathbb{R})$ be the strict upper triangular group. It can be identified with $\mathbb{R}^{n(n-1)/2}$. The measure μ defined on G by

$$\int_G f(x)\mu(dx) = \int_{\mathbb{R}^{n(n-1)/2}} f(x) \prod_{i<j} dx_{ij}$$

is a left and right Haar measure. The group G is unimodular.

(5) Let $G = T(n, \mathbb{R})$ be the upper triangular group. It can be identified with

$$(\mathbb{R}^*)^n \times \mathbb{R}^{n(n-1)/2} \subset \mathbb{R}^{n(n+1)/2}.$$

The measure μ_ℓ defined on G by

$$\int_G f(x)\mu_\ell(dx) = \int \int_{(\mathbb{R}^*)^n \times \mathbb{R}^{n(n-1)/2}} f(x) \prod_{i=1}^{n} |x_{ii}|^{i-n-1} dx_{ii} \prod_{i<j} dx_{ij}$$

is a left Haar measure, and the measure μ_r defined by

$$\int_G f(x)\mu_r(dx) = \int_{(\mathbb{R}^*)^n \times \mathbb{R}^{n(n-1)/2}} f(x) \prod_{i=1}^{n} |x_{ii}|^{-i} dx_{ii} \prod_{i<j} dx_{ij}$$

is a right Haar measure. The module function is given by

$$\Delta(x) = \prod_{i=1}^{n} |x_{ii}|^{2i-n-1}.$$

5.3 Haar measure on a product

The following theorem has numerous applications.

Theorem 5.3.1 *Let G be a locally compact group, P and Q two closed subgroups of G such that $G = PQ$. More precisely one assumes that the* *map*

$$P \times Q \to G, \ (x, y) \mapsto xy$$

is a homeomorphism. Let Δ be the module of G. Let α denote a left Haar measure on P, and β a right Haar measure on Q. Then the measure μ defined on G by

$$\int_G f(g)\mu(dg) = \int_{P \times Q} f(xy)\Delta(y)\alpha(dx)\beta(dy),$$

is a left Haar measure on G

Proof. For $g = xy$ ($g \in G$, $x \in P$, $y \in Q$) let us write $x = \varphi_1(g)$, $y = \varphi_2(g)$. Let μ be a left Haar measure on G. For $f_1 \in \mathcal{C}_c(P)$, $f_2 \in \mathcal{C}_c(Q)$, let us consider the integral

$$I(f_1, f_2) = \int_G f_1\big(\varphi_1(g)\big) f_2\big(\varphi_2(g)\big) \Delta\big(\varphi_2(g)^{-1}\big) \mu(dg).$$

One can check that, for f_2 fixed, the map $f_1 \mapsto I(f_1, f_2)$ defines a left invariant measure on P. Hence one can write

$$I(f_1, f_2) = B(f_2) \int_P f_1(x)\alpha(dx),$$

where B is a positive linear form on the space $\mathcal{C}_c(Q)$ of continuous functions on Q with compact support. Similarly, for f_1 fixed, the map $f_2 \mapsto I(f_1, f_2)$ defines a right invariant measure on Q, and therefore

$$I(f_1, f_2) = A(f_1) \int_Q f_2(y)\beta(dy),$$

where A is a positive linear form on $\mathcal{C}_c(P)$. It follows that there is a positive constant C such that, for $f_1 \in \mathcal{C}_c(P)$, $f_2 \in \mathcal{C}_c(Q)$,

$$I(f_1, f_2) = C \int_{P \times Q} f_1(x)f_2(y)\alpha(dx)\beta(dy).$$

Therefore, if f is the function defined on G by

$$f(xy) = f_1(x)f_2(y) \quad (x \in P, \ y \in Q),$$

then

$$\int_G f(g)\mu(dg) = I(f_1, f_2\Delta)$$

$$= C \iint_{P \times Q} f_1(x)f_2(y)\Delta(y)\alpha(dx)\beta(dy).$$

The statement is then proven for a function which can be written

$$f(g) = f_1(x)f_2(y)\Delta(y) \quad (g = xy, \ x \in P, \ y \in Q),$$

where $f_1 \in \mathcal{C}_c(P)$, $f_2 \in \mathcal{C}_c(Q)$ and, by linearity, for a finite sum of such functions,

$$f(g) = \sum_{i=1}^{N} f_1^i(x)f_2^j(y)\Delta(y).$$

Since every function in $\mathcal{C}_c(G)$ can be approximated by such functions for the topology of $\mathcal{C}_c(G)$, the statement is now proven.   □

Let us give a first application of this theorem. Let $G = GL(n, \mathbb{R})$, $K = O(n)$, and $T = \mathrm{T}(n, \mathbb{R})_+$ the group of upper triangular matrices with positive diagonal entries. Let us recall the Gram decomposition (Theorem 1.6.1). Every element g in G can be written

$$g = kt,$$

with $k \in K, t \in T$. The decomposition is unique, and the map

$$\varphi : K \times T \to G, \quad (k, t) \mapsto kt,$$

is a homeomorphism.

Proposition 5.3.2 *Let $K = O(n)$ and $T = \mathrm{T}(n, \mathbb{R})_+$, and let α denote the normalised Haar measure of K. There exists a constant $c_n > 0$ such that, for every function f, which is integrable on $G = GL(n, \mathbb{R})$,*

$$\int_G f(x)|\det(x)|^{-n} \prod_{i,j=1}^{n} dx_{ij} = c_n \int_{K \times T} f(kt)\alpha(dk) \prod_{i=1}^{n} t_{ii}^{-i} \prod_{i \leq j} dt_{ij}.$$

Proof. We saw that the group $G = GL(n, \mathbb{R})$ is unimodular (Example 3), and that the measure defined on T by

$$\beta_r(dt) = \prod_{i=1}^{n} t_{ii}^{-i} \prod_{i \leq j} dt_{ij},$$

is a right Haar measure (example 5). Hence this proposition is a direct consequence of Theorem 5.3.1.   □

For the evaluation of the constant c_n see Exercise 4.
One can also consider the decomposition

$$\psi : T \times K \to G, \quad (t, k) \mapsto tk,$$

and, similarly, there exists a constant $d_n > 0$ such that

$$\int_G f(x)|\det(x)|^{-n} \prod_{i,j=1}^n dx_{ij} = d_n \int_{T \times K} f(tk) \prod_{i=1}^n t_{ii}^{i-n-1} \prod_{i \le j} dt_{ij} \, \alpha(dk).$$

In fact,

$$\beta_\ell(dt) = \prod_{i=1}^n t_{ii}^{i-n-1} \prod_{i \le j} dt_{ij}$$

is a left Haar measure on the group T. One can show that $c_n = d_n$ (see Exercise 4).

5.4 Some facts about differential calculus

We saw how it is possible to determine a Haar measure on a group G which can be realised as an open set in $\mathbb{R}^m$, and when the transformations $L(g)$ and $R(g)$ are restrictions to G of affine linear maps. This method does not apply to groups whose geometry is less simple, such as the orthogonal group $O(n)$ or the unitary group $U(n)$. We will see in Section 5.5 how it is possible to determine a Haar measure on a linear Lie group by using differential forms. For that we will first recall some facts in differential calculus.

Let V be a submanifold in $\mathbb{R}^N$, and x_0 a point of V. A tangent vector X at x_0 can be written

$$X = \gamma'(t_0),$$

where γ is a C^1 curve drawn on V such that $\gamma(t_0) = x_0$. The tangent vectors at x_0 form a vector subspace in $\mathbb{R}^N$ which is called the tangent vector space of V at x_0 and is denoted by $T_{x_0}(V)$.

Let V and W be two submanifolds in $\mathbb{R}^N$, and φ a differential map from V into W. The image under φ of a C^1 curve γ which is drawn on V running through x_0 is a curve $\varphi \circ \gamma$ which is drawn on W running through $y_0 = \varphi(x_0)$, and

$$\frac{d}{dt}\varphi \circ \gamma(t)\Big|_{t=t_0} = D\varphi_{x_0}\big(\gamma'(t_0)\big).$$

If V and W have the same dimension and if φ is a diffeomorphism, then the differential $(D\varphi)_x$ of φ at every point $x \in V$ is an isomorphism from $T_x(V)$ onto $T_y(W)$, where $y = \varphi(x)$. If V and W have the same dimension and if, for every $x \in V$, the differential $(D\varphi)_x$ is an isomorphism, then (V, φ) is a covering of W.

A *vector field* ξ on V is the prescription at each point x of V of a tangent vector $\xi(x)$ in $T_x(V)$. It is said to be C^k if $x \mapsto \xi(x)$ is C^k.

Let $\varphi : V \to W$ be a diffeomorphism from V onto W, and ξ a vector field on V. The image of ξ under φ is denoted by $\varphi_* \xi$:

$$(\varphi_* \xi)(\varphi(x)) = D\varphi_x(\xi(x)).$$

To every vector field ξ on V one associates the differential operator $\tilde{\xi}$ of order one defined by

$$\tilde{\xi} f(x) = Df_x(\xi(x)) \quad (f \in C^\infty(V)).$$

If ξ is a vector field on V, then the map $f \mapsto \tilde{\xi}(f)$ is a derivation of the algebra $C^\infty(V)$:

$$\tilde{\xi}(fg) = \tilde{\xi}(f)g + f\tilde{\xi}(g) \quad (f, g \in C^\infty(V)),$$

and one can show that every derivation of $C^\infty(V)$ is obtained in that way. The space of C^∞ vector fields on V will be denoted by $\Xi(V)$. If ξ and η are two vector fields in $\Xi(V)$ their bracket $[\xi, \eta]$ is defined by

$$\widetilde{[\xi, \eta]} = [\tilde{\xi}, \tilde{\eta}] = \tilde{\xi} \circ \tilde{\eta} - \tilde{\eta} \circ \tilde{\xi}.$$

Hence $\Xi(V)$ is equipped with a Lie algebra structure.

If the vector fields ξ and η are written in local coordinates

$$\xi(x) = (\xi_1(x), \ldots, \xi_m(x)), \quad \eta(x) = (\eta_1(x), \ldots, \eta_m(x)),$$

($m = \dim V$) then

$$\tilde{\xi} f = \sum_{i=1}^m \xi_i \frac{\partial f}{\partial x_i}, \quad \tilde{\eta} f = \sum_{i=1}^m \eta_i \frac{\partial f}{\partial x_i},$$

and

$$\tilde{\xi} \circ \tilde{\eta} f - \tilde{\eta} \circ \tilde{\xi} f = \sum_{i=1}^m \xi_i \frac{\partial}{\partial x_i} \left(\sum_{j=1}^m \eta_j \frac{\partial f}{\partial x_j} \right) - \sum_{i=1}^m \eta_i \frac{\partial}{\partial x_i} \left(\sum_{j=1}^m \xi_j \frac{\partial f}{\partial x_j} \right)$$

$$= \sum_{i=1}^m \zeta_i \frac{\partial f}{\partial x_i},$$

with

$$\zeta_i = \sum_{j=1}^m \left(\xi_j \frac{\partial \eta_i}{\partial x_j} - \eta_j \frac{\partial \xi_i}{\partial x_j} \right).$$

A *differential form* of degree one is a map

$$\alpha : \Xi(V) \to C^\infty(V)$$

which is $\mathcal{C}^\infty(V)$-linear:

$$\alpha(f\xi) = f\alpha(\xi) \quad (f \in \mathcal{C}^\infty(V)).$$

The function $\alpha(\xi)$ can be written

$$x \mapsto \alpha_x\big(\xi(x)\big),$$

where α_x is a linear form on $T_x(V)$.

Let $u \in \mathcal{C}^\infty(V)$, then the map

$$\alpha : \xi \mapsto Du(\xi)$$

defines a differential form. One writes $\alpha = du$. Let φ be a $\mathcal{C}^\infty$ map from a manifold V into a manifold W. If α is a differential form of degree one on W, one denotes by $\varphi^*\alpha$ the differential form defined on V by

$$\varphi^*\alpha(\xi) = \alpha\big(D\varphi(\xi)\big).$$

In a system of local coordinates a differential form α of degree one can be written as a linear combination with coefficients in $\mathcal{C}^\infty(V)$ of the differential dx_i of the coordinates:

$$\alpha = \sum_{i=1}^m \alpha_i(x)dx_i.$$

A *differential form* of degree k on V is a map

$$\Xi(V) \times \cdots \times \Xi(V) \to \mathcal{C}^\infty(V)$$

which is k-$\mathcal{C}^\infty(V)$-linear and alternate. If ω is a differential form of degree k on V, and if $\xi_1, \ldots, \xi_k$ are k $\mathcal{C}^\infty$ vector fields on V, then $\omega(\xi_1, \ldots, \xi_k)$ is a function on V which can be written

$$x \mapsto \omega_x\big(\xi_1(x), \ldots, \xi_k(x)\big)$$

where ω_x is a k-skewlinear form on $T_x(V)$.

The wedge product $\alpha_1 \wedge \cdots \wedge \alpha_k$ of k linear forms $\alpha_1, \ldots, \alpha_k$ of degree one is the differential form of degree k defined by

$$\alpha_1 \wedge \cdots \wedge \alpha_k(\xi_1, \ldots, \xi_k) = \det\big(\alpha_i(\xi_j)\big)_{1 \le i,j \le k}.$$

If φ is a differential map from V into W, and if ω is a differential form of degree k on W, one denotes by $\varphi^*\omega$ the differential form of degree k defined on V by

$$\varphi^*\omega(\xi_1, \ldots, \xi_k) = \omega\big(D\varphi(\xi_1), \ldots, D\varphi(\xi_k)\big).$$

If $X_1, \ldots, X_k$ are k tangent vectors at $x \in V$,

$$(\varphi^*\omega)_x(X_1, \ldots, X_k) = \omega_{\varphi(x)}\big((D\varphi)_x X_1, \ldots, (D\varphi)_x X_k\big).$$

This is an important formula that we will use several times in the following.

A differential form ω of degree m on an open set V in $\mathbb{R}^m$ can be written

$$\omega = a(x)dx_1 \wedge \cdots \wedge dx_m,$$

where a is a function defined on V. Let φ be a diffeomorphism from V onto W, where V and W are two open sets in $\mathbb{R}^m$, and ω a differential form of degree m on W,

$$\omega = a(y)dy_1 \wedge \cdots \wedge dy_m.$$

Then

$$\varphi^*\omega = a\big(\varphi(x)\big)J\varphi(x)dx_1 \wedge \cdots \wedge dx_m,$$

where $J\varphi$ is the Jacobian determinant of φ,

$$J\varphi = \det\left(\frac{\partial\varphi_j}{\partial x_i}\right)_{1 \le i,j \le m}.$$

To every differential form ω of degree m on a manifold V of dimension m one associates a positive measure which is called the *modulus* of ω and denoted by $|\omega|$. Let V_0 be an open set where local coordinates are available. In V_0 the form ω can be written

$$\omega = a(x)dx_1 \wedge \cdots \wedge dx_m,$$

and the measure $|\omega|$ has the density $|a(x)|$ with respect to the Lebesgue measure:

$$|\omega|(dx) = |a(x)|dx_1 \ldots dx_m.$$

If φ is a diffeomorphism from V onto W, and if ω is a differential form of degree m on W, then

$$|\varphi^*\omega| = \varphi^{-1}(|\omega|),$$

that is, if f is a continuous function with compact support on W,

$$\int_W f(y)|\omega|(dy) = \int_V (f \circ \varphi)|\varphi^*\omega|(dx).$$

In terms of local coordinates this relation is nothing but the change of variable formula for multiple integrals:

$$\int_{W_0} f(y)|a(y)|dy_1 \ldots dy_m$$
$$= \int_{V_0} f(\varphi(x))|a(\varphi(x))||J\varphi(x)|dx_1 \ldots dx_m.$$

More generally, if φ is a covering of order k,

$$\int_V (f \circ \varphi)|\varphi^*\omega|(dx) = k \int_W f(y)|\omega|(dy).$$

If φ is a diffeomorphism from V onto V, and if ω is a differential form of degee m on V, which is invariant under φ up to a sign, that is $\varphi^*\omega = \pm\omega$, then the measure $|\omega|$ is invariant under φ, that is, if f is a continuous function on V with compact support,

$$\int_V (f \circ \varphi)|\omega|(dx) = \int_V f|\omega|(dx).$$

Examples. (1) Let us consider on $\mathbb{R}^n$ the differential form

$$\omega = dx_1 \wedge \cdots \wedge dx_n$$

of degree n. If $X_1, \ldots, X_n$ are n vectors,

$$\omega(X_1, \ldots, X_n) = \det(X_1, \ldots, X_n)$$

(the determinant being relative to the canonical basis). Then the associated measure $\lambda = |\omega|$ is the Lebesgue measure.
  (2) Let us consider on $\mathbb{R}^n$ the differential form

$$\omega = \sum_{i=1}^{n}(-1)^{i-1}x_i dx_1 \wedge \cdots \wedge \widehat{dx_i} \wedge \cdots dx_n$$

of degree $n - 1$ (the notation $\widehat{dx_i}$ means that the factor dx_i is omitted). At $x \in \mathbb{R}^n$, if $X_1, \ldots, X_{n-1}$ are $n - 1$ vectors,

$$\omega_x(X_1, \ldots, X_{n-1}) = \det(x, X_1, \ldots, X_{n-1}).$$

The form ω is invariant under every linear transformation in $SL(n, \mathbb{R})$. The restriction ω_0 of the form ω to the unit sphere S in $\mathbb{R}^n$ is invariant under $SO(n)$. The associated measure $\Sigma = |\omega_0|$ is a measure on S which is invariant under $O(n)$. One can show that every measure on S which is invariant under $O(n)$ is equal, up to a factor, to Σ. We will see that $\varpi_n = \Sigma(S)$ equals

$$\varpi_n = 2\frac{\pi^{n/2}}{\Gamma\left(\frac{n}{2}\right)}.$$

The normalised invariant measure will be denoted by σ:

$$\sigma = \frac{1}{\varpi_n}\Sigma.$$

5.5 Invariant vector fields and Haar measure on a linear Lie group

Let G be a linear Lie group, that is a closed subgroup in $GL(n, \mathbb{R})$. It is a submanifold in $M(n, \mathbb{R})$ (Corollary 3.3.5).

Proposition 5.5.1 *The tangent vector space to G at the identity element $e = I$ is the Lie algebra $\mathfrak{g} = Lie(G)$ of G.*

Proof. (a) Let $X \in \mathfrak{g}$. Then $\gamma(t) = \exp tX$ is a curve drawn on G running through e for $t = 0$, and $\gamma'(0) = X$, hence $X \in T_e(G)$ and $\mathfrak{g} \subset T_e(G)$.

(b) Conversely let $\gamma(t)$ be a curve drawn on G running through e for $t = t_0$. For t close to t_0, $X(t) = \log \gamma(t)$ is well defined and $t \mapsto X(t)$ is a curve in $\mathfrak{g}$; furthermore

$$\gamma'(t_0) = (D\exp)_0\big(X'(t_0)\big).$$

Since $(D\exp)_0 = Id$, $\gamma'(t_0) = X'(t_0) \in \mathfrak{g}$. This shows that $T_e(G) \subset \mathfrak{g}$. $\square$

To $X \in \mathfrak{g}$ one associates the vector field ξ_X on G defined by

$$\xi_X(g) = \big(DL(g)\big)_e X = g \cdot X.$$

This is a left invariant vector field: it is invariant under the diffeomorphisms $L(g) : x \mapsto gx$,

$$L(g)_*\xi_X = \xi_X.$$

To this vector field one associates the left invariant differential operator

$$\big(\tilde{\xi}_X f\big)(g) = (Df)_g(gX) = \frac{d}{dt}\bigg|_{t=0} f(g \exp tX).$$

Proposition 5.5.2 *The map $X \mapsto \xi_X$ is an isomorphism of the Lie algebra $\mathfrak{g}$ onto the Lie algebra made of the left invariant vector fields on G.*

Proof. The map

$$\mathfrak{g} \to \Xi(G), \quad X \mapsto \xi_X,$$

is injective, and every left invariant vector field on G is of that form. For $X, Y \in \mathfrak{g}$,

$$([\tilde{\xi}_X, \tilde{\xi}_Y] f)(g) = Df_g(g[X, Y]) = (\tilde{\xi}_{[X,Y]} f)(g).$$

In fact,

$$\tilde{\xi}_X \cdot \tilde{\xi}_Y f(g) = \frac{d}{ds}\bigg|_{s=0} \frac{d}{dt}\bigg|_{t=0} f(g \exp sX \exp tY)$$
$$= (D^2 f)_g(gX, gY) + (Df)_g(gXY). \qquad \square$$

Let ω be a left invariant differential form of degree k on G. Then, if $X_1, \ldots, X_k \in \mathfrak{g} = T_e(G)$,

$$\omega_g(gX_1, \ldots, gX_k) = \omega_e(X_1, \ldots, X_k).$$

Hence, the form ω is determined by ω_e, which is a k-skewlinear form on $\mathfrak{g}$. Conversely, given a k-skewlinear form ω_o on $\mathfrak{g}$, there is a unique left invariant differential form ω of degree k on G such that $\omega_e = \omega_0$.

Proposition 5.5.3 *Let ω be a (non-zero) left invariant differential form of degree $m = \dim G$ on G. Then $|\omega|$ is a left Haar measure on G.*

Proof. In fact, if $\varphi = L(g)$, then $\varphi^*\omega = \omega$ and, for every continuous function f on G with compact support,

$$\int_G f(gx)|\omega|(dx) = \int_G f(x)|\omega|(dx). \qquad \square$$

In Section 5.2 we considered the case of a group G which can be identified with an open set in $\mathbb{R}^m$. This means that there exists on G a system of global coordinates. We can rephrase what was said. Let ω be a differential form on G of degree m:

$$\omega = a(x)dx_1 \wedge \cdots \wedge dx_m.$$

If $J_g(x)$ denotes the Jacobian determinant of $L(g)$ at x,

$$\varphi^*\omega = a(g \cdot x)J_g(x)dx_1 \wedge \cdots \wedge dx_m.$$

The measure $|\omega|$ is left invariant if

$$a(g \cdot x)J_g(x) = \pm a(x).$$

Therefore, the measure μ defined by

$$\mu(dx) = \frac{C}{|J_x(e)|} dx_1 \ldots dx_m,$$

where C is a positive constant, is a left Haar measure.

Proposition 5.5.4

$$\Delta(g) = |\det \mathrm{Ad}(g^{-1})|.$$

Proof. Let ω be a left invariant differential form of degree m on G. For $g \in G$ the inner automorphism

$$x \mapsto \varphi(x) = gxg^{-1}$$

is a diffeomorphism of G. Let us show that

$$\varphi^* \omega = \det \mathrm{Ad}(g)\omega.$$

For $X_1, \ldots, X_m \in \mathfrak{g}$,

$$\begin{aligned}
(\varphi^*\omega)_x(xX_1, \ldots, xX_m) &= \omega_{gxg^{-1}}\big(g(xX_1)g^{-1}, \ldots, g(xX_m)g^{-1}\big) \\
&= \omega_{gxg^{-1}}\big(gxg^{-1}\mathrm{Ad}(g)X_1, \ldots, gxg^{-1}\mathrm{Ad}(g)X_m\big) \\
&= \omega_e\big(\mathrm{Ad}(g)X_1, \ldots, \mathrm{Ad}(g)X_m\big) \\
&= \det \mathrm{Ad}(g)\omega_e(X_1, \ldots, X_m) \\
&= \det \mathrm{Ad}(g)\omega_x(xX_1, \ldots, xX_m).
\end{aligned}$$

It follows that, if μ denotes the left Haar measure associated to ω, then

$$\int_G f(gxg^{-1})\mu(dx) = |\det \mathrm{Ad}(g)|^{-1} \int_G f(x)\mu(dx),$$

and

$$\Delta(g) = |\det \mathrm{Ad}(g)|^{-1}. \qquad \square$$

Corollary 5.5.5 *In the three following cases the group G is unimodular:*

(i) $\mathrm{Ad}(G)$ *is compact,*
(ii) $\mathfrak{g} = Lie(G)$ *is semi-simple,*
(iii) $\mathfrak{g} = Lie(G)$ *is nilpotent and G is connected.*

We already saw that a compact group is unimodular (Proposition 5.1.3).

In (iii) it is necessary to assume that G is connected. In fact $\mathfrak{g}$ could be nilpotent and G non-unimodular. (See Exercise 3.)

Proof. (i) The map

$$g \mapsto |\det \mathrm{Ad}(g)|, \quad g \to \mathbb{R}_+^*$$

is a continuous morphism. If Ad(G) is compact, then this map is bounded, hence constant and equal to 1.

(ii) Let B be the Killing form of $\mathfrak{g}$. From the relation

$$\mathrm{ad}\big(\mathrm{Ad}(g)X\big) = \mathrm{Ad}(g)\,\mathrm{ad}\,X\,\mathrm{Ad}(g^{-1}) \quad (g \in G,\ X \in \mathfrak{g}),$$

it follows that

$$B\big(\mathrm{Ad}(g)X, \mathrm{Ad}(g)Y\big) = B(X, Y) \quad (X, Y \in \mathfrak{g}).$$

This means that Ad(g) belongs to the orthogonal group of B. Therefore

$$|\det \mathrm{Ad}(g)| = 1.$$

(iii) For every $X \in \mathfrak{g}$, ad X is nilpotent and

$$\det \mathrm{Ad}(\exp X) = \det \mathrm{Exp}\,\mathrm{ad}\,X = e^{\mathrm{tr}(\mathrm{ad}\,X)} = 1.$$

Therefore, for every g in a neighbourhood of e,

$$\det \mathrm{Ad}(g) = 1.$$

Hence, the subgroup

$$H = \{g \in G \mid \det \mathrm{Ad}(g) = 1\}$$

is open and closed and, since G is connected, $H = G$. $\qquad\square$

Let φ be the diffeomorphism of G defined by

$$x \mapsto \varphi(x) = x^{-1}.$$

One can show that, if ω is a left invariant differential form of degree m, then

$$\varphi^*\omega = \det\big(-\mathrm{Ad}(x)\big)\omega.$$

(See Exercise 6.)

In the following proposition we express left Haar measures in the exponential chart.

Proposition 5.5.6 *Let U be a connected neighbourhood of 0 in $\mathfrak{g} = Lie(G)$ such that the exponential map is a diffeomorphism of U onto $V = \exp U$. Let μ be a left Haar measure on G and λ a Lebesgue measure on $\mathfrak{g}$. Let f be an integrable function on G supported in V. Then*

$$\int_G f(g)\mu(dg) = c \int_{\mathfrak{g}} f(\exp X)\det A(X)\lambda(dX),$$

where

$$A(X) = \frac{I - \exp(-\operatorname{ad} X)}{\operatorname{ad} X},$$

and c is a positive constant.

Observe that, if $\mathfrak{g}$ is nilpotent then, for every X, $\operatorname{ad} X$ is nilpotent and $\det A(X) = 1$.

Proof. Let ω be a left invariant differential form of degree m on G. For $\varphi = \exp$, the exponential map,

$$(\varphi^*\omega)_X(Y_1, \ldots, Y_m) = \omega_{\exp X}\big((D\exp)_X Y_1, \ldots, (D\exp)_X Y_m\big),$$

and, by Theorem 2.1.4,

$$\begin{aligned}
(\varphi^*\omega)_X(Y_1, \ldots, Y_m) &= \omega_{\exp X}\big(\exp X A(X) Y_1, \ldots, \exp X A(X) Y_m\big) \\
&= \omega_e\big(A(X) Y_1, \ldots, A(X) Y_m\big) \\
&= \det A(X)\omega_e(Y_1, \ldots, Y_m).
\end{aligned}$$

Since the exponential map is a diffeomorphism from U onto V it follows that $\det A(X) \neq 0$. Therefore, since U is connected, and $A(0) = Id$, then $\det A(X) > 0$ on U. $\square$

5.6 Exercises

1. Let G be the motion group of the plane. It can be identified with the group consisting of the matrices

$$\begin{pmatrix} \cos\theta & -\sin\theta & a \\ \sin\theta & \cos\theta & b \\ 0 & 0 & 1 \end{pmatrix} \quad (\theta \in \mathbb{R}/2\pi\mathbb{Z}, a, b \in \mathbb{R}).$$

   (a) Show that G is unimodular
   (b) Determine a Haar measure on G.

2. (a) Let $V = M(p, q; \mathbb{R})$ be the vector space of $p \times q$ real matrices. Let $A \in GL(p, \mathbb{R})$, $B \in GL(q, \mathbb{R})$. Consider the endomorphism T of V given by:

$$T : X \mapsto AXB.$$

   Determine the determinant of T.

   (b) Let G be the group consisting of the $n \times n$ matrices:

$$g = \begin{pmatrix} A & C \\ 0 & B \end{pmatrix}, \quad A \in GL(p, \mathbb{R}), \quad B \in GL(q, \mathbb{R}), \quad C \in M(p, q, \mathbb{R}),$$

   with $n = p + q$.

Describe the adjoint representation of G on its Lie algebra, and determine the module function of G.

Determine a left Haar measure and a right Haar measure on G.

3. Fix $q > 1$, and consider the group G consisting of the matrices

$$\begin{pmatrix} q^k & x \\ 0 & 1 \end{pmatrix}, \quad k \in \mathbb{Z}, \ x \in \mathbb{R}.$$

Show that G is a closed subgroup in $GL(2, \mathbb{R})$. Determine its Lie algebra. Is it nilpotent?

Show that the linear form

$$f \mapsto \sum_{k=-\infty}^{\infty} \frac{1}{q^k} \int_{-\infty}^{\infty} f(k, x)\,dx$$

defines a left Haar measure, and that

$$f \mapsto \sum_{k=-\infty}^{\infty} \int_{-\infty}^{\infty} f(k, x)\,dx$$

defines a right Haar measure. (The group G has been identified with $\mathbb{Z} \times \mathbb{R}$.) Determine the module function of G.

4. Let μ denote the Haar measure on $G = GL(n, \mathbb{R})$ given by

$$\int_G f(x)\mu(dx) = \int_{M(n,\mathbb{R})} f(x)|\det x|^{-n} \prod_{i,j=1}^{n} dx_{ij},$$

and let α denote the normalised Haar measure of $K = O(n)$. By Proposition 5.3.2 there exists a constant c_n such that

$$\int_G f(x)\mu(dx) = c_n \int_{K \times T} f(kt)\alpha(dk) \prod_{i=1}^{n} t_{ii}^{-i} \prod_{i \leq j} dt_{ij}.$$

(a) By considering the function f:

$$f(x) = |\det x|^n \exp(-\operatorname{tr} x^T x),$$

show that

$$c_n = \frac{2^n \pi^{n(n+1)/4}}{\prod_{i=1}^{n} \Gamma\left(\frac{i}{2}\right)}.$$

Use the formulae:

$$\int_0^{\infty} e^{-t^2} t^\alpha\,dt = \tfrac{1}{2}\Gamma\left(\frac{\alpha+1}{2}\right), \qquad \int_{-\infty}^{\infty} e^{-t^2}\,dt = \sqrt{\pi}.$$

(b) Show that the constant d_n which occurs in the integration formula

$$\int_G f(x)|\det(x)|^{-n} \prod_{i,j=1}^n dx_{ij} = d_n \int_{T \times K} f(tk) \prod_{i=1}^n t_{ii}^{i-n-1} \prod_{i \leq j} dt_{ij}\alpha(dk)$$

is equal to c_n.

5. (a) Let $G = GL(n, \mathbb{C})$. Show that the measure μ defined on G by

$$\int_G f(x)\mu(dx) = \int_G f(x)|\det(x)|^{-2n}\lambda(dx),$$

where λ is a Lebesgue measure on $M(n, \mathbb{C})$, seen as a real vector space $M(n, \mathbb{C}) \simeq \mathbb{R}^{2n^2}$, is a left and right Haar measure.

(b) Let T be the subgroup of G consisting of upper triangular matrices with positive diagonal entries. Show that the measure defined on T by

$$\int_T f(t) \prod_{i=1}^n t_{ii}^{-2(n-i)-1}\lambda(dt),$$

where λ is a Lebesgue measure on the real vector space consisting of complex upper triangular matrices with real diagonal entries, is a left Haar measure.

Show that the measure defined on T by

$$\int_T f(t) \prod_{i=1}^n t_{ii}^{-2(i-1)-1}d\lambda(t),$$

is a right Haar measure, and that the module function is given by:

$$\Delta(t) = \prod_{i=1}^n t_{ii}^{4i-2n-2}.$$

(c) Let α be a Haar measure on the unitary group $U = U(n)$. Show that there is a constant C such that, for every function f on G which is integrable with respect to the measure μ,

$$\int_G f(x)\mu(dx) = C \int_{U \times T} f(ut)\alpha(du) \prod_{i=1}^n t_{ii}^{-2(i-1)-1}\lambda(dt).$$

6. Let G be a linear Lie group and φ the diffeomorphism of G given by $\varphi(x) = x^{-1}$.

(a) Show that the differential of φ at the identity element e in G equals $-Id$:

$$(D\varphi)_e X = -X \quad (X \in Lie(G)).$$

Show then that, for every $g \in G$,

$$(D\varphi)_g(gX) = g^{-1}(-\operatorname{Ad}(g)X).$$

(b) Let ω be a differential form on G of degree $m = \dim G$ which is left invariant. Show that

$$\varphi^*\omega = \det(-\operatorname{Ad}(x))\omega.$$

(c) Show that this result provides an alternative proof of Proposition 5.5.4:

$$\Delta(x) = |\det \operatorname{Ad}(x^{-1})|.$$

7. (a) The *Cayley transform* φ is the bijection of $\mathbb{C} \setminus \{-1\}$ given by

$$\varphi(z) = \frac{1-z}{1+z}.$$

Check that φ is involutive: $\varphi \circ \varphi = Id$. Determine the image of $i\mathbb{R}$. Let $\mathbb{U}$ be the unit circle, and let μ be the measure on $\mathbb{U}$ defined by

$$\int_{\mathbb{U}} f(u)\mu(du) = \frac{1}{2\pi} \int_0^{2\pi} f(e^{i\theta})d\theta.$$

Show that the image of the measure μ through the map φ is the Cauchy measure:

$$\frac{1}{2\pi} \int_{-\pi}^{\pi} f(\varphi(e^{i\theta}))d\theta = \frac{1}{\pi} \int_{-\infty}^{\infty} f(it)\frac{dt}{1+t^2}.$$

(b) Let $X \in M(n, \mathbb{R})$ be such that $\det(I + X) \neq 0$. The Cayley transform of X is defined by

$$\varphi(X) = (I - X)(I + X)^{-1}.$$

Show that φ is a diffeomorphism of

$$\mathcal{D}(\varphi) = \{X \in M(n, \mathbb{R}) \mid \det(I + X) \neq 0\},$$

and that its differential is given by:

$$(D\varphi)_X Y = -2(I + X)^{-1}Y(I + X)^{-1}.$$

(c) Let $V = Skew(n, \mathbb{R})$ be the space of real skewsymmetric $n \times n$ matrices. To a matrix $A \in M(n, \mathbb{R})$ one associates the endomorphism $T(A)$ of V defined by

$$T(A)X = AXA^T.$$

Show that

$$\operatorname{Det} T(A) = (\det A)^{n-1}.$$

(d) Determine the image through φ of V.

(e) Let μ be the normalised Haar measure of the orthogonal group $K = SO(n)$. Establish the following integration formula: there is a constant $a_n > 0$ such that, if f is an integrable function on K, then

$$\int_K f(k)\mu(dk) = a_n \int_V f\big(\varphi(X)\big) \det(I + X)^{-(n-1)} \prod_{i<j} dx_{ij}.$$

Hint. Consider a differential form ω on K of degree $m = \dim K$ which is left invariant. Let $X, Y_1, \ldots, Y_m \in V$. Show that

$$(\varphi^*\omega)_X(Y_1, \ldots, Y_m) = \omega_{\varphi(X)}\big(\varphi(X)A(X)Y_1, \ldots, \varphi(X)A(X)Y_m\big),$$

where $A(X)$ is an endomorphism of V to be determined.

(f) Show that

$$a_2 = \frac{1}{\pi}, \quad a_3 = \frac{1}{\pi^2}.$$

Hint. Apply the integration formula to the function $f = 1$. Show that

$$\frac{1}{a_3} = \int_{\mathbb{R}^3} \big(1 + x_{12}^2 + x_{13}^2 + x_{23}^2\big)^{-2} dx_{12}dx_{13}dx_{23}.$$

(g) Consider the same questions for $M(n, \mathbb{C})$ instead of $M(n, \mathbb{R})$, *SkewHerm*$(n, \mathbb{C}) = i$*Herm*$(n, \mathbb{C})$ instead of *Skew*$(n, \mathbb{R})$, and $U(n)$ instead of $SO(n)$.

6

Representations of compact groups

In this chapter we present the Peter–Weyl theory for compact groups. By using spectral theory for compact operators we will see that an irreducible representation of a compact group is finite dimensional. Using the Peter–Weyl theory, classical Fourier analysis is extended to compact groups.

6.1 Unitary representations

Let G be a topological group and $\mathcal{V}$ a normed vector space over $\mathbb{R}$ or $\mathbb{C}$ ($\mathcal{V} \neq \{0\}$). Let $\mathcal{L}(\mathcal{V})$ denote the algebra of bounded operators on $\mathcal{V}$. A *representation* of G on $\mathcal{V}$ is a map

$$\pi : G \to \mathcal{L}(\mathcal{V}),$$
$$g \mapsto \pi(g),$$

such that

1. $\pi(g_1 g_2) = \pi(g_1)\pi(g_2)$, $\pi(e) = I$,
2. for every $v \in \mathcal{V}$, the map

$$G \to \mathcal{V},$$
$$g \mapsto \pi(g)v,$$

is continuous.

The definition, we give here, differs slightly from that given in Section 4.1, where we only considered the case of a finite dimensional vector space $\mathcal{V}$.

A subspace $\mathcal{W} \subset \mathcal{V}$ is said to be *invariant* if, for every $g \in G$, $\pi(g)\mathcal{W} = \mathcal{W}$. Putting $\pi_0(g) = \pi(g)|_{\mathcal{W}}$, the restriction of $\pi(g)$ to $\mathcal{W}$, we get a representation of G on $\mathcal{W}$. One says that π_0 is a *subrepresentation* of π. Assume $\mathcal{W}$ to be closed. The representation π_1 of G on the quotient space $\mathcal{V}/\mathcal{W}$ is called the *quotient representation*. The representation π is said to be *irreducible* if the only invariant closed subspaces are $\{0\}$ and $\mathcal{V}$. Observe that, by definition, a one dimensional representation is irreducible.

Let $(\pi_1, \mathcal{V}_1)$ and $(\pi_2, \mathcal{V}_2)$ be two representations of G. If a continuous linear map A from $\mathcal{V}_1$ into $\mathcal{V}_2$ satisfies the relation

$$A\pi_1(g) = \pi_2(g)A,$$

for every $g \in G$, one says that A is an *intertwinning operator* or that A *intertwins* the representations π_1 and π_2. The representations $(\pi_1, \mathcal{V}_1)$ and $(\pi_2, \mathcal{V}_2)$ are said to be *equivalent* if there exists an isomorphism $A : \mathcal{V}_1 \to \mathcal{V}_2$ which intertwins the representations π_1 and π_2.

Let $\mathcal{H}$ be a Hilbert space. Recall that an operator A on $\mathcal{H}$ is said to be *unitary* if it is invertible and $A^{-1} = A^*$. A representation π of G on $\mathcal{H}$ is said to be *unitary* if, for every $g \in G$, $\pi(g)$ is a unitary operator; this can be written

$$\forall g \in G, \ \forall v \in \mathcal{H}, \quad \|\pi(g)v\| = \|v\|.$$

If the representation π is unitary, and if $\mathcal{W}$ is an invariant subspace, then the orthogonal subspace $\mathcal{W}^\perp$ is invariant as well. If $\mathcal{W}$ is closed, the quotient representation on $\mathcal{H}/\mathcal{W}$ is equivalent to the subrepresentation on $\mathcal{W}^\perp$.

Proposition 6.1.1 *Let π be a representation of a compact group G on a finite dimensional vector space $\mathcal{V}$. There exists on $\mathcal{V}$ a Euclidean inner product for which π is unitary.*

Proof. Let us choose arbitrarily on $\mathcal{V}$ a Euclidean inner product $(\cdot|\cdot)_0$ and put

$$(u|v) = \int_G \big(\pi(g)u|\pi(g)v\big)_0 \, \mu(dg),$$

where μ is a Haar measure on G. One can check easily that $(\cdot|\cdot)$ is a Euclidean inner product on $\mathcal{V}$, and that the representation π is unitary with respect to this Euclidean inner porduct. $\qquad\square$

Corollary 6.1.2 *Let π be a representation of a compact group G on a finite dimensional vector space $\mathcal{V}$.*

(i) For every invariant subspace there is an invariant complementary subspace.

(ii) The vector space $\mathcal{V}$ can be decomposed into a direct sum of irreducible invariant subspaces:

$$\mathcal{V} = \mathcal{V}_1 \oplus \cdots \oplus \mathcal{V}_N.$$

Proof. By Proposition 6.1.1 there exists on $\mathcal{V}$ a Euclidean inner product for which the representation π is unitary. If $\mathcal{W}$ is an invariant subspace, then the orthogonal subspace $\mathcal{W}^\perp$ is invariant and complementary to $\mathcal{W}$.

Let $\mathcal{V}_1$ be a non-zero invariant subspace with minimal dimension. Then

$$\mathcal{V} = \mathcal{V}_1 \oplus \mathcal{V}_1^\perp.$$

If $\mathcal{V}_1^\perp \neq \{0\}$, let $\mathcal{V}_2$ be a non-zero invariant subspace in $\mathcal{V}_1^\perp$ with minimal dimension. One continues the process as long as the subspace $\mathcal{V}_k^\perp$ is not zero. Since the dimension of $\mathcal{V}$ is finite, the process stops necessarily. □

Theorem 6.1.3 (Schur's Lemma) (i) *Let $(\pi_1, \mathcal{V}_1)$ and $(\pi_2, \mathcal{V}_2)$ be two finite dimensional irreducible representations of a topological group G. Let $A : \mathcal{V}_1 \to \mathcal{V}_2$ be a linear map which intertwins the representations π_1 and π_2:*

$$A\pi_1(g) = \pi_2(g)A$$

for every $g \in G$. Then either $A = 0$, or A is an isomorphism.

(ii) *Let π be an irreducible $\mathbb{C}$-linear representation of a topological group G on a finite dimensional complex vector space $\mathcal{V}$. Let $A : \mathcal{V} \to \mathcal{V}$ be a $\mathbb{C}$-linear map which commutes to the representation π:*

$$A\pi(g) = \pi(g)A$$

for every $g \in G$. Then there exists $\lambda \in \mathbb{C}$ such that

$$A = \lambda I.$$

Proof. (i) In fact, the kernel $\ker(A)$ and the image $\mathrm{Im}(A)$ are two invariant subspaces. The statement follows immediately.

(ii) There exists $\lambda \in \mathbb{C}$ such that $A - \lambda I$ is not invertible. It follows from (i) that

$$A - \lambda I = 0. \qquad\qquad □$$

If the group G is commutative, by Schur's Lemma an irreducible $\mathbb{C}$-linear representation is one dimensional. It is a character of G. In this setting a *character* is defined as a continuous function $\chi : G \to \mathbb{C}$ satisfying

$$\chi(xy) = \chi(x)\chi(y).$$

For instance the characters of the group $G = SO(2) \simeq U(1) \simeq \mathbb{R}/2\pi\mathbb{Z}$ are the functions

$$\chi_m(\theta) = e^{im\theta} \quad (m \in \mathbb{Z}).$$

In part (ii) of Theorem 6.1.3, the assumption that the representation π is $\mathbb{C}$-linear cannot be dropped. For the $\mathbb{R}$-linear representations the situation is quite different. Consider for instance the representation π of the group $G = SO(2) \simeq \mathbb{R}/2\pi\mathbb{Z}$ on $\mathbb{R}^2$ defined by

$$\pi(\theta) = \begin{pmatrix} \cos\theta & -\sin\theta \\ \sin\theta & \cos\theta \end{pmatrix}.$$

This representation is irreducible. But the matrices

$$A = \begin{pmatrix} a & b \\ -b & a \end{pmatrix} \quad (a, b \in \mathbb{R}),$$

commute with the matrices $\pi(\theta)$. (See Exercise 4 about irreducible $\mathbb{R}$-linear representations.)

In the same way one can establish similar statements for representations of Lie algebras.

Proposition 6.1.4 (i) *Let $(\rho_1, \mathcal{V}_1)$ and $(\rho_2, \mathcal{V}_2)$ be two finite dimensional irreducible representations of a Lie algebra $\mathfrak{g}$. Let $A : \mathcal{V}_1 \to \mathcal{V}_2$ be a linear map which intertwins the representations ρ_1 and ρ_2:*

$$A\rho_1(X) = \rho_2(X)A$$

for every $X \in \mathfrak{g}$. Then either $A = 0$, or A is an isomorphism.

(ii) *Let ρ be a $\mathbb{C}$-linear representation of a complex Lie algebra $\mathfrak{g}$ on a finite dimensional complex vector space $\mathcal{V}$. Let $A : \mathcal{V} \to \mathcal{V}$ be a $\mathbb{C}$-linear map which commutes with the representation ρ:*

$$A\rho(X) = \rho(X)A$$

for every $X \in \mathfrak{g}$. Then there exists $\lambda \in \mathbb{C}$ such that

$$A = \lambda I.$$

6.2 Compact self-adjoint operators

Let A be a bounded operator on a Hilbert space $\mathcal{H}$. Its norm $\|A\|$ is defined by

$$\|A\| = \sup_{\|u\| \leq 1} \|Au\|.$$

For v fixed, the map

$$u \mapsto (Au|v)$$

is a continuous linear form on $\mathcal{H}$. By the Riesz representation theorem there exists a unique $w \in \mathcal{H}$ such that

$$(Au|v) = (u|w)$$

for every $u \in \mathcal{H}$. The map $v \mapsto w$ is linear, it is denoted by A^* and is called the *adjoint operator* of A. It is defined by the relation

$$(Au|v) = (u|A^*v).$$

One can show that $\|A^*\| = \|A\|$ and that $(A^*)^* = A$. If $A^* = A$, one says that the operator A is *self-adjoint*, that is

$$(Au|v) = (u|Av)$$

for every $u, v \in \mathcal{H}$.

Proposition 6.2.1 *Let A be a self-adjoint operator.*

(i) *The eigenvalues of A are real.*

(ii) *If λ and μ are distinct eigenvalues of A, the corresponding eigenspaces are orthogonal.*

Proof. (a) Let λ be an eigenvalue of A, and u an associated eigenvector:

$$Au = \lambda u, \ u \neq 0.$$

Then

$$(Au|u) = (u|Au) \quad \text{and} \quad \lambda\|u\|^2 = \bar{\lambda}\|u\|^2.$$

(b) Let λ and μ be two eigenvalues of A, $\lambda \neq \mu$, u and v associated eigenvectors:

$$Au = \lambda u, \ Av = \mu v.$$

Then

$$(Au|v) = (u|Av) \quad \text{and} \quad (\lambda - \mu)(u|v) = 0. \qquad \square$$

Proposition 6.2.2 *Let A be a self-adjoint operator. Then*

$$\|A\| = \sup_{\|u\| \leq 1} |(Au|u)|.$$

Proof. Put

$$M = \sup_{\|u\| \leq 1} |(Au|u)|.$$

Observe first that, by the Schwarz inequality, $M \leq \|A\|$. On the other hand,

$$\|A\| = \sup_{\|u\| \leq 1, \|v\| \leq 1} |\operatorname{Re}(Au|v)|.$$

In fact, for $w \in \mathcal{H}$,

$$\|w\| = \sup_{\|v\| \leq 1} |\operatorname{Re}(w|v)|,$$

and, by definition of the norm of an operator,

$$\|A\| = \sup_{\|u\| \le 1} \|Au\|.$$

From the identity

$$4\operatorname{Re}(Au|v) = \big(A(u+v)|u+v\big) - \big(A(u-v)|u-v\big),$$

it follows that

$$|\operatorname{Re}(Au|v)| \le \frac{M}{4}(\|u+v\|^2 + \|u-v\|^2) = \frac{M}{2}(\|u\|^2 + \|v\|^2).$$

Hence, if $\|u\| \le 1$, $\|v\| \le 1$,

$$|\operatorname{Re}(Au|v)| \le M,$$

therefore $\|A\| \le M$.   □

Let A be an operator acting on $\mathcal{H}$. The operator A is said to be *compact* if the following property holds:

the image under A of a bounded set is relatively compact.

This property is equivalent to each of the two following:

the image under A of the unit ball is relatively compact;
if (u_n) is a bounded sequence, there is a subsequence (u_{n_k}) such that the sequence (Au_{n_k}) converges.

A finite rank operator is compact. If A is a compact operator and B a bounded operator, then AB and BA are compact operators.

Proposition 6.2.3 *If (A_n) is a sequence of compact operators with limit A,*

$$\lim_{n \to \infty} \|A_n - A\| = 0,$$

then the operator A is compact.

Proof. Let (u_k) be a sequence in $\mathcal{H}$ such that $\|u_k\| \le 1$. Since the operator A_1 is compact, there is a subsequence $(u_k^{(1)})$ such that the sequence $(A_1 u_k^{(1)})$ converges. Since the operator A_2 is compact, one can extract from the subsequence $(u_k^{(1)})$ a subsequence $(u_k^{(2)})$ such that $(A_2 u_k^{(2)})$ converges, and so on. Then one considers the sequence $(u_k') = (u_k^{(k)})$. For every n the sequence $k \mapsto A_n u_k'$ converges. Let us show that (Au_k') is a Cauchy sequence. Let $\varepsilon > 0$. There exists n such that

$$\|A_n - A\| \le \frac{\varepsilon}{3}.$$

Since $(A_n u'_k)$ is a Cauchy sequence, there exists $K > 0$ such that, if $k, \ell \geq K$,

$$\| A_n u'_k - A_n u'_\ell \| \leq \frac{\varepsilon}{3}.$$

Hence, if $k, \ell \geq K$,

$$\| A u'_k - A u'_\ell \| \leq \| A u'_k - A_n u'_k \| + \| A_n u'_k - A_n u'_\ell \| + \| A_n u'_\ell - A u'_\ell \| \leq \varepsilon.$$

$\square$

Finally we can state that the set of compact operators is a closed two-sided ideal in $\mathcal{L}(\mathcal{H})$.

Example. Let $\mathcal{H} = \ell^2(\mathbb{N})$. Let (λ_n) be a sequence of complex numbers with limit 0, and let $A \in \mathcal{L}(\mathcal{H})$ be defined by

$$A(u_n) = (\lambda_n u_n).$$

The operator A is compact. In fact, let A_N be the operator defined as follows: if $v = A_N u$,

$$v_n = \lambda_n u_n \quad \text{if } n \leq N,$$
$$v_n = 0 \qquad \text{if } n > N.$$

The operator A_N has finite rank and

$$\| A - A_N \| = \sup_{n > N} |\lambda_n|.$$

Theorem 6.2.4 *Let A be a compact self-adjoint operator. Then, either $\| A \|$ or $-\| A \|$ is an eigenvalue of A.*

Hence, a non-zero compact self-adjoint operator has a non-zero eigenvalue.

Proof. Since the operator A is self-adjoint,

$$\| A \| = \sup_{\| u \| \leq 1} |(A u | u)|$$

(Proposition 6.2.2). Observe that the numbers $(Au|u)$ are real; one may assume, by taking $-A$ instead of A if necessary, that

$$\| A \| = \sup_{\| u \| \leq 1} (A u | u).$$

Let us then show that $\lambda = \| A \|$ is an eigenvalue of A. There is a sequence (u_n)

such that

$$\|u_n\| = 1, \quad \lim_{n \to \infty} (Au_n | u_n) = \lambda.$$

Since the operator A is compact, there is a subsequence (u_{n_k}) such that the sequence (Au_{n_k}) converges:

$$\lim_{k \to \infty} Au_{n_k} = v.$$

From the expansion

$$\|Au_{n_k} - \lambda u_{n_k}\|^2 = \|Au_{n_k}\|^2 - 2\lambda(Au_{n_k}|u_{n_k}) + \lambda^2$$

it follows that

$$\lim_{k \to \infty} \|Au_{n_k} - \lambda u_{n_k}\|^2 = \|v\|^2 - \lambda^2.$$

On the other hand, since $\|A\| = \lambda$,

$$\|v\| = \lim_{k \to \infty} \|Au_{n_k}\| \le \lambda,$$

hence

$$\lim_{k \to \infty} \|Au_{n_k} - \lambda u_{n_k}\| = 0.$$

Therefore the sequence (u_{n_k}) converges:

$$\lim_{k \to \infty} u_{n_k} = u.$$

Furthermore $Au = v$ and $Au = \lambda u$. $\qquad\qquad\qquad\qquad\qquad\qquad$ $\square$

Theorem 6.2.5 (Spectral theorem) *Let A be a compact self-adjoint operator. The non-zero eigenvalues of A form a sequence (λ_n) which is finite or converges to 0. Let $\mathcal{H}_n$ be the eigenspace associated to λ_n and let P_n be the orthogonal projection onto $\mathcal{H}_n$. The dimension of $\mathcal{H}_n$ is finite and*

$$A = \sum_{n=1}^{N} \lambda_n P_n,$$

if the number N of non-zero eigenvalues is finite, otherwise

$$A = \sum_{n=0}^{\infty} \lambda_n P_n,$$

as a convergent series in the norm topology.

Lemma 6.2.6 *Let $\mathcal{H}$ be a Hilbert space. If the unit ball in $\mathcal{H}$ is compact, then $\mathcal{H}$ is finite dimensional.*

Proof. If $\mathcal{H}$ were not finite dimensional, there would be in $\mathcal{H}$ an infinite orthonormal sequence (e_n). Since

$$\|e_p - e_q\| = \sqrt{2}$$

for $p \neq q$, there cannot be a converging subsequence. $\qquad\square$

Let A be a self-adjoint operator, and λ a non-zero eigenvalue of A. From this lemma it follows that the associated eigenspace is finite dimensional.

Proof of Theorem 6.2.5 By Theorem 6.2.4 there exists an eigenvalue λ_1 of A such that $|\lambda_1| = \|A\|$. Let $\mathcal{H}_1$ be the associated eigenspace. From Lemma 6.2.6 it follows that $\mathcal{H}_1$ is finite dimensional. Put $A_1 = A - \lambda_1 P_1$. The operator A_1 is self-adjoint and compact, and $\|A_1\| \leq \|A\|$. By continuing, either one finds an integer N such that $A_N = 0$, and then

$$A = \sum_{n=1}^{N} \lambda_n P_n,$$

or the sequence (λ_n) is infinite. Observe that the sequence $(|\lambda_n|)$ is decreasing by construction. Let us show that, when infinite, the sequence (λ_n) goes to 0. Let us assume the opposite, that $|\lambda_n| \geq \alpha > 0$. For every n let us take $v_n \in \mathcal{H}_n$, $\|v_n\| = 1$. Since A is compact, one can extract from the sequence (Av_n) a converging subsequence. But this is impossible since

$$\|Av_p - Av_q\|^2 = \|\lambda_p v_p - \lambda_q v_q\|^2 = \lambda_p^2 + \lambda_q^2 \geq 2\alpha^2.$$

It follows that

$$A = \sum_{n=1}^{\infty} \lambda_n P_n.$$

In fact,

$$A = \sum_{n=1}^{N} \lambda_n P_n + A_{N+1},$$

and $\|A_{N+1}\| = |\lambda_{N+1}|$. Finally, the dimension of $\mathcal{H}_n$ is finite since the unit ball of $\mathcal{H}_n$ is compact. $\qquad\square$

6.3 Schur orthogonality relations

Let G be a compact group, and μ the normalised Haar measure of G. Let $(\pi, \mathcal{H})$ be a unitary representation of G. For $v \in \mathcal{H}$ one considers the operator K_v of

$\mathcal{H}$ defined by

$$K_v w = \int_G (w|\pi(g)v)\pi(g)v\, \mu(dg).$$

This can also be written

$$(K_v w|w') = \int_G (w|\pi(g)v)\overline{(w'|\pi(g)v)}\, \mu(dg).$$

Proposition 6.3.1 (i) K_v *is bounded,* $\|K_v\| \leq \|v\|^2$.
(ii) K_v *is self-adjoint:* $K_v^* = K_v$.
(iii) K_v *commutes with the representation* π: *for every* $g \in G$,

$$K_v\,\pi(g) = \pi(g)\,K_v.$$

(iv) K_v *is a compact operator.*

Proof.
(i) $$\|K_v w\| \leq \|v\|^2\|w\|.$$

(ii) $$(K_v^* w|w') = (w|K_v w') = (K_v w|w').$$

(iii) Let $g_0 \in G$,

$$K_v\big(\pi(g_0)w\big) = \int_G (w|\pi(g_0^{-1}g)v)\pi(g)v\, \mu(dg),$$

and, by the invariance of the measure μ,

$$K_v\big(\pi(g_0)w\big) = \int_G (w|\pi(g)v)\pi(g_0 g)v\, \mu(dg) = \pi(g_0)K_v w.$$

(iv) For $v \in \mathcal{H}$ let P_v be the rank one operator defined by

$$P_v w = (w|v)v.$$

It is a compact operator and, for v fixed, the map

$$G \to \mathcal{L}(\mathcal{H}),$$

$$g \mapsto P_{\pi(g)v},$$

is continuous for the norm topology. The operator K_v can be written

$$K_v = \int_G P_{\pi(g)v}\mu(dg).$$

Since the space of compact operators is closed for the norm topology (Proposition 6.2.3), the operator K_v is compact. $\qquad\square$

Observe that

$$(K_v w|w) = \int_G |(\pi(g)v|w)|^2\mu(dg) \geq 0,$$

and that, if $v \neq 0$,

$$(K_v v | v) > 0,$$

hence $K_v \neq 0$.

Theorem 6.3.2 (i) *Every unitary representation of a compact group contains a finite dimensional subrepresentation.*

(ii) *Every irreducible unitary representation of a compact group is finite dimensional.*

Proof. Let $(\pi, \mathcal{H})$ be a unitary representation of a compact group. The operator K_v is self-adjoint, compact (Proposition 6.3.1), and non-zero if $v \neq 0$. By Theorem 6.2.4 it has a non-zero eigenvalue, and the corresponding eigenspace is finite dimensional. This subspace is invariant under the representation π. □

Theorem 6.3.3 *Let π be an irreducible unitary $\mathbb{C}$-linear representation of a compact group G on a complex Euclidean vector space $\mathcal{H}$ with dimension d_π. Then, for $u, v \in \mathcal{H}$,*

$$\int_G |(\pi(g)u|v)|^2 \mu(dg) = \frac{1}{d_\pi} \|u\|^2 \|v\|^2,$$

and, by polarisation, for $u, v, u', v' \in \mathcal{H}$,

$$\int_G (\pi(g)u|v)\overline{(\pi(g)u'|v')} \mu(dg) = \frac{1}{d_\pi}(u|u')\overline{(v|v')}.$$

Proof. For $v \in \mathcal{H}$, the operator K_v commutes with the representation π. By Schur's Lemma (Theorem 6.1.3) there is $\lambda(v) \in \mathbb{C}$ such that

$$K_v = \lambda(v)I.$$

Hence,

$$\int_G |(\pi(g)u|v)|^2 \mu(dg) = \lambda(v)\|u\|^2.$$

By permuting u and v we get

$$\lambda(u)\|v\|^2 = \lambda(v)\|u\|^2,$$

hence $\lambda(u) = \lambda_0 \|u\|^2$, where λ_0 is a constant. Let $\{e_1, \ldots, e_n\}$ be an orthonormal basis of $\mathcal{H}$ ($n = d_\pi$):

$$\sum_{i=1}^n |(\pi(g)u|e_i)|^2 = \|u\|^2.$$

By integration over G we get

$$\|u\|^2 = \sum_{i=1}^{n} \int_G |(\pi(g)u|e_i)|^2 \mu(dg) = n\lambda_0 \|u\|^2,$$

hence $\lambda_0 = 1/n$. Finally

$$\int_G |(\pi(g)u|v)|^2 \mu(dg) = \frac{1}{n} \|u\|^2 \|v\|^2. \qquad \square$$

Let $\pi_{ij}(g)$ denote the entries of the matrix $\pi(g)$ with respect to the basis $\{e_i\}$,

$$\pi_{ij}(g) = (\pi(g)e_j|e_i).$$

From the preceding theorem one obtains *Schur's orthogonality relations*:

$$\int_G \pi_{ij}(g)\overline{\pi_{k\ell}(g)}\mu(dg) = \frac{1}{d_\pi}\delta_{ik}\delta_{j\ell}.$$

This can be written in the following alternative form: if A and B are two endomorphisms of $\mathcal{H}$, then

$$\int_G \mathrm{tr}\big(A\pi(g)\big)\overline{\mathrm{tr}\big(B\pi(g)\big)}\mu(dg) = \frac{1}{d_\pi}\,\mathrm{tr}(AB^*).$$

In fact one can check that, if A and B are two rank one endomorphisms, the above formula is precisely the second formula of the preceding theorem.

Let $\mathcal{M}_\pi$ denote the subspace of $L^2(G)$ generated by the entries of the representation π, that is by the functions of the following form:

$$g \mapsto (\pi(g)u|v) \ (u, v \in \mathcal{H}).$$

Theorem 6.3.4 *Let* $(\pi, \mathcal{H})$ *and* $(\pi', \mathcal{H}')$ *be two irreducible unitary representations of a compact group* G *which are not equivalent. Then* $\mathcal{M}_\pi$ *and* $\mathcal{M}_{\pi'}$ *are two orthogonal subspaces of* $L^2(G)$:

$$\int_G (\pi(g)u|v)\overline{(\pi'(g)u'|v')}\mu(dg) = 0 \quad (u, v \in \mathcal{H}, u', v' \in \mathcal{H}').$$

Proof. Let A be a linear map from $\mathcal{H}$ into $\mathcal{H}'$ and put

$$\tilde{A} = \int_G \pi'(g^{-1})A\pi(g)\mu(dg).$$

Then $\tilde{A}$ is a linear map from $\mathcal{H}$ into $\mathcal{H}'$ which intertwins the representations π and π',

$$\tilde{A} \circ \pi(g) = \pi'(g) \circ \tilde{A}.$$

By Schur's Lemma (Theorem 6.1.3), $\tilde{A} = 0$. Hence

$$(\tilde{A}u|u') = \int_G (A\pi(g)u|\pi'(g)u')\mu(dg) = 0.$$

Take for A the rank one operator defined by

$$Au = (u|v)v' \quad (v \in \mathcal{H}, v' \in \mathcal{H}'),$$

then

$$A\pi(g)u = (\pi(g)u|v)v',$$

and

$$\int_G (\pi(g)u|v)\overline{(\pi(g)u'|v')}\mu(dg) = 0. \qquad \square$$

It follows that two irreducible representations π_1 and π_2 of a compact group G are equivalent if and only if the spaces $\mathcal{M}_{\pi_1}$ and $\mathcal{M}_{\pi_2}$ agree.

6.4 Peter–Weyl Theorem

Let G be a compact group, and let R denote the right regular representation of G on $L^2(G)$:

$$(R(g)f)(x) = f(xg).$$

Let $(\pi, \mathcal{H})$ be an irreducible representation of G, and let $\{e_1, \ldots, e_n\}$ be an orthonormal basis of $\mathcal{H}$ $(n = d_\pi)$. One puts

$$\pi_{ij}(x) = (\pi(x)e_j|e_i).$$

Let $\mathcal{M}_\pi^{(1)}$ be the subspace of $\mathcal{M}_\pi$ generated by the entries of the first row, that is by the functions $x \mapsto \pi_{1j}(x)$, for $j = 1, \ldots, n$. Observe that

$$\pi_{1j}(xg) = \sum_{k=1}^n \pi_{1k}(x)\pi_{kj}(g).$$

This shows that the subspace $\mathcal{M}_\pi^{(1)}$ is invariant under R. Furthermore, the map

$$A : \sum_{j=1}^n c_j e_j \mapsto \sum_{j=1}^n c_j \pi_{1j}(x)$$

from $\mathcal{H}$ into $\mathcal{M}_\pi^{(1)}$ is an isomorphism, and intertwins the representations π and R. In fact, if $u = \sum_{j=1}^{n} c_j e_j$, then

$$A\pi(g)u = A \sum_{j=1}^{n} c_j \pi(g)e_j = A \sum_{j=1}^{n} c_j \left(\sum_{i=1}^{n} \pi_{ij}(g)e_i \right)$$

$$= \sum_{i=1}^{n} \left(\sum_{j=1}^{n} \pi_{ij}(g)c_j \right) \pi_{1i}(x) = \sum_{j=1}^{n} c_j \pi_{1j}(xg) = R(g)Au.$$

Furthermore

$$\|Au\|^2 = \frac{1}{n} \|u\|^2.$$

Let $\mathcal{M}_\pi^{(i)}$ denote the subspace of $\mathcal{M}_\pi$ generated by the coefficients of the ith line. Then

$$\mathcal{M}_\pi = \mathcal{M}_\pi^{(1)} \oplus \cdots \oplus \mathcal{M}_\pi^{(n)},$$

and the restriction to $\mathcal{M}_\pi$ of the representation R is equivalent to

$$\pi \oplus \cdots \oplus \pi = n\pi.$$

By considering the columns instead of the rows one gets the same statement with, instead of the representation R, the regular left representation L:

$$\big(L(g)f\big)(x) = f(g^{-1}x).$$

Theorem 6.4.1 (Peter–Weyl Theorem) *Let $\hat{G}$ be the set of equivalence classes of irreducible unitary representations of the compact group G and, for each $\lambda \in \hat{G}$, let $\mathcal{M}_\lambda$ be the space generated by the coefficients of a representation in the class λ. Then*

$$L^2(G) = \widehat{\bigoplus_{\lambda \in \hat{G}}} \mathcal{M}_\lambda.$$

Recall that

$$\widehat{\bigoplus_{\lambda \in \hat{G}}} \mathcal{M}_\lambda$$

denotes the closure in $L^2(G)$ of

$$\mathcal{M} = \bigoplus_{\lambda \in \hat{G}} \mathcal{M}_\lambda,$$

which is the space of finite linear combinations of coefficients of finite dimensional representations of G.

Proof. We saw that the subspaces $\mathcal{M}_\lambda$ are two by two orthogonal (Theorem 6.3.4). Put

$$\mathcal{H} = \widehat{\bigoplus_{\lambda \in \hat{G}}} \mathcal{M}_\lambda,$$

and

$$\mathcal{H}_0 = \mathcal{H}^\perp.$$

We will show that $\mathcal{H}_0 = \{0\}$. Let us assume the opposite, that $\mathcal{H}_0 \neq \{0\}$. The space $\mathcal{H}_0$ is invariant under the representation R and closed. By Theorem 6.3.2 it contains a closed subspace $\mathcal{Y} \neq \{0\}$ which is invariant under R and irreducible. The restriction of R to $\mathcal{Y}$ belongs to one of the classes λ. Let $f \in \mathcal{Y}$, $f \neq 0$, and put

$$F(g) = \int_G f(xg)\overline{f(x)}\mu(dx) = (R(g)f|f).$$

The function F belongs to $\mathcal{M}_\lambda$. We will see that it is orthogonal to $\mathcal{M}_\lambda$. Let $(\pi, \mathcal{V})$ be a representation of the class λ, and $u, v \in \mathcal{V}$. Then

$$\int_G F(g)\overline{(\pi(g)u|v)}\mu(dg) = \int_G \int_G f(xg)\overline{f(x)}\,\overline{(\pi(g)u|v)}\mu(dg)\mu(dx),$$

and, by putting $xg = g'$,

$$\int_G F(g)\overline{(\pi(g)u|v)}\mu(dg) = \int_G \overline{f(x)}\left(\int_G f(g')\overline{(\pi(g')u|\pi(x)v)}\mu(dg')\right)\mu(dx) = 0.$$

Therefore $F = 0$, and, since

$$F(e) = \int_G |f(x)|^2 \mu(dx),$$

it follows that $f = 0$. This yields a contradiction. $\qquad\square$

Let $\mathcal{H}$ be a finite dimensional Hilbert space and $A \in \mathcal{L}(\mathcal{H})$. The *Hilbert–Schmidt norm* of A is defined by

$$\|\|A\|\|^2 = \mathrm{tr}(AA^*).$$

If $\{e_1, \ldots, e_n\}$ is an orthonormal basis of $\mathcal{H}$, and if (a_{ij}) is the matrix of A with respect to this basis,

$$\|\|A\|\|^2 = \sum_{i,j=1}^{n} |a_{ij}|^2.$$

For every $\lambda \in \hat{G}$ one chooses a representative $(\pi_\lambda, \mathcal{H}_\lambda)$. Let d_λ denote the dimension of $\mathcal{H}_\lambda$. If f is an integrable function on G, its Fourier coefficient

$\hat{f}(\lambda)$ is the operator acting on the space $\mathcal{H}_\lambda$ defined by

$$\hat{f}(\lambda) = \int_G f(g)\pi_\lambda(g^{-1})\mu(dg).$$

The following theorem follows directly from the Peter–Weyl Theorem and from Schur's orthogonality relations.

Theorem 6.4.2 (Plancherel's Theorem) *Let $f \in L^2(G)$. Then f is equal to the sum of its Fourier series:*

(i) $$f(g) = \sum_{\lambda \in \hat{G}} d_\lambda \operatorname{tr}\big(\hat{f}(\lambda)\pi_\lambda(g)\big).$$

This holds in the L^2 sense.

(ii) $$\int_G |f(g)|^2 \mu(dg) = \sum_{\lambda \in \hat{G}} d_\lambda \|\|\hat{f}(\lambda)\|\|^2.$$

And, if $f_1, f_2 \in L^2(G)$,

$$\int_G f_1(g)\overline{f_2(g)}\mu(dg) = \sum_{\lambda \in \hat{G}} d_\lambda \operatorname{tr}\big(\hat{f}_1(\lambda)\hat{f}_2(\lambda)^*\big).$$

(iii) *The map $f \mapsto \hat{f}$ is a unitary isomorphism from $L^2(G)$ onto the space of sequences of operators $A = (A_\lambda)\ \big(A_\lambda \in \mathcal{L}(\mathcal{H}_\lambda)\big)$, for which*

$$\|A\|^2 = \sum_{\lambda \in \hat{G}} d_\lambda \|\|A_\lambda\|\|^2 < \infty,$$

and equipped with this norm.

If the compact group G is commutative then a $\mathbb{C}$-linear irreducible representation is one dimensional, and $\hat{G}$ is the set of continuous characters. Recall that, in this setting, a continuous *character* is a continuous function

$$\chi : G \to \mathbb{C}^*,$$

satisfying

$$\chi(xy) = \chi(x)\chi(y).$$

Since G is compact, the set $\chi(G)$ is a compact subgroup of $\mathbb{C}^*$, hence consists of modulus one complex numbers. Therefore

$$\chi : G \to \{z \in \mathbb{C} \mid |z| = 1\}.$$

The set $\hat{G}$ is a commutative group for the ordinary product of the characters which is called the dual group of G, and the continuous characters form a Hilbert basis of $L^2(G)$.

The Fourier coefficient $\hat{f}(\chi)$ of a square integrable function f on G is given by

$$\hat{f}(\chi) = \int_G f(x)\overline{\chi(x)}\mu(dx).$$

The Fourier series of f is written as:

$$\sum_{\chi \in \hat{G}} \hat{f}(\chi)\chi(x),$$

and the Plancherel formula:

$$\int_G |f(x)|^2\mu(dx) = \sum_{\chi \in \hat{G}} |\hat{f}(\chi)|^2.$$

For instance, if $G = SO(2) \simeq U(1) \simeq \mathbb{R}/2\pi\mathbb{Z}$, then a character χ has the form

$$\chi(\theta) = e^{im\theta},$$

where $m \in \mathbb{Z}$. Hence $\hat{G} \simeq \mathbb{Z}$. In this case one obtains the classical formulae. If f is an integrable function on $\mathbb{R}/2\pi\mathbb{Z}$, the Fourier coefficients of f are given by

$$\hat{f}(m) = \frac{1}{2\pi} \int_0^{2\pi} f(\theta)e^{-im\theta}d\theta.$$

The Fourier series of f is written as

$$\sum_{m \in \mathbb{Z}} \hat{f}(m)e^{im\theta},$$

and the Plancherel formula, if f is square integrable,

$$\frac{1}{2\pi} \int_0^{2\pi} |f(\theta)|^2 d\theta = \sum_{m \in \mathbb{Z}} |\hat{f}(m)|^2.$$

Recall that $\mathcal{M}$ denotes the space of finite linear combinations of coefficients of finite dimensional representations of G,

$$\mathcal{M} = \bigoplus_{\lambda \in \hat{G}} \mathcal{M}_\lambda.$$

We will show that $\mathcal{M}$ is dense in the space of continuous complex valued functions on G. For that we will apply the Stone–Weierstrass Theorem which we recall below.

Theorem 6.4.3 (Stone–Weierstrass Theorem) *Let X be a compact topological space, and $C(X)$ the space of complex valued continuous functions on X,* *equipped with the topology of uniform convergence. Let $\mathcal{A}$ be a subspace of $C(X)$ with the following properties:*

(i) *$\mathcal{A}$ is an algebra (for the ordinary product of functions),*
(ii) *$\mathcal{A}$ separates the points of X, and constant functions belong to $\mathcal{A}$,*
(iii) *if f belongs to $\mathcal{A}$, then $\bar{f}$ also belongs to $\mathcal{A}$.*

Then $\mathcal{A}$ is dense in $C(X)$.

See for instance: K. Yosida (1968). *Functional Analysis.* Springer (Corollary 2, p. 10).

Let $(\pi_1, \mathcal{H}_1)$ and $(\pi_2, \mathcal{H}_2)$ be two finite dimensional representations of G. The tensor product $\pi_1 \otimes \pi_2$ is the representation of G on $\mathcal{H}_1 \otimes \mathcal{H}_2$ such that

$$(\pi_1 \otimes \pi_2)(g)(u_1 \otimes u_2) = \pi_1(g)u_1 \otimes \pi_2(g)u_2.$$

If $\mathcal{H}_1$ and $\mathcal{H}_2$ are finite dimensional Hilbert spaces, then $\mathcal{H}_1 \otimes \mathcal{H}_2$ is equipped with an inner product such that

$$(u_1 \otimes u_2 | v_1 \otimes v_2) = (u_1 | v_1)(u_2 | v_2),$$

and

$$\big((\pi_1 \otimes \pi_2)(g)(u_1 \otimes u_2) | (v_1 \otimes v_2)\big) = (\pi_1(g)u_1 | v_1)(\pi_2(g)u_2 | v_2).$$

Therefore the product of a coefficient of π_1 and a coefficient of π_2 is a coefficient of $\pi_1 \otimes \pi_2$.

For $\lambda, \mu \in \hat{G}$ the representation $\pi_\lambda \otimes \pi_\mu$ can be decomposed into a sum of irreducible representations:

$$\pi_\lambda \otimes \pi_\mu = \bigoplus_{\nu \in E(\lambda, \mu)} c(\lambda, \mu; \nu)\pi_\nu,$$

where $E(\lambda, \mu)$ is a finite subset of $\hat{G}$. The numbers $c(\lambda, \mu; \nu)$, which are positive integers, are called *Clebsch–Gordan coefficients*. This shows that the space $\mathcal{A}$ of finite linear combinations of coefficients of finite dimensional representations of G is an algebra.

Let $\mathcal{V}$ be a normed vector space, and $\mathcal{V}'$ its topological dual. Let π be a representation of G on $\mathcal{V}$. The *contragredient representation* of π is the representation π' of G on $\mathcal{V}'$ defined by

$$\langle \pi'(g)f, u \rangle = \langle f, \pi(g^{-1})u \rangle \quad (f \in \mathcal{V}', u \in \mathcal{V}).$$

Assume now that $\mathcal{V} = \mathcal{H}$ is a Hilbert space and that π is unitary. There is an antilinear isomorphism T from $\mathcal{H}$ onto $\mathcal{H}'$ defined by

$$\langle Tv, u \rangle = (u | v).$$

We will write $\bar{v} = Tv$, and $\bar{\mathcal{H}} = \mathcal{H}'$. The inner product on $\bar{\mathcal{H}}$ is defined by

$$(\bar{u}|\bar{v}) = (v|u).$$

The contragredient representation is then called the *conjugate representation*. One writes $\pi' = \bar{\pi}$,

$$\bar{\pi}(g) = T\pi(g)T^{-1}.$$

Hence

$$(\bar{\pi}(g)\bar{u}|\bar{v}) = (\overline{\pi(g)u}|\bar{v}) = (v|\pi(g)u) = \overline{(\pi(g)u|v)}.$$

If $\lambda \in \hat{G}$ is the class of π, the class of $\bar{\pi}$ will be denoted by $\bar{\lambda}$. From the preceding relation it follows that

$$\mathcal{M}_{\bar{\lambda}} = \overline{\mathcal{M}_\lambda},$$

and that

$$\overline{\mathcal{M}} = \mathcal{M}.$$

Lemma 6.4.4 *Let G be a compact group. If $g \neq e$, there exists a finite dimensional representation π of G such that $\pi(g) \neq I$.*

Proof. Let $g_0 \in G$ such that $\pi_\lambda(g_0) = I$ for every $\lambda \in \hat{G}$. Let f be a continuous function on G, and put

$$\varphi(x) = f(xg_0) - f(x).$$

Then

$$\hat{\varphi}(\lambda) = \pi_\lambda(g_0)\hat{f}(\lambda) - \hat{f}(\lambda) = 0.$$

By Theorem 6.4.2 it follows that

$$\int_G |f(xg_0) - f(x)|^2 \mu(dx) = 0,$$

hence $f(xg_0) = f(x)$, $f(g_0) = f(e)$, and therefore $g_0 = e$. □

From this lemma it follows that the space $\mathcal{M}$ separates the points of G. All assumptions of the Stone–Weierstrass Theorem hold, and we can state the following.

Theorem 6.4.5 *The space $\mathcal{M}$ is dense in $\mathcal{C}(G)$.*

Let $(\pi_1, \mathcal{H}_1)$ and $(\pi_2, \mathcal{H}_2)$ be two irreducible representations of G. The representation τ_1 of $G \times G$ on $\mathcal{H}_1 \otimes \mathcal{H}_2$ such that

$$\tau_1(g_1, g_2)(u_1 \otimes u_2) = \pi_1(g_1)u_1 \otimes \pi_2(g_2)u_2$$

is irreducible.

In particular we can take $\pi_1 = \pi$, and $\pi_2 = \bar{\pi}$. To an element $u \otimes \bar{v}$ in $\mathcal{H} \otimes \bar{\mathcal{H}}$ one associates the coefficient f of the representation π defined by

$$f(x) = (\pi(x)u|v),$$

and this map extends as an isomorphism from $\mathcal{H} \otimes \bar{\mathcal{H}}$ onto $\mathcal{M}_\pi$. The action of $G \times G$ on $\mathcal{M}_\pi$ can be written

$$\big(\tau_2(g_1, g_2)f\big)(x) = f(g_2^{-1}xg_1).$$

In fact, to $\pi(g)u \otimes \bar{v}$ corresponds

$$\big(R(g)f\big)(x) = (\pi(x)\pi(g)u|v),$$

and, to $u \otimes \bar{\pi}(g)\bar{v}$,

$$\big(L(g)f\big)(x) = (\pi(x)u|\pi(g)v).$$

To an element $u \otimes \bar{v}$ of $\mathcal{H} \otimes \bar{\mathcal{H}}$ one associates also the rank one operator $A \in \mathcal{L}(\mathcal{H})$ defined by

$$Aw = (w|v)u,$$

and this map extends as an isomorphism from $\mathcal{H} \otimes \bar{\mathcal{H}}$ onto $\mathcal{L}(\mathcal{H})$. The action of $G \times G$ on $\mathcal{L}(\mathcal{H})$ can be written as

$$\tau_3(g_1, g_2)A = \pi(g_1)A\pi(g_2^{-1}).$$

The map from $\mathcal{M}_\pi$ into $\mathcal{L}(\mathcal{H})$ which maps f to

$$\hat{f}(\pi) = \int_G f(x)\pi(x^{-1})\mu(dx)$$

is an isomorphism which intertwins the representations τ_2 and τ_3 of $G \times G$. In fact,

$$\widehat{R(g)}f(\pi) = \pi(g)\hat{f}(\pi)$$
$$\widehat{L(g)}f(\pi) = \hat{f}(\pi)\pi(g^{-1}).$$

Observe that, if $f \in \mathcal{M}_\pi$, then

$$f(g) = d_\pi \operatorname{tr}\big(\hat{f}(\pi)\pi(g)\big).$$

We can restate the Peter–Weyl Theorem as follows:

$$L^2(G) = \widehat{\bigoplus_{\lambda \in \hat{G}}} (\mathcal{H}_\lambda \otimes \overline{\mathcal{H}_\lambda}),$$

and this corresponds to the decomposition of the representation of $G \times G$ on $L^2(G)$ (G acting on the right and on the left) into a sum of irreducible representations of $G \times G$.

6.5 Characters and central functions

Let G be a compact group. A function f which is defined on G is said to be *central* if

$$f(gxg^{-1}) = f(x) \quad (g, x \in G).$$

Let π be a representation of G on a finite dimensional complex vector space $\mathcal{V}$. The *character* of π is the function χ_π defined on G by

$$\chi_\pi(g) = \operatorname{tr}\pi(g).$$

It is a central function which only depends on the equivalence class of π. One can establish easily the following properties:

$$\chi_\pi(e) = \dim \mathcal{V},$$
$$\chi_{\pi_1 \oplus \pi_2}(g) = \chi_{\pi_1}(g) + \chi_{\pi_2}(g),$$
$$\chi_{\pi_1 \otimes \pi_2}(g) = \chi_{\pi_1}(g) \cdot \chi_{\pi_2}(g),$$
$$\chi_{\bar\pi}(g) = \chi_\pi(g^{-1}) = \overline{\chi_\pi(g)}.$$

Let us denote by $\mathcal{V}^G$ the subspace of invariant vectors:

$$\mathcal{V}^G = \{v \in \mathcal{V} \mid \forall g \in G,\ \pi(g)v = v\}.$$

The operator P, defined by

$$Pv = \int_G \pi(g)v\, \mu(dg),$$

where μ is the normalised Haar measure of G, is a projection onto $\mathcal{V}^G$. Since $\operatorname{tr} P = \dim \mathcal{V}^G$, it follows that

$$\int_G \chi_\pi(g)d\mu(g) = \dim \mathcal{V}^G.$$

If $(\pi_1, \mathcal{V}_1)$ and $(\pi_2, \mathcal{V}_2)$ are two finite dimensional representations of G one puts

$$\mathcal{E}(\pi_1, \pi_2) = \{A \in \mathcal{L}(\mathcal{V}_1, \mathcal{V}_2) \mid \forall g \in G,\ A\pi_1(g) = \pi_2(g)A\}.$$

This is the space of operators which intertwin the representations π_1 and π_2. The group G acts on the space $\mathcal{L}(\mathcal{V}_1, \mathcal{V}_2)$ by the representation T defined by

$$T(g)A = \pi_2(g)A\pi_1(g^{-1}).$$

This representation is equivalent to $\pi_2 \otimes \bar\pi_1$. Observe that

$$\mathcal{E}(\pi_1, \pi_2) = \mathcal{L}(\mathcal{V}_1, \mathcal{V}_2)^G,$$

and that the character of T is equal to

$$\chi_T(g) = \chi_{\pi_2}(g)\overline{\chi_{\pi_1}(g)}.$$

This yields the following statement.

Proposition 6.5.1 *Let π_1 and π_2 be two finite dimensional representations of G. Then*

$$\int_G \chi_{\pi_1}(g)\overline{\chi_{\pi_2}(g)}\mu(dg) = \dim \mathcal{E}(\pi_1, \pi_2).$$

Assume that π_1 and π_2 are irreducible. They are equivalent if and only if they have the same character:

$$\chi_{\pi_1}(g) = \chi_{\pi_2}(g) \quad (g \in G).$$

A finite dimensional representation π of G is irreducible if and only if

$$\int_G |\chi_\pi(g)|^2 \mu(dg) = 1.$$

Let π be an irreducible representation of G on a vector space $\mathcal{V}$ with finite dimension d_π. We saw at the end of Section 6.4 that the space $\mathcal{M}_\pi$ generated by the coefficients of π is isomorphic to the space $\mathcal{L}(\mathcal{V})$ of the endomorphisms of $\mathcal{V}$, and that this isomorphism is $G \times G$-equivariant. By Schur's Lemma (Theorem 6.1.3) it follows that the central functions in $\mathcal{M}_\pi$ are proportional to the character χ_π of π. It follows that, for $f \in \mathcal{M}_\pi$,

$$\int_G f(gxg^{-1})\mu(dg) = \frac{1}{d_\pi}f(e)\chi_\pi(x).$$

In fact, the left-hand side is a central function of x which belongs to $\mathcal{M}_\pi$, and is hence proportional to χ_π: equal to a factor times χ_π. One determines the factor by evaluating both sides at $x = e$. Furthermore, one obtains the relation

$$\int_G f(xgyg^{-1})\mu(dg) = \frac{1}{d_\pi}f(x)\chi_\pi(y),$$

by observing that the left-hand side is a central function of y which belongs to $\mathcal{M}_\pi$ and by evaluating both sides at $y = e$. In particular, if $f = \chi_\pi$, we get the remarkable following relation.

Proposition 6.5.2 *If π is an irreducible representation of the compact group G, then*

$$\int_G \chi_\pi(xgyg^{-1})\mu(dg) = \frac{1}{d_\pi}\chi_\pi(x)\chi_\pi(y).$$

For $\lambda \in \hat{G}$, an equivalence class of irreducible representations of G, one denotes by χ_λ the character of a representation of this class.

Proposition 6.5.3 *The system $\{\chi_\lambda \mid \lambda \in \hat{G}\}$ is a Hilbert basis for the subspace of $L^2(G)$ consisting of square integrable central functions.*

Proof. From what has been said, or from Schur's orthogonality relations, one deduces that

$$\int_G |\chi_\lambda(g)|^2 \mu(dg) = 1,$$

and, if $\lambda \neq \lambda'$,

$$\int_G \chi_\lambda(g)\overline{\chi_{\lambda'}(g)} \, d\mu(g) = 0.$$

Let f be a central function and π_λ be a representation of the class λ. The operator $\hat{f}(\lambda)$ commutes with the representation π_λ. Hence, by Schur's Lemma (Theorem 6.1.3), it is a scalar multiple of the identity. In fact

$$\hat{f}(\lambda) = \int_G f(g)\pi_\lambda(g^{-1})\mu(dg) = \frac{1}{d_\lambda}(f|\chi_\lambda)I,$$

where d_λ is the dimension of the representation space, and $(f|\chi_\lambda)$ is the inner product of f and χ_λ in $L^2(G)$. Therefore

$$\|\hat{f}(\lambda)\|^2 = \frac{1}{d_\lambda}|(f|\chi_\lambda)|^2.$$

By the Plancherel Theorem (Theorem 6.4.2),

$$\int_G |f(g)|^2 d\mu(g) = \sum_{\lambda \in \hat{G}} |(f|\chi_\lambda)|^2.$$

It follows that $\{\chi_\lambda \mid \lambda \in \hat{G}\}$ is a Hilbert basis of the space of square integrable central functions. $\square$

6.6 Absolute convergence of Fourier series

We saw in the preceding section that the Fourier series of a function f in $L^2(G)$ converges to f in L^2-norm. We will now study the uniform convergence of a Fourier series.

Proposition 6.6.1 (i) *Let (A_λ) be a family of operators with $A_\lambda \in \mathcal{L}(\mathcal{H}_\lambda)$. If*

$$\sum_{\lambda \in \hat{G}} d_\lambda^{3/2} \|A_\lambda\| < \infty,$$

then the Fourier series

$$\sum_{\lambda \in \hat{G}} d_\lambda \operatorname{tr}\big(A_\lambda \pi_\lambda(g)\big)$$

converges absolutly and uniformly on G.

   (ii) *Let f be a continuous function on G such that*

$$\sum_{\lambda \in \hat{G}} d_\lambda^{3/2} \|\|\hat{f}(\lambda)\|\| < \infty,$$

then

$$f(g) = \sum_{\lambda \in \hat{G}} d_\lambda \operatorname{tr}\big(\hat{f}(\lambda)\pi_\lambda(g)\big);$$

the convergence is absolute and uniform on G.

Proof. (i) By the inequality

$$|\operatorname{tr}(AB)| \leq \|\|A\|\|\,\|\|B\|\|,$$

and from the relation

$$\|\|\pi_\lambda(g)\|\| = \sqrt{d_\lambda},$$

it follows that

$$d_\lambda |\operatorname{tr}\big(A_\lambda \pi_\lambda(g)\big)| \leq d_\lambda^{3/2} \|\|A_\lambda\|\|,$$

and this yields the statement.

   (ii) Put

$$h(g) = \sum_{\lambda \in \hat{G}} d_\lambda \operatorname{tr}\big(\hat{f}(\lambda)\pi_\lambda(g)\big).$$

Since the convergence is uniform, the function h is continuous. Let us compute $\hat{h}(\lambda)$,

$$\hat{h}(\lambda) = \int_G h(g)\pi_\lambda(g^{-1})d\mu(g)$$

$$= \int_G \left(\sum_{\lambda' \in \hat{G}} d_{\lambda'} \operatorname{tr}\big(\hat{f}(\lambda')\pi_{\lambda'}(g)\big)\right) \pi_\lambda(g^{-1}d\mu(g)).$$

We can integrate termwise. From Schur's orthogonality relations (Theorem 6.3.3) it follows that

$$\hat{h}(\lambda) = d_\lambda \int_G \operatorname{tr}\big(\hat{f}(\lambda)\pi_\lambda(g)\big)\pi_\lambda(g^{-1})d\mu(g) = \hat{f}(\lambda).$$

By the Plancherel Theorem (Theorem 6.4.2),

$$\int_G |f(g) - h(g)|^2 d\mu(g) = 0.$$

Therefore $f(g) = h(g)$ for every $g \in G$. □

Let G be a compact linear Lie group. One can show that, if $f \in C^k(G)$ with $k > \frac{1}{2} \dim G$, then

$$\sum_{\lambda \in \hat{G}} d_\lambda^{3/2} \|\hat{f}(\lambda)\| < \infty,$$

therefore the Fourier series of f converges to f uniformly. We will prove this in the case of the group $G = SU(2)$, and of the groups $U(n)$. For that we will use the Casimir operator, which we introduce in the following section.

6.7 Casimir operator

A symmetric bilinear form β on a Lie algebra $\mathfrak{g}$ is said to be invariant if

$$\beta([X, Y], Z) = -\beta(Y, [X, Z]) \quad (X, Y, Z \in \mathfrak{g}),$$

that is, for $X \in \mathfrak{g}$, the endomorphism $\mathrm{ad}\, X$ is skewsymmetric with respect to β. If $\mathfrak{g}$ is the Lie algebra of a connected linear Lie group G, it is equivalent to saying that β is invariant under the adjoint representation:

$$\beta\big(\mathrm{Ad}(g)X, \mathrm{Ad}(g)Y\big) = \beta(X, Y) \quad (g \in G, \ X, Y \in \mathfrak{g}).$$

Let $\mathfrak{g}$ be a Lie algebra, and assume that there exists on $\mathfrak{g}$ an invariant non-degenerate symmetric bilinear form β. For instance, if $\mathfrak{g}$ is semi-simple, we can take $\beta = B$, the Killing form of $\mathfrak{g}$. If $\mathfrak{g}$ is the Lie algebra of a compact linear Lie group, there exists on $\mathfrak{g}$ a Euclidean inner product which is invariant under the adjoint representation (Proposition 6.1.1), and we can take

$$\beta(X, Y) = (X|Y).$$

If $\mathfrak{g}$ is a subalgebra of $M(n, \mathbb{R})$ such that, if $X \in \mathfrak{g}$, then the transpose $X^T \in \mathfrak{g}$ also, and we can take

$$\beta(X, Y) = \mathrm{tr}(XY).$$

This bilinear form β is invariant, and non-degenerate. In fact let $X \in \mathfrak{g}$ be such that, for every $Y \in \mathfrak{g}$,

$$\mathrm{tr}(XY) = 0.$$

Then in particular, for $Y = X^T$,

$$\mathrm{tr}(XX^T) = 0,$$

therefore $X = 0$.

Let $\{X_1, \ldots, X_n\}$ be a basis of $\mathfrak{g}$, and put

$$g_{ij} = \beta(X_i, X_j), \quad (g^{ij}) = (g_{ij})^{-1}.$$

Let ρ be a representation of $\mathfrak{g}$ on a vector space $\mathcal{V}$. The *Casimir operator* Ω_ρ of the representation ρ is defined by

$$\Omega_\rho = \sum_{ij=1}^{n} g^{ij} \rho(X_i)\rho(X_j).$$

In particular, if β is positive definite, and if $\{X_1, \ldots, X_n\}$ is an orthonormal basis for the Euclidean inner product defined by β, then

$$\beta(X_i, X_j) = \delta_{ij},$$

and

$$\Omega_\rho = \sum_{i=1}^{n} \rho(X_i)^2.$$

Proposition 6.7.1 *The preceding definition does not depend on the choice of basis. The operator Ω_ρ commutes with the representation ρ.*

Proof. (a) Let $\{Y_1, \ldots, Y_n\}$ be another basis, and put

$$Y_i = \sum_{j=1}^{n} a_{ij} X_j, \quad (a^{ij}) = (a_{ij})^{-1},$$

$$h_{ij} = \beta(Y_i, Y_j), \quad (h^{ij}) = (h_{ij})^{-1}.$$

Then

$$X_k = \sum_{i=1}^{n} a^{ki} Y_i,$$

$$h_{ij} = \sum_{k,\ell=1}^{n} a_{ik} g_{k\ell} a_{j\ell};$$

this can be written in terms of matrices as $H = AGA^T$, and

$$h^{ij} = \sum_{k,\ell=1}^{n} a^{ki} g^{k\ell} a^{\ell j},$$

or $H^{-1} = A^{T-1} G^{-1} A^{-1}$. Hence

$$\sum_{i,j=1}^{n} h^{ij} \rho(Y_i)\rho(Y_j) = \sum_{i,j,k,\ell=1}^{n} a^{ki} g^{k\ell} a^{\ell j} \rho(Y_i)\rho(Y_j)$$

$$= \sum_{k,\ell=1}^{n} g^{k\ell} \rho(X_k)\rho(X_\ell).$$

(b) Let $\{X^1, \ldots, X^n\}$ be the dual basis:

$$X^i = \sum_{j=1}^{n} g^{ij} X_j.$$

Then

$$\beta(X^i, X_j) = \delta^i_j,$$

and

$$\Omega_\rho = \sum_{i=1}^{n} \rho(X_i)\rho(X^i).$$

Let $X \in \mathfrak{g}$, and put

$$[X, X_i] = \sum_{j=1}^{n} c_{ij}(X) X_j,$$

then

$$\beta([X, X_i], X^j) = c_{ij}(X).$$

Similarly,

$$[X, X^i] = \sum_{j=1}^{n} d_{ij}(X) X^j,$$

and

$$\beta([X, X^i], X_j) = d_{ij}(X).$$

Hence

$$c_{ij}(X) = -d_{ji}(X).$$

From the identity

$$[A, BC] = [A, B]C + B[A, C],$$

it follows that

$$[\rho(X), \Omega_\rho] = \sum_{i=1}^{n} \big([\rho(X), \rho(X_i)]\rho(X^i) + \rho(X_i)[\rho(X), \rho(X^i)]\big)$$

$$= \sum_{i,j}^{n} \big(c_{ij}(X)\rho(X_j)\rho(X^i) + d_{ij}(X)\rho(X_i)\rho(X^j)\big) = 0. \qquad \square$$

By using Schur's Lemma, one deduces the following from Proposition 6.7.1.

Corollary 6.7.2 *If ρ is a $\mathbb{C}$-linear irreducible representation of $\mathfrak{g}$ on a finite dimensional complex vector space V, then there exists $\kappa_\rho \in \mathbb{C}$ such that*

$$\Omega_\rho = -\kappa_\rho I.$$

Proof. Since the Casimir operator Ω_ρ commutes with the representation ρ, the statement follows from Schur's Lemma (Proposition 6.1.4). $\qquad \square$

Assume that $\mathfrak{g}$ is the Lie algebra of a connected compact linear Lie group G. There exists on $\mathfrak{g}$ a Euclidean inner product for which the adjoint representation is unitary. Let π be a representation of G on a finite dimensional complex vector space V, and let $d\pi$ be the derived representation. Put

$$\Omega_\pi = \sum_{i=1}^{n} \big(d\pi(X_i)\big)^2,$$

where $\{X_i\}$ is an orthonormal basis of $\mathfrak{g}$. The operator Ω_π commutes with the representation π. If the representation π is irreducible, then there exists a number κ_π such that

$$\Omega_\pi = -\kappa_\pi I.$$

Proposition 6.7.3 *If the representation π is not trivial, then $\kappa_\pi > 0$.*

Proof. In fact, there exists on V a Euclidean inner product for which π is unitary and then

$$d\pi(X)^* = -d\pi(X),$$

and, if $v \neq 0$,

$$(\Omega_\pi v | v) = -\sum_{i=1}^{n} \|d\pi(X_i)v\|^2 < 0. \qquad \square$$

6.8 Exercises

1. Let $\mathcal{H}$ be a separable Hilbert space, and $\{e_i\}$ a Hilbert basis of $\mathcal{H}$. One says that an operator A acting on $\mathcal{H}$ is *Hilbert–Schmidt* if

$$\|\|A\|\|^2 = \sum_{i=1}^{\infty} \|Ae_i\|^2 < \infty.$$

(a) Show that this definition does not depend on the choice of basis. Show that A is Hilbert–Schmidt if and only if its adjoint A^* is such, and that

$$\|\|A\|\| = \|\|A^*\|\|.$$

(b) Show that a Hilbert–Schmidt operator is compact.
   Hint. Consider the finite rank operators A_N defined by

$$A_N = \sum_{j=1}^{N} (Ax|e_j)e_j,$$

and show that

$$\|A - A_N\|^2 \leq \sum_{i=N+1}^{\infty} \sum_{j=1}^{\infty} |a_{ij}|^2,$$

with $a_{ij} = (Ae_j|e_i)$.
(c) Assume that A is a Hilbert–Schmidt self-adjoint operator. Show that

$$\|\|A\|\|^2 = \sum_{n=1}^{\infty} d_n \lambda_n^2,$$

where λ_n is the sequence of non-zero eigenvalues of A, and d_n is the dimension of the eigensubspace corresponding to the eigenvalue λ_n.

2. Let λ_n be a sequence of real numbers >0 with limit $+\infty$. Put

$$E = \left\{ x \in \ell^2(\mathbb{N}) \mid \sum_{n=0}^{\infty} \lambda_n |x_n|^2 \leq 1 \right\}.$$

   Show that the set E is compact.

3. Let G be a finite group. Show that the number of conjugacy classes in G is equal to the number of elements in $\hat{G}$. Show that

$$\#(G) = \sum_{\lambda \in \hat{G}} d_\lambda^2,$$

where $\#(G)$ denotes the number of elements in G.
(These statements form one of the Burnside Theorems.)

4. Let G be a compact group. For two finite dimensional $\mathbb{K}$-linear representations $(\pi_1, \mathcal{V}_1)$ and $(\pi_2, \mathcal{V}_2)$ of G ($\mathbb{K} = \mathbb{R}$ or $\mathbb{C}$), one defines

$$\mathcal{E}(\pi_1, \pi_2) = \{A \in \mathcal{L}(\mathcal{V}_1, \mathcal{V}_2) \mid \forall g \in G, \ A\pi_1(g) = \pi_2(g)A\}.$$

If $\mathcal{E}(\pi_1, \pi_2) = \{0\}$, one says that π_1 et π_2 are *disjoint*.

(a) Show that two irreducible representations are either equivalent or disjoint.

(b) For a finite dimensional representation π one defines $\mathcal{E}(\pi) = \mathcal{E}(\pi, \pi)$. Show that $\mathcal{E}(\pi)$ is an algebra, and that, if π is irreducible, then $\mathcal{E}(\pi)$ is a field.

(c) Assume that $\mathbb{K} = \mathbb{C}$. Show that, if π is irreducible, then $\mathcal{E}(\pi)$ is isomorphic to $\mathbb{C}$.

Up to now we have assumed that $\mathbb{K} = \mathbb{R}$. The following result (due to Frobenius) will be used. *Let $\mathcal{A}$ be a finite dimensional associative algebra over $\mathbb{R}$. If $\mathcal{A}$ is a field, then $\mathcal{A}$ is isomorphic to $\mathbb{R}$, $\mathbb{C}$ or $\mathbb{H}$, the quaternionic field.* Let π be an irreducible $\mathbb{R}$-linear representation. If $\mathcal{E}(\pi)$ is isomorphic to $\mathbb{R}$ (respectively to $\mathbb{C}$, to $\mathbb{H}$), one says that π is *of real type* (respectively *of complex type*, *of quaternion type*).

(d) Let π be the $\mathbb{R}$-linear representation of $G = \mathbb{R}/2\pi\mathbb{Z}$ on $\mathcal{V} = \mathbb{R}^2$ defined by

$$\pi(\theta) = \begin{pmatrix} \cos\theta & \sin\theta \\ -\sin\theta & \cos\theta \end{pmatrix}.$$

Show that π is irreducible. Of which type is π?

(e) Let $\mathcal{V}$ be the real vector space of dimension four consisting of the matrices

$$v = \begin{pmatrix} a & b \\ -\bar{b} & \bar{a} \end{pmatrix} \quad (a, b \in \mathbb{C}),$$

and let π be the $\mathbb{R}$-linear representation of $SU(2)$ on $\mathcal{V}$ defined by

$$\pi(g)v = gv.$$

Show that π is irreducible. Of which type is π?

Let $\mathcal{V}$ be a finite dimensional real vector space, and let A be an endomorphism of $\mathcal{V}$. Let $A_\mathbb{C}$ denote the $\mathbb{C}$-linear endomorphism of $\mathcal{V}_\mathbb{C} = \mathcal{V} + i\mathcal{V}$ defined by

$$A_\mathbb{C}(u + iv) = Au + iAv.$$

Observe that

$$\mathrm{tr}_\mathbb{C} A_\mathbb{C} = \mathrm{tr}_\mathbb{R} A.$$

If (π, V) is a $\mathbb{R}$-linear representation of G one will denote by $\tilde{\pi}$ the $\mathbb{C}$-linear representation of G on $V_{\mathbb{C}}$ defined by $\tilde{\pi}(g) = \pi(g)_{\mathbb{C}}$.

(f) Let (π, V) be an irreducible $\mathbb{R}$-linear representation of G, and W a nontrivial complex subspace of $V_{\mathbb{C}}$ which is invariant under $\tilde{\pi}$. Show that W is of the form

$$W = \{x + iAx \mid x \in V\},$$

where $A \in \mathcal{E}(\pi)$ with $A^2 = -I$, and that W is irreducible.

(g) Assume that (π, V) is an irreducible $\mathbb{R}$-linear representation of real type. Show that $\tilde{\pi}$ is irreducible.

(h) Assume that (π, V) is an irreducible $\mathbb{R}$-linear representation of complex or quaternion type. Let $A \in \mathcal{E}(\pi)$ be such that $A^2 = -I$. Put

$$W_{\pm} = \{x \pm iAx \mid x \in V\}.$$

Denote by $\pi_{\pm}$ the restriction of $\tilde{\pi}$ to $W_{\pm}$. Show that $\mathcal{E}_{\mathbb{C}}(\pi_+, \pi_-)$ is isomorphic to

$$\{B \in \mathcal{E}(\pi) \mid BA = -AB\}$$

as a real vector space.

Deduce that, if π is of complex type, then π_+ and π_- are not equivalent, and that, if π is of quaternion type, then π_+ and π_- are equivalent.

(i) Let χ_{π} denote the character of π:

$$\chi_{\pi}(g) = \mathrm{tr}_{\mathbb{R}} \, \pi(g).$$

Show that

$$\int_G \chi_{\pi}(g)^2 d\mu(g) = \dim_{\mathbb{R}} \mathcal{E}(\pi),$$

where μ denotes the normalised Haar measure of G.

See: A. A. Kirillov (1976). *Elements of the Theory of Representations.* Springer (Section 8.2, p. 119).

5. Let G be a compact group and μ the normalised Haar measure of G. The convolution product of two integrable functions f_1 and f_2 is defined by

$$f_1 * f_2(x) = \int_G f_1(xy^{-1}) f_2(y) \mu(dy).$$

(a) Show that

$$\|f_1 * f_2\|_1 \le \|f_1\|_1 \|f_2\|_1.$$

Hence $L^1(G)$ is endowed with a Banach algebra structure.

(b) Let $L^1(G)^\sharp$ denote the subspace of (classes of) central integrable functions. Show that $L^1(G)^\sharp$ is a commutative subalgebra of $L^1(G)$. Show that $L^1(G)^\sharp$ is equal to the centre of $L^1(G)$.

(c) Denote by $\hat{G}$ the set of equivalence classes of irreducible representations of G. For $\lambda \in \hat{G}$, one denotes by χ_λ the character of a representation of the class λ. Show that

$$\chi_\lambda * \chi_\lambda = \frac{1}{d_\lambda} \chi_\lambda,$$

where d_λ is the dimension of a representation of the class λ, and show that, if $\lambda \neq \lambda'$,

$$\chi_\lambda * \chi_{\lambda'} = 0.$$

(d) Denote by $\mathcal{M}_\lambda$ the space generated by the coefficients of a representation of the class λ. Prove that the orthogonal projection $P_\lambda : L^2(G) \to \mathcal{M}_\lambda$ can be written

$$P_\lambda f = d_\lambda \chi_\lambda * f.$$

6. Let G be a compact linear Lie group and $\mathfrak{g} = Lie(G)$ its Lie algebra. The aim of this exercise is to show that

$$\mathfrak{g} = \mathfrak{z} \oplus \mathfrak{g}',$$

where $\mathfrak{z}$ is the centre of $\mathfrak{g}$ and $\mathfrak{g}'$ is a semi-simple Lie algebra.

(a) Show that there exists on $\mathfrak{g}$ a Euclidean inner product such that, for $X, Y, Z \in \mathfrak{g}$,

$$([X, Y]|Z) = (X|[Y, Z]).$$

Fix such an inner product.

(b) Show that the orthogonal $\mathfrak{g}'$ of the centre $\mathfrak{z}$ of $\mathfrak{g}$ is an ideal.

(c) Show that there is a constant $C > 0$ such that, for every $X \in \mathfrak{g}$ and every $t \in \mathbb{R}$,

$$\| Exp(t \, ad \, X) \| \le C.$$

(d) Let B be the Killing form of $\mathfrak{g}$. Deduce from (c), using Exercise 12 of Chapter 2, that, for every $X \in \mathfrak{g}$,

$$B(X, X) \le 0,$$

and that $B(X, X) = 0$ if and only if $X \in \mathfrak{z}$.

(e) Show that $\mathfrak{g}'$ is a semi-simple Lie algebra.

7

The groups $SU(2)$ and $SO(3)$, Haar measures, and irreducible representations

The special orthogonal group $SO(3)$ and its simply connected covering $SU(2)$ are the simplest non-commutative compact linear Lie groups. In this chapter we study the irreducible representations of these groups. The irreducible representations of $SO(3)$ can be realised on spaces of harmonic homogeneous polynomials in three variables.

7.1 Adjoint representation of $SU(2)$

The group $SU(2)$ consists of the matrices

$$g = \begin{pmatrix} \alpha & \beta \\ -\bar{\beta} & \bar{\alpha} \end{pmatrix}, \quad \alpha, \beta \in \mathbb{C}, \ |\alpha|^2 + |\beta|^2 = 1.$$

The inverse of this matrix is

$$g^{-1} = \begin{pmatrix} \bar{\alpha} & -\beta \\ \bar{\beta} & \alpha \end{pmatrix}.$$

The group $SU(2)$ is homeomorphic to the unit sphere of $\mathbb{C}^2$, therefore compact, connected, and simply connected.

The matrices

$$X_1 = \begin{pmatrix} i & 0 \\ 0 & -i \end{pmatrix}, \quad X_2 = \begin{pmatrix} 0 & 1 \\ -1 & 0 \end{pmatrix}, \quad X_3 = \begin{pmatrix} 0 & i \\ i & 0 \end{pmatrix},$$

form a basis of its Lie algebra $\mathfrak{su}(2)$: every matrix T in $\mathfrak{su}(2)$ can be written uniquely as

$$T = t_1 X_1 + t_2 X_2 + t_3 X_3 = \begin{pmatrix} it_1 & t_2 + it_3 \\ -t_2 + it_3 & -it_1 \end{pmatrix}, \quad t_1, t_2, t_3 \in \mathbb{R}.$$

The commutation relations are the following:

$$[X_1, X_2] = 2X_3, \quad [X_2, X_3] = 2X_1, \quad [X_3, X_1] = 2X_2.$$

With respect to this basis the adjoint representation can be written, if $T = t_1 X_1 + t_2 X_2 + t_3 X_3$, as

$$\mathrm{ad}(T) = \begin{pmatrix} 0 & -2t_3 & 2t_2 \\ 2t_3 & 0 & -2t_1 \\ -2t_2 & 2t_1 & 0 \end{pmatrix}.$$

One can deduce easily the following formula for the Killing form:

$$B(T, T) = \mathrm{tr}(\mathrm{ad}\, T)^2 = -8(t_1^2 + t_2^2 + t_3^2).$$

Since the Killing form is invariant under $\mathrm{Ad}(g)$ ($g \in SU(2)$), the adjoint representation Ad is a morphism from $SU(2)$ into $O(3)$. From the above formula for $\mathrm{ad}(T)$ it follows that that the adjoint representation ad is an isomorphism from $\mathfrak{su}(2)$ onto $\mathfrak{so}(3)$. Since $SU(2)$ is connected, the image of the map Ad is contained in $SO(3)$, the identity component of $SO(3)$. By Proposition 4.1.2 this image is equal to $SO(3)$. The kernel of Ad is the centre of $SU(2)$, that is $\{\pm e\}$ (e is the identity). We can state the following.

Proposition 7.1.1 *The map* Ad *is a surjective morphism from* $SU(2)$ *onto* $SO(3)$. *Its kernel, which is the centre of* $SU(2)$, *is equal to* $\{\pm e\}$.

Hence $(SU(2), \mathrm{Ad})$ is a covering of order 2 of $SO(3)$.

Every element $x \in SU(2)$ is conjugate to a diagonal matrix of the form

$$a(\theta) = \exp(\theta X_1) = \begin{pmatrix} e^{i\theta} & 0 \\ 0 & e^{-i\theta} \end{pmatrix},$$

that is, $x = g a(\theta) g^{-1}$ with $g \in SU(2)$, $\theta \in \mathbb{R}$. In fact a unitary matrix x is normal: $xx^* = x^*x$, hence diagonalisable in an orthogonal basis. This means that

$$x = g \begin{pmatrix} e^{i\theta_1} & 0 \\ 0 & e^{i\theta_2} \end{pmatrix} g^{-1} \quad (g \in U(2), \ \theta_1, \theta_2 \in \mathbb{R}).$$

One can choose g with determinant equal to one and, since $\det x = 1$, one can choose $\theta_2 = -\theta_1$:

$$x = g \begin{pmatrix} e^{i\theta} & 0 \\ 0 & e^{-i\theta} \end{pmatrix} g^{-1} \quad (g \in SU(2), \ \theta \in \mathbb{R}).$$

The matrix x can also be written as

$$x = \exp X, \quad \text{with } X = g \begin{pmatrix} i\theta & 0 \\ 0 & -i\theta \end{pmatrix} g^{-1}.$$

Hence the exponential map $\exp : \mathfrak{su}(2) \to SU(2)$ is surjective.

The *quaternion field* $\mathbb{H}$ can be described as the set of matrices in $M(2, \mathbb{C})$ of the form

$$q = \begin{pmatrix} \alpha & \beta \\ -\bar{\beta} & \bar{\alpha} \end{pmatrix}, \quad \alpha, \beta \in \mathbb{C}.$$

The absolute value $|q|$ of q is given by

$$|q| = \sqrt{|\alpha|^2 + |\beta|^2}.$$

Observe that

$$|q|^2 = |\alpha|^2 + |\beta|^2 = \det \begin{pmatrix} \alpha & \beta \\ -\bar{\beta} & \bar{\alpha} \end{pmatrix}.$$

The quaternion

$$\begin{pmatrix} x & 0 \\ 0 & x \end{pmatrix} \quad (x \in \mathbb{R})$$

is identified with the real number x. The set $\mathbb{H}$ is also a vector space over $\mathbb{R}$ with dimension four: $\mathbb{H} \simeq \mathbb{R}^4$. A basis is formed by the following matrices

$$\underline{1} = \begin{pmatrix} 1 & 0 \\ 0 & 1 \end{pmatrix}, \quad \underline{i} = \begin{pmatrix} 0 & 1 \\ -1 & 0 \end{pmatrix}, \quad \underline{j} = \begin{pmatrix} 0 & i \\ i & 0 \end{pmatrix}, \quad \underline{k} = \begin{pmatrix} i & 0 \\ 0 & -i \end{pmatrix}.$$

These elements satisfy the following relations:

$$\underline{i}^2 = -\underline{1}, \quad \underline{j}^2 = -\underline{1}, \quad \underline{k}^2 = -\underline{1}, \quad \underline{i}\,\underline{j} = \underline{k}, \quad \underline{j}\,\underline{k} = \underline{i}, \quad \underline{k}\,\underline{i} = \underline{j}.$$

The group $SU(2)$ can be identified with the group of quaternions with modulus one.

Let the group $SU(2) \times SU(2)$ act on $\mathbb{H}$ by

$$q \mapsto uqv^{-1} \quad (u, v \in SU(2)).$$

Since $\det(uqv^{-1}) = \det q$, and $|uqv^{-1}| = |q|$, this action defines a group morphism

$$\phi : SU(2) \times SU(2) \to O(4).$$

Proposition 7.1.2 *The morphism* $\phi : SU(2) \times SU(2) \to SO(4)$ *is surjective, with kernel* $\ker\phi = \big((e, e), (-e, -e)\big)$.

Hence $\big(SU(2) \times SU(2), \phi\big)$ is a covering of order two of $SO(4)$.

Proof. Since $SU(2) \times SU(2)$ is connected, the image is contained in $SO(4)$ which is the identity component of $O(4)$.

Let $(u, v) \in \ker\phi$:

$$\forall q \in \mathbb{H}, \ uqv^{-1} = q.$$

Taking $q = 1$ it follows that $u = v$, and that u is in the centre of $\mathbb{H}$ which equals $\mathbb{R}1$. Therefore $u = e$, or $u = -e$.

The differential $d\phi = (D\phi)_{(e,e)}$ is a Lie algebra morphism:

$$d\phi : \mathfrak{su}(2) \times \mathfrak{su}(2) \to \mathfrak{so}(4).$$

If $(S, T) \in \mathfrak{su}(2) \times \mathfrak{su}(2)$, then $X = d\phi(S, T)$ is defined by

$$X : q \mapsto Sq - qT.$$

Therefore $d\phi$ is injective. Since $\dim\big(\mathfrak{su}(2) \times \mathfrak{su}(2)\big) = 6 = \dim \mathfrak{so}(4)$, then $d\phi$ is an isomorphism. And since $SO(4)$ is connected, it follows that ϕ is surjective (Proposition 4.1.2). $\qquad\qquad\square$

We have also proved the following.

Corollary 7.1.3

$$\mathfrak{so}(4) \simeq \mathfrak{su}(2) \oplus \mathfrak{su}(2).$$

Hence, the Lie algebra $\mathfrak{so}(4)$, which is semi-simple, is not simple.

7.2 Haar measure on $SU(2)$

A measure on the unit sphere S^3 of $\mathbb{R}^4 \simeq \mathbb{H}$, which is invariant under $SO(4)$ is a Haar measure on $SU(2) \simeq S^3$.

Let ω be the differential form of degree $n - 1$ on $\mathbb{R}^n$ defined by

$$\omega = \sum_{i=1}^{n}(-1)^{i-1}x_i dx_1 \wedge \cdots \wedge \widehat{dx_i} \wedge \cdots \wedge dx_n.$$

At every point x, for vectors $\xi_1, \ldots, \xi_{n-1} \in \mathbb{R}^n$,

$$\omega_x(\xi_1, \ldots, \xi_{n-1}) = \det(x, \xi_1, \ldots, \xi_{n-1}).$$

The differential form ω is invariant under $SL(n, \mathbb{R})$. Its restriction to the unit sphere S^{n-1} of $\mathbb{R}^n$ is invariant under $SO(n)$. It defines a measure on S^{n-1}, which is invariant under $SO(n)$. For $n = 4$ we get in this way a Haar measure on $SU(2) \simeq S^3$. Let μ be the corresponding normalised Haar measure:

$$\int_{SU(2)} f(x)\mu(dx) = \frac{1}{\varpi_4} \int_{SU(2)} f\omega,$$

with

$$\varpi_4 = \int_{SU(2)} \omega.$$

We will see below that $\varpi_4 = 2\pi^2$.

Let us write an element in $SU(2)$ as

$$x = \begin{pmatrix} x_1 + ix_2 & x_3 + ix_4 \\ -x_3 + ix_4 & x_1 - ix_2 \end{pmatrix},$$

and put

$$x_1 = \cos\theta,$$
$$x_2 = \sin\theta\cos\varphi,$$
$$x_3 = \sin\theta\sin\varphi\cos\psi,$$
$$x_4 = \sin\theta\sin\varphi\sin\psi.$$

Let Φ denote the map $(\theta, \varphi, \psi) \mapsto x = (x_1, x_2, x_3, x_4)$.

Proposition 7.2.1 *If f is an integrable function on $SU(2)$, then*

$$\int_{SU(2)} f(x)\mu(dx) = \frac{1}{2\pi^2} \int_0^\pi d\theta \int_0^\pi d\varphi \int_0^{2\pi} d\psi \, f \circ \Phi(\theta, \varphi, \psi) \sin^2\theta \sin\varphi.$$

Proof. We will see that

$$\Phi^*\omega = \sin^2\theta \sin\varphi \, d\theta \wedge d\varphi \wedge d\psi.$$

The relations

$$dx_1 = -\sin\theta\, d\theta,$$
$$dx_2 = \cos\theta\cos\varphi\, d\theta - \sin\theta\sin\varphi\, d\varphi,$$
$$dx_3 = \cos\theta\sin\varphi\cos\psi\, d\theta + \sin\theta\cos\varphi\cos\psi\, d\varphi - \sin\theta\sin\varphi\sin\psi\, d\psi,$$
$$dx_4 = \cos\theta\sin\varphi\sin\psi\, d\theta + \sin\theta\cos\varphi\sin\psi\, d\varphi + \sin\theta\sin\varphi\cos\psi\, d\psi,$$

can be written as

$$\begin{pmatrix} dx_1 \\ dx_2 \\ dx_3 \\ dx_4 \end{pmatrix} = \begin{pmatrix} -\sin\theta \\ \cos\theta\cos\varphi \\ \cos\theta\sin\varphi\cos\psi \\ \cos\theta\sin\varphi\sin\psi \end{pmatrix} d\theta$$
$$+ \begin{pmatrix} 0 \\ -\sin\varphi \\ \cos\varphi\cos\psi \\ \cos\varphi\sin\psi \end{pmatrix} \sin\theta\, d\varphi + \begin{pmatrix} 0 \\ 0 \\ -\sin\psi \\ \cos\psi \end{pmatrix} \sin\theta\sin\varphi\, d\psi.$$

The vector $x = \Phi(\theta, \varphi, \psi)$ and the three columns of the above right-hand side are unit vectors and orthogonal, and form an orthonormal basis, which is direct (one can check this for $\theta = \varphi = \psi = 0$, and the statement follows since the determinant of these vectors is equal to ± 1, and it depends continuously on

θ, φ, ψ). Therefore

$$\det\left(\Phi, \frac{\partial \Phi}{\partial \theta}, \frac{\partial \Phi}{\partial \varphi}, \frac{\partial \Phi}{\partial \psi}\right) = \sin^2\theta \sin\varphi.$$

The statement follows since, for $f = 1$, one has

$$\varpi_4 = \int_{SU(2)} \omega = 2\pi^2. \qquad \square$$

Let us recall that a function f on a group G is said to be central if it is invariant under inner automorphisms:

$$f(gxg^{-1}) = f(x) \quad (g, x \in G).$$

For every matrix $x \in SU(2)$, there exists $g \in SU(2)$ such that

$$gxg^{-1} = \begin{pmatrix} e^{i\theta} & 0 \\ 0 & e^{-i\theta} \end{pmatrix}.$$

Furthermore, if

$$g = \begin{pmatrix} 0 & i \\ i & 0 \end{pmatrix},$$

then

$$g\begin{pmatrix} e^{i\theta} & 0 \\ 0 & e^{-i\theta} \end{pmatrix}g^{-1} = \begin{pmatrix} e^{-i\theta} & 0 \\ 0 & e^{i\theta} \end{pmatrix}.$$

Hence a central function f on $SU(2)$ only depends on the trace: there is a function F on $[-1, 1]$ such that

$$f(x) = F\left(\frac{1}{2}\operatorname{tr}x\right) = F(\operatorname{Re}\alpha) = F(x_1),$$

if

$$x = \begin{pmatrix} \alpha & \beta \\ -\bar{\beta} & \bar{\alpha} \end{pmatrix} = \begin{pmatrix} x_1 + ix_2 & x_3 + ix_4 \\ -x_3 + ix_4 & x_1 - ix_2 \end{pmatrix}.$$

Corollary 7.2.2 *If f is an integrable central function on $SU(2)$:*

$$f(x) = F\left(\frac{1}{2}\operatorname{tr}x\right),$$

then

$$\int_{SU(2)} f(x)d\mu(x) = \frac{2}{\pi}\int_0^\pi F(\cos\theta)\sin^2\theta\,d\theta$$

$$= \frac{2}{\pi}\int_{-1}^1 F(t)\sqrt{1-t^2}\,dt.$$

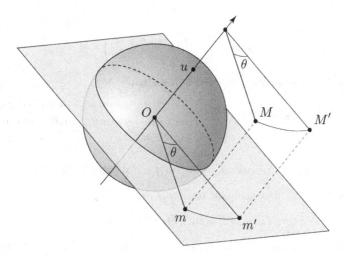

Figure 2

7.3 The group $SO(3)$

We saw that the adjoint representation Ad is a surjective morphism from $SU(2)$ onto $SO(3)$ with kernel $\{\pm e\}$. It is a covering of $SO(3)$ of order two. Hence

$$SO(3) \simeq SU(2)/\{\pm e\}.$$

The group $SU(2)$ is homeomorphic to the unit sphere S^3 of $\mathbb{R}^4$, therefore the group $SO(3)$ is homeomorphic to the projective space $P^3(\mathbb{R})$.

The Lie algebra $\mathfrak{so}(3)$ of $SO(3)$ is made of the real 3×3 skewsymmetric matrices. The exponential map

$$\exp : \mathfrak{so}(3) \to SO(3),$$

is surjective. Let us consider polar coordinates in $\mathfrak{so}(3)$: $X = \theta T(u)$, where θ is a real number, $u = (a, b, c)$ is a point of the unit sphere S^2 of $\mathbb{R}^3$ ($a^2 + b^2 + c^2 = 1$), and

$$T(u) = \begin{pmatrix} 0 & -c & b \\ c & 0 & -a \\ -b & a & 0 \end{pmatrix}.$$

These coordinates have the following meaning: $g = \exp X = \exp\big(\theta T(u)\big)$ is the rotation of angle θ around the axis determined by the unit vetor $u = (a, b, c)$. Therefore one may see $SO(3)$ as the ball in $\mathbb{R}^3$ with radius π, two extremities of each diameter being identified. Note that

$$T(u) = \tfrac{1}{2} \operatorname{ad} S(u),$$

with

$$S(u) = \begin{pmatrix} ia & b+ic \\ -b+ic & -ia \end{pmatrix}.$$

From the relation $S(u)^2 = -I$ it follows that

$$\exp\big(\theta S(u)\big) = \cos\theta\, I + \sin\theta\, S(u) = \begin{pmatrix} \cos\theta + ia\sin\theta & (b+ic)\sin\theta \\ -(b-ic)\sin\theta & \cos\theta - ia\sin\theta \end{pmatrix}.$$

Recall that

$$\exp\big(\theta T(u)\big) = \exp\tfrac{1}{2}\theta\big(\mathrm{ad}\, S(u)\big) = \mathrm{Ad}\,\big(\exp\big(\tfrac{1}{2}\theta S(u)\big)\big).$$

One can establish also, using the relation $T(u)^3 = -T(u)$, that

$$\exp\big(\theta T(u)\big) = I + \sin\theta\, T(u) + (1 - \cos\theta)T(u)^2.$$

This formula has the following meaning: $I + T(u)^2$ is the projection onto the axis defined by u, $-T(u)^2$ is the projection onto the plane orthogonal to u, and the expression $\sin\theta\, T(u) - \cos\theta\, T(u)^2$ means a projection onto this plane, followed by a rotation of angle θ in this plane.

Since the image by the map Ad of the Haar measure μ of $SU(2)$ is equal to the Haar measure ν of $SO(3)$, from Proposition 6.2.1 we have the following result.

Proposition 7.3.1 *For an integrable function f on $SO(3)$,*

$$\int_{SO(3)} f(g)\nu(dg) = \frac{2}{\pi}\int_0^\pi d\theta \int_{S^2} \sigma(du) f\Big(\exp\big(\theta T(u)\big)\Big)\sin^2\frac{\theta}{2},$$

where σ is the normalised uniform measure on the unit sphere S^2 in $\mathbb{R}^3$.

If the function f is central, it only depends on the rotation angle θ:

$$f\Big(\exp\big(\theta T(u)\big)\Big) = F(\theta),$$

where F is a function on $\mathbb{R}$, even and 2π-periodic. In this case the integration formula simplifies to

$$\int_{SO(3)} f(g)\nu(dg) = \frac{2}{\pi}\int_0^\pi F(\theta)\sin^2\frac{\theta}{2}d\theta.$$

7.4 Euler angles

Every matrix $x \in SU(2)$ decomposes as

$$x = \begin{pmatrix} e^{i\psi} & 0 \\ 0 & e^{-i\psi} \end{pmatrix}\begin{pmatrix} \cos\theta & \sin\theta \\ -\sin\theta & \cos\theta \end{pmatrix}\begin{pmatrix} e^{i\varphi} & 0 \\ 0 & e^{-i\varphi} \end{pmatrix}$$
$$= \exp(\psi X_1)\exp(\theta X_2)\exp(\varphi X_1)$$

with $0 \leq \theta \leq \frac{\pi}{2}, 0 \leq \varphi \leq \pi, -\pi \leq \psi \leq \pi$. If

$$x = \begin{pmatrix} \alpha & \beta \\ -\bar{\beta} & \bar{\alpha} \end{pmatrix}, \quad |\alpha|^2 + |\beta|^2 = 1,$$

this decomposition can be written

$$\alpha = e^{i(\psi+\varphi)} \cos\theta, \quad \beta = e^{i(\psi-\varphi)} \sin\theta.$$

The numbers θ, φ, ψ are called the *Euler angles* of the matrix x. We will establish an integration formula corresponding to this decomposition. Let us denote by Φ the map $(\theta, \varphi, \psi) \mapsto x$.

Proposition 7.4.1 *Let μ be the normalised Haar measure of $SU(2)$, and f an integrable function on $SU(2)$. Then*

$$\int_{SU(2)} f(x) d\mu(x) = \frac{1}{2\pi^2} \int_0^{\pi/2} \sin 2\theta \, d\theta \int_0^\pi d\varphi \int_{-\pi}^\pi d\psi \, f \circ \Phi(\theta, \varphi, \psi).$$

Proof. If

$$x = \begin{pmatrix} x_1 + ix_2 & x_3 + ix_4 \\ -x_3 + ix_4 & x_1 - ix_2 \end{pmatrix},$$

this decomposition can be written

$$x_1 = \cos\theta \cos s,$$
$$x_2 = \cos\theta \sin s,$$
$$x_3 = \sin\theta \cos t,$$
$$x_4 = \sin\theta \sin t,$$

with $s = \psi + \varphi, t = \psi - \varphi$. The differential of Φ can be written

$$\begin{pmatrix} dx_1 \\ dx_2 \\ dx_3 \\ dx_4 \end{pmatrix} = \begin{pmatrix} -\sin\theta\cos s \\ -\sin\theta\sin s \\ \cos\theta\cos t \\ \cos\theta\sin t \end{pmatrix} d\theta + \begin{pmatrix} -\sin s \\ \cos s \\ 0 \\ 0 \end{pmatrix} \cos\theta \, ds + \begin{pmatrix} 0 \\ 0 \\ -\sin t \\ \cos t \end{pmatrix} \sin\theta \, dt.$$

The vector x and the three columns above are orthogonal unit vectors, and form a direct orthonormal basis. As in Section 7.3, we get

$$\Phi^* \omega = \cos\theta \sin\theta \, d\theta \wedge ds \wedge dt = 2\cos\theta \sin\theta \, d\theta \wedge d\varphi \wedge d\psi. \qquad \square$$

One can prove also that every matrix $x \in SU(2)$ can be written

$$x = \exp\left(\tfrac{1}{2}\psi X_3\right) \exp\left(\tfrac{1}{2}\theta X_1\right) \exp\left(\tfrac{1}{2}\varphi X_3\right),$$

with $0 \leq \theta \leq \pi, 0 \leq \varphi \leq 2\pi, -2\pi \leq \psi \leq 2\pi$.

Let us consider the following basis of $\mathfrak{so}(3)$:

$$Y_1 = \tfrac{1}{2} \operatorname{ad} X_1 = \begin{pmatrix} 0 & 0 & 0 \\ 0 & 0 & -1 \\ 0 & 1 & 0 \end{pmatrix},$$

$$Y_2 = \tfrac{1}{2} \operatorname{ad} X_2 = \begin{pmatrix} 0 & 0 & 1 \\ 0 & 0 & 0 \\ -1 & 0 & 0 \end{pmatrix},$$

$$Y_3 = \tfrac{1}{2} \operatorname{ad} X_3 = \begin{pmatrix} 0 & -1 & 0 \\ 1 & 0 & 0 \\ 0 & 0 & 0 \end{pmatrix}.$$

The commutation relations are

$$[Y_1, Y_2] = Y_3, \quad [Y_2, Y_3] = Y_1, \quad [Y_3, Y_1] = Y_2.$$

Using the surjective morphism

$$\mathrm{Ad} : SU(2) \to SO(3),$$

it follows that every element $g \in SO(3)$ can be written

$$g = \exp(\psi Y_3) \exp(\theta Y_1) \exp(\varphi Y_3)$$

$$= \begin{pmatrix} \cos\psi & -\sin\psi & 0 \\ \sin\psi & \cos\psi & 0 \\ 0 & 0 & 1 \end{pmatrix} \begin{pmatrix} 1 & 0 & 0 \\ 0 & \cos\theta & -\sin\theta \\ 0 & \sin\theta & \cos\theta \end{pmatrix} \begin{pmatrix} \cos\varphi & -\sin\varphi & 0 \\ \sin\varphi & \cos\varphi & 0 \\ 0 & 0 & 1 \end{pmatrix}.$$

One obtains the following integration formula.

Proposition 7.4.2 *Let ν be the normalised Haar measure of $SO(3)$, and f an integrable function on $SO(3)$. Then*

$$\int_{SO(3)} f(g)\nu(dg)$$

$$= \frac{1}{8\pi^2} \int_0^\pi \sin\theta \, d\theta \int_0^{2\pi} d\varphi \int_0^{2\pi} d\psi f\big(\exp(\psi Y_3)\exp(\theta Y_1)\exp(\varphi Y_3)\big).$$

The numbers θ, φ, ψ are called the *Euler angles* of the rotation g, ψ is the *precession angle*, θ the *nutation angle*, and φ the *angle of proper rotation*.

7.5 Irreducible representations of $SU(2)$

In order to study the irreducible representations of $SU(2)$, we will first consider the irreducible finite dimensional representations of the complex Lie algebra $\mathfrak{g} = \mathfrak{sl}(2, \mathbb{C})$. The reason for this is that $\mathfrak{sl}(2, \mathbb{C})$ is the complexification of the real Lie algebra $\mathfrak{su}(2)$. In fact every $Z \in \mathfrak{sl}(2, \mathbb{C})$ can be written uniquely as $Z = X + iY$, with $X, Y \in \mathfrak{su}(2)$.

Let $\mathcal{P}_m$ be the space of polynomials in two variables, with complex coefficients, and homogeneous of degree m. Note that $\dim \mathcal{P}_m = m + 1$. Let π_m be the representation of $SL(2, \mathbb{C})$ on $\mathcal{P}_m$ defined by

$$\big(\pi_m(g)f\big)(u, v) = f(au + cv, bu + dv),$$

if

$$g = \begin{pmatrix} a & b \\ c & d \end{pmatrix}.$$

Note that

$$(au + cv \quad bu + dv) = (u \quad v)\begin{pmatrix} a & b \\ c & d \end{pmatrix}.$$

In order to study the derived representation $\rho_m = d\pi_m$ of $\mathfrak{sl}(2, \mathbb{C})$ on $\mathcal{P}_m$ we will use the following basis of $\mathfrak{sl}(2, \mathbb{C})$:

$$H = \begin{pmatrix} 1 & 0 \\ 0 & -1 \end{pmatrix}, \quad E = \begin{pmatrix} 0 & 1 \\ 0 & 0 \end{pmatrix}, \quad F = \begin{pmatrix} 0 & 0 \\ 1 & 0 \end{pmatrix},$$

for which the commutation relations are

$$[H, E] = 2E, \quad [H, F] = -2F, \quad [E, F] = H.$$

The derived representation is obtained as follows:

$$\big(\pi_m(\exp t H)f\big)(u, v) = f(e^t u, e^{-t} v),$$

$$\rho_m(H)f = u\frac{\partial f}{\partial u} - v\frac{\partial f}{\partial v},$$

$$\big(\pi_m(\exp t E)f\big)(u, v) = f(u, tu + v),$$

$$\rho_m(E)f = u\frac{\partial f}{\partial v},$$

$$\big(\pi_m(\exp t F)f\big)(u, v) = f(u + tv, v),$$

$$\rho_m(F)f = v\frac{\partial f}{\partial u}.$$

The monomials f_j,

$$f_j(u, v) = u^j v^{m-j}, \quad j = 0, \ldots, m,$$

form a basis of $\mathcal{P}_m$, and

$$\rho_m(H)f_j = (2j - m)f_j,$$
$$\rho_m(E)f_j = (m - j)f_{j+1},$$
$$\rho_m(F)f_j = jf_{j-1}.$$

The matrices of $\rho_m(H)$, $\rho_m(E)$ and $\rho_m(F)$ with respect to the basis $\{f_j\}$ are

$$\rho_m(H) = \begin{pmatrix} -m & & & & \\ & -m+2 & & & \\ & & \ddots & & \\ & & & m-2 & \\ & & & & m \end{pmatrix},$$

$$\rho_m(E) = \begin{pmatrix} 0 & & & & \\ m & 0 & & & \\ & \ddots & \ddots & & \\ & & 2 & 0 & \\ & & & 1 & 0 \end{pmatrix},$$

$$\rho_m(F) = \begin{pmatrix} 0 & 1 & & & \\ & 0 & 2 & & \\ & & \ddots & \ddots & \\ & & & 0 & m \\ & & & & 0 \end{pmatrix}.$$

Proposition 7.5.1 *The representation ρ_m is irreducible.*

Proof. Let $\mathcal{W}$ be a non-zero invariant subspace in $\mathcal{P}_m$. The restriction to $\mathcal{W}$ of the operator $\rho_m(H)$ admits at least one eigenvalue. Therefore one of the vectors f_j belongs to $\mathcal{W}$. Letting the powers of $\rho_m(E)$ and of $\rho_m(F)$ act on this vector one sees that all the vectors f_j belong to $\mathcal{W}$. Hence $\mathcal{W} = \mathcal{P}_m$. □

Theorem 7.5.2 *Every irreducible finite dimensional $\mathbb{C}$-linear representation of $\mathfrak{sl}(2, \mathbb{C})$ is equivalent to one of the representations ρ_m.*

Proof. Let ρ be an irreducible $\mathbb{C}$-linear representation of $\mathfrak{sl}(2, \mathbb{C})$ on a finite dimensional complex vector space $\mathcal{V}$. Let λ_0 be an eigenvalue of $\rho(H)$ with minimal real part, and φ_0 an associated eigenvector:

$$\rho(H)\varphi_0 = \lambda_0\varphi_0.$$

We will show that the vector $\varphi_1 = \rho(E)\varphi_0$, if non-zero, is an eigenvector of $\rho(H)$:

$$\begin{aligned}
\rho(H)\varphi_1 &= \rho(H)\rho(E)\varphi_0 \\
&= \rho(E)\rho(H)\varphi_0 + \rho([H, E])\varphi_0 \\
&= \lambda_0\rho(E)\varphi_0 + 2\rho(E)\varphi_0 = (\lambda_0 + 2)\varphi_1.
\end{aligned}$$

Hence one constructs a sequence of vectors $\varphi_k = \rho(E)^k\varphi_0$, and

$$\rho(H)\varphi_k = (\lambda_0 + 2k)\varphi_k.$$

If these vectors are non-zero, since they are eigenvectors of $\rho(H)$ for distinct eigenvalues, they are linearly independent. There exists an integer m such that

$$\varphi_k \neq 0 \text{ if } k \leq m, \text{ and } \varphi_{m+1} = 0.$$

Let us show that the space $\mathcal{W}$ of dimension $m + 1$ generated by $\varphi_0, \ldots, \varphi_m$ is invariant. It is invariant under $\rho(H)$ and $\rho(E)$. In fact

$$\begin{aligned}
\rho(H)\varphi_k &= (\lambda_0 + 2k)\varphi_k, \\
\rho(E)\varphi_k &= \varphi_{k+1}.
\end{aligned}$$

Let us determine the action of $\rho(F)$ on the vectors φ_k. We show first that $\rho(F)\varphi_0 = 0$:

$$\begin{aligned}
\rho(H)\rho(F)\varphi_0 &= \rho(F)\rho(H)\varphi_0 + \rho([H, F])\varphi_0 \\
&= \lambda_0\rho(F)\varphi_0 - 2\rho(F)\varphi_0 = (\lambda_0 - 2)\rho(F)\varphi_0.
\end{aligned}$$

Since λ_0 is an eigenvalue with minimal real part, $\rho(F)\varphi_0 = 0$. Let us show recursively that

$$\rho(F)\varphi_k = \alpha_k\varphi_{k-1},$$

with

$$\alpha_k = -k(\lambda_0 + k - 1).$$

For $k = 1$,

$$\begin{aligned}
\rho(F)\varphi_1 &= \rho(F)\rho(E)\varphi_0 \\
&= \rho(E)\rho(F)\varphi_0 + \rho([F, E])\varphi_0 \\
&= -\rho(H)\varphi_0 = -\lambda_0\varphi_0.
\end{aligned}$$

Assume that

$$\rho(F)\varphi_k = -k(\lambda_0 + k - 1)\varphi_{k-1},$$

then

$$\begin{aligned}
\rho(F)\varphi_{k+1} &= \rho(F)\rho(E)\varphi_k \\
&= \rho(E)\rho(F)\varphi_k + \rho([F, E])\varphi_k \\
&= -k(\lambda_0 + k - 1)\rho(E)\varphi_{k-1} - \rho(H)\varphi_k \\
&= \big(-k(\lambda_0 + k - 1) - (\lambda_0 + 2k)\big)\varphi_k = -(k + 1)(\lambda_0 + k)\varphi_k.
\end{aligned}$$

Let us show now that $\lambda_0 = -m$. For that one observes that

$$\operatorname{tr}\rho(H) = \operatorname{tr}[\rho(E), \rho(F)] = 0.$$

But,

$$\begin{aligned}
\operatorname{tr}\rho(H) &= \lambda_0 + (\lambda_0 + 2) + \cdots + (\lambda_0 + 2m) \\
&= (m + 1)\lambda_0 + m(m + 1) \\
&= (m + 1)(\lambda_0 + m).
\end{aligned}$$

Finally

$$\begin{aligned}
\rho(H)\varphi_k &= (2k - m)\varphi_k, \\
\rho(E)\varphi_k &= \varphi_{k+1}, \\
\rho(F)\varphi_k &= k(m - k + 1)\varphi_{k-1}.
\end{aligned}$$

It follows that the representation ρ is equivalent to ρ_m. In fact the linear map A from $\mathcal{V}$ onto $\mathcal{P}_m$ defined by

$$A\varphi_k = c_k f_k,$$

with

$$c_0 = 1, \quad c_k = m(m - 1)\ldots(m - k + 1),$$

intertwins the representations ρ and ρ_m:

$$A \circ \rho(X) = \rho_m(X) \circ A,$$

for every $X \in \mathfrak{sl}(2, \mathbb{C})$. $\qquad\square$

The representation π_m (or more precisely its restriction to $SU(2)$) is an irreducible representation of $SU(2)$. In fact, if a subspace is invariant under π_m, it is also invariant under the derived representation ρ_m.

Theorem 7.5.3 *Let π be an irreducible representation of $SU(2)$ on a finite dimensional complex vector space $\mathcal{V}$. Then π is equivalent to one of the representations π_m.*

Proof. The derived representation $d\pi$ extends linearly as a $\mathbb{C}$-linear representation ρ of $\mathfrak{sl}(2, \mathbb{C})$ on $\mathcal{V}$. The representation ρ is irreducible. In fact let $\mathcal{W} \neq \{0\}$ be a subspace of $\mathcal{V}$, invariant under ρ. Then $\mathcal{W}$ is invariant under $\operatorname{Exp}\rho(X) = \pi(\exp X)$, for $X \in \mathfrak{su}(2)$. Since the group $SU(2)$ is connected, it is generated by $\exp(\mathfrak{su}(2))$, therefore $\mathcal{W}$ is invariant under $SU(2)$, and $\mathcal{W} = \mathcal{V}$ since π is irreducible. By Theorem 7.5.2 the representation ρ is equivalent to one of the representations ρ_m. Hence there exists an isomorphism A from $\mathcal{V}$ onto $\mathcal{P}_m$ such that

$$A\rho(X) = \rho_m(X)A,$$

for every $X \in \mathfrak{sl}(2, \mathbb{C})$. It follows that, for every $X \in \mathfrak{su}(2)$,

$$A \operatorname{Exp}\rho(X) = \operatorname{Exp}\rho_m(X)A;$$

this can be written

$$A\pi(\exp X) = \pi_m(\exp X)A,$$

and, since $\exp(\mathfrak{su}(2))$ generates $SU(2)$, for every $g \in SU(2)$,

$$A\pi(g) = \pi_m(g)A. \qquad \square$$

Note that the natural representation of $SU(2)$ on $\mathbb{C}^2$ is equivalent to the representation π_1, and that the adjoint representation is equivalent to π_2.

Let us recall that the character of a finite dimensional representation π of a group G is the function χ_π defined on G by

$$\chi_\pi(g) = \operatorname{tr}\pi(g).$$

It is a central function, that is, invariant under inner automorphisms:

$$\chi_\pi(gxg^{-1}) = \chi_\pi(x).$$

Recall also that every element in $SU(2)$ is conjugate to a matrix of the form

$$a(\theta) = \exp(\theta X_1) = \begin{pmatrix} e^{i\theta} & 0 \\ 0 & e^{-i\theta} \end{pmatrix} \quad (\theta \in \mathbb{R}).$$

Proposition 7.5.4 *Let χ_m denote the character of the representation π_m. Then*

$$\chi_m(a(\theta)) = \frac{\sin(m+1)\theta}{\sin\theta}.$$

Proof. The eigenvalues of $\pi_m(a(\theta))$ are the numbers

$$e^{i(2j-m)\theta}, \quad j = 0, \dots, m,$$

hence

$$\chi_m\big(a(\theta)\big) = e^{-im\theta}\left(1 + e^{2i\theta} + \cdots + e^{2im\theta}\right)$$

$$= e^{-im\theta}\frac{e^{2i(m+1)\theta} - 1}{e^{2i\theta} - 1}$$

$$= \frac{e^{i(m+1)\theta} - e^{-i(m+1)\theta}}{e^{i\theta} - e^{-i\theta}}. \qquad \Box$$

7.6 Irreducible representations of $SO(3)$

Let T be an irreducible representation of $SO(3)$ on a finite dimensional vector space $\mathcal{V}$. If Ad denotes the adjoint representation which maps $SU(2)$ onto $SO(3)$, then $\pi = T \circ \mathrm{Ad}$ is an irreducible representation of $SU(2)$ on $\mathcal{V}$, hence π is equivalent to π_m for some $m \in \mathbb{N}$. Since $\mathrm{Ad}(-e) = I$, necessarily $\pi_m(-e) = I$. This happens if and only if m is even. Conversely, if $m = 2\ell$, then the representation $\pi_{2\ell}$ of $SU(2)$ factors to the quotient $SU(2)/\{\pm e\} \simeq SO(3)$, and there exists a representation $\tilde{\pi}_{2\ell}$ of $SO(3)$ such that

$$\tilde{\pi}_{2\ell} \circ \mathrm{Ad} = \pi_{2\ell}.$$

Finally we can state the following.

Proposition 7.6.1 *Every finite dimensional irreducible representation of $SO(3)$ is equivalent to one of the representations $\tilde{\pi}_{2\ell}$.*

Recall that

$$X_1 = \begin{pmatrix} i & 0 \\ 0 & -i \end{pmatrix},$$

and that

$$\mathrm{Ad}(\exp\theta X_1) = \exp(\theta\, \mathrm{ad}\, X_1) = \begin{pmatrix} 1 & 0 & 0 \\ 0 & \cos 2\theta & -\sin 2\theta \\ 0 & \sin 2\theta & \cos 2\theta \end{pmatrix}.$$

The eigenvalues of $\tilde{\pi}_{2\ell}\big(\mathrm{Ad}(\exp\theta X_1)\big)$ are the numbers

$$e^{2ij\theta}, \quad -\ell \le j \le \ell.$$

We will show that the representations $\tilde{\pi}_{2\ell}$ can be realised on the space of harmonic polynomials in three variables, homogeneous of degree ℓ.

A C^2 function f defined in an open set in $\mathbb{R}^n$ is said to be *harmonic* if

$$\Delta f = 0,$$

where Δ is the Laplace operator,

$$\Delta = \sum_{i=1}^{n} \frac{\partial^2 f}{\partial x_i^2}.$$

A transformation g of $GL(n, \mathbb{R})$ acts on the functions as follows

$$\big(r(g)f\big)(x) = f(xg)$$

(a vector $x \in \mathbb{R}^n$ is identified with a $1 \times n$ matrix). Let $D = P(\frac{\partial}{\partial x})$ be a differential operator with constant coefficients.

Proposition 7.6.2

$$r(g)P\left(\frac{\partial}{\partial x}\right)r(g^{-1}) = Q\left(\frac{\partial}{\partial x}\right),$$

where Q is the polynomial defined by

$$Q(\xi) = P(\xi(g^{-1})^T).$$

Proof. The operator $r(g)P(\frac{\partial}{\partial x})r(g^{-1})$ is a constant coefficient differential operator:

$$\left(r(g)P\left(\frac{\partial}{\partial x}\right)r(g^{-1})f\right) = Q\left(\frac{\partial}{\partial x}\right)f.$$

Take $f(x) = e^{(x|\xi)}$ ($\xi \in \mathbb{R}^n$):

$$\begin{aligned}
\left(r(g)P\left(\frac{\partial}{\partial x}\right)r(g^{-1})f\right)(x) &= r(g)P\left(\frac{\partial}{\partial x}\right)e^{(xg^{-1}|\xi)} \\
&= r(g)P\left(\frac{\partial}{\partial x}\right)e^{(x|\xi(g^{-1})^T)} \\
&= r(g)P\big(\xi(g^{-1})^T\big)e^{(x|\xi(g^{-1})^T)} \\
&= P\big(\xi(g^{-1})^T\big).
\end{aligned}$$

The statement follows. $\qquad\square$

Corollary 7.6.3 *If $g \in O(n)$,*

$$r(g)\Delta = \Delta r(g),$$

and, if f is harmonic, then $r(g)f$ is harmonic too.

Let $\mathcal{H}_\ell$ be the space of harmonic polynomials in three variables, homogeneous of degree ℓ ($\ell \in \mathbb{N}$).

Proposition 7.6.4

$$\dim \mathcal{H}_\ell = 2\ell + 1.$$

Proof. A polynomial f in $\mathcal{H}_\ell$ can be written

$$f(x_1, x_2, x_3) = \sum_{k=0}^{\ell} \frac{x_1^k}{k!} f_k(x_2, x_3),$$

and

$$\Delta f = \sum_{k=0}^{\ell-2} \frac{x_1^k}{k!} f_{k+2} + \sum_{k=0}^{\ell} \frac{x_1^k}{k!} \left(\frac{\partial^2 f_k}{\partial x_2^2} + \frac{\partial^2 f_k}{\partial x_3^2} \right),$$

hence

$$f_{k+2} = - \left(\frac{\partial^2 f_k}{\partial x_2^2} + \frac{\partial^2 f_k}{\partial x_3^2} \right).$$

Therefore f is determined by the polynomials f_0 and f_1; f_0 is an arbitrary polynomial in two variables of degree ℓ, and f_1 is arbitrary of degree $\ell - 1$. Therefore

$$\dim \mathcal{H}_\ell = (\ell + 1) + \ell = 2\ell + 1. \qquad \square$$

Let T_ℓ be the representation of $SO(3)$ on $\mathcal{H}_\ell$ defined by

$$\big(T_\ell(g) f \big)(x) = f(xg).$$

Theorem 7.6.5 *The representation T_ℓ is irreducible. It is equivalent to $\tilde{\pi}_{2\ell}$.*

Proof. By Corollary 6.1.2, the space $\mathcal{H}_\ell$ can be decomposed as a direct sum of irreducible subspaces

$$\mathcal{H}_\ell = \mathcal{H}_\ell^{(1)} \oplus \cdots \oplus \mathcal{H}_\ell^{(N)};$$

$\mathcal{H}_\ell^{(k)}$ is invariant under T_ℓ and the restriction $T_\ell^{(k)}$ of T_ℓ to $\mathcal{H}_\ell^{(k)}$ is an irreducible representation of $SO(3)$. Hence there exists an integer ℓ_k such that $T_\ell^{(k)}$ is equivalent to $\tilde{\pi}_{2\ell_k}$. Therefore

$$\dim \mathcal{H}_\ell^{(k)} = 2\ell_k + 1,$$

and necessarily $\ell_k \leq \ell$. The eigenvalues of $T_\ell^{(k)} \, (\mathrm{Ad}(\exp \theta X_1))$ are the numbers

$$e^{2ij\theta}, \quad -\ell_k \leq j \leq \ell_k, \; 1 \leq k \leq N.$$

Let us consider the polynomial

$$f(x_1, x_2, x_3) = (x_2 + ix_3)^\ell.$$

It belongs to $\mathcal{H}_\ell$ and

$$\left(T_\ell\left(\mathrm{Ad}(\exp\theta X_1)\right)f\right)(x_1, x_2, x_3)$$
$$= f(x_1, x_2\cos 2\theta + x_3\sin 2\theta, -x_2\sin 2\theta + x_3\cos 2\theta)$$
$$= (x_2\cos 2\theta + x_3\sin 2\theta - ix_2\sin 2\theta + ix_3\cos 2\theta)^\ell$$
$$= e^{-2i\ell\theta} f(x_1, x_2, x_3).$$

Hence $e^{-2i\ell\theta}$ is one of the eigenvalues of $T_\ell\left(\mathrm{Ad}(\exp\theta X_1)\right)$. It follows that one of the numbers ℓ_k is equal to ℓ, and that $\mathcal{H}_\ell = \mathcal{H}_\ell^{(k)}$. Therefore T_ℓ is irreducible and is equivalent to $\tilde{\pi}_{2\ell}$. $\qquad\square$

We will determine a basis of $\mathcal{H}_\ell$ by using the derived representation τ_ℓ of T_ℓ. This representation extends linearly as a $\mathbb{C}$-linear representation of the Lie algebra $\mathfrak{so}(3, \mathbb{C})$, whose elements are complex 3×3 skewsymmetric matrices.

Let τ be the representation of $\mathfrak{so}(3, \mathbb{C})$ on the space $\mathcal{P}$ of polynomials in three variables, for which $Y \in \mathfrak{so}(3, \mathbb{C})$ maps to the differential operator $\tau(Y)$ defined by

$$\left(\tau(Y)f\right)(x) = \frac{d}{dt}f(x\exp tY)\Big|_{t=0}.$$

We have

$$\tau(Y_1) = x_3\frac{\partial}{\partial x_2} - x_2\frac{\partial}{\partial x_3},$$
$$\tau(Y_2) = x_1\frac{\partial}{\partial x_3} - x_3\frac{\partial}{\partial x_1},$$
$$\tau(Y_3) = x_2\frac{\partial}{\partial x_1} - x_1\frac{\partial}{\partial x_2}.$$

For $Y \in \mathfrak{so}(3, \mathbb{C})$, $\tau(Y)$ commutes with the Laplacian Δ,

$$\tau(Y) \circ \Delta = \Delta \circ \tau(Y),$$

and with the dilations. Hence $\tau(Y)$ leaves $\mathcal{H}_\ell$ invariant. One has a representation τ_ℓ of $\mathfrak{so}(3, \mathbb{C})$ on $\mathcal{H}_\ell$, whose restriction to $\mathfrak{so}(3)$ is the derived representation of T_ℓ.

Put

$$H_0 = 2iY_3, \quad E_0 = Y_1 + iY_2, \quad F_0 = -Y_1 + iY_2.$$

Then

$$[H_0, E_0] = 2E_0, \quad [H_0, F_0] = -2F_0, \quad [E_0, F_0] = H_0,$$

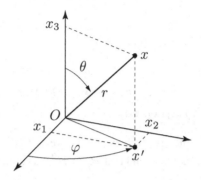

Figure 3

and the map from $\mathfrak{sl}(2, \mathbb{C})$ into $\mathfrak{so}(3, \mathbb{C})$ defined by

$$\begin{pmatrix} 1 & 0 \\ 0 & -1 \end{pmatrix} \mapsto H_0, \quad \begin{pmatrix} 0 & 1 \\ 0 & 0 \end{pmatrix} \mapsto E_0, \quad \begin{pmatrix} 0 & 0 \\ 1 & 0 \end{pmatrix} \mapsto F_0,$$

is a Lie algebra isomorphism.

Let us now introduce the *spherical coordinates* (r, θ, φ) such that

$$x_1 = r \sin \theta \cos \varphi,$$
$$x_2 = r \sin \theta \sin \varphi,$$
$$x_3 = r \cos \theta.$$

We will express the operators $\tau(H_0)$, $\tau(E_0)$, and $\tau(F_0)$ in these coordinates. Express first the derivatives with respect to x_1, x_2, x_3 in terms of the derivatives with respect to r, θ, φ:

$$\begin{pmatrix} \frac{\partial}{\partial r} \\ \frac{\partial}{\partial \theta} \\ \frac{\partial}{\partial \varphi} \end{pmatrix} = \begin{pmatrix} \sin \theta \cos \varphi & \sin \theta \sin \varphi & \cos \theta \\ r \cos \theta \cos \varphi & r \cos \theta \sin \varphi & -\sin \theta \\ -r \sin \theta \sin \varphi & r \sin \theta \cos \varphi & 0 \end{pmatrix} \begin{pmatrix} \frac{\partial}{\partial x_1} \\ \frac{\partial}{\partial x_2} \\ \frac{\partial}{\partial x_3} \end{pmatrix};$$

this can be written

$$\begin{pmatrix} \frac{\partial}{\partial r} \\ \frac{1}{r} \frac{\partial}{\partial \theta} \\ \frac{1}{r \sin \theta} \frac{\partial}{\partial \varphi} \end{pmatrix} = \begin{pmatrix} \sin \theta \cos \varphi & \sin \theta \sin \varphi & \cos \theta \\ \cos \theta \cos \varphi & \cos \theta \sin \varphi & -\sin \theta \\ -\sin \varphi & \cos \varphi & 0 \end{pmatrix} \begin{pmatrix} \frac{\partial}{\partial x_1} \\ \frac{\partial}{\partial x_2} \\ \frac{\partial}{\partial x_3} \end{pmatrix}.$$

The above matrix is orthogonal, hence its inverse equals its transpose:

$$\begin{pmatrix} \dfrac{\partial}{\partial x_1} \\[6pt] \dfrac{\partial}{\partial x_2} \\[6pt] \dfrac{\partial}{\partial x_3} \end{pmatrix} = \begin{pmatrix} \sin\theta\cos\varphi & \cos\theta\cos\varphi & -\sin\varphi \\ \sin\theta\sin\varphi & \cos\theta\sin\varphi & \cos\varphi \\ \cos\theta & -\sin\theta & 0 \end{pmatrix} \begin{pmatrix} \dfrac{\partial}{\partial r} \\[6pt] \dfrac{1}{r}\dfrac{\partial}{\partial \theta} \\[6pt] \dfrac{1}{r\sin\theta}\dfrac{\partial}{\partial \varphi} \end{pmatrix}.$$

It follows that

$$\tau(Y_1) = x_3\frac{\partial}{\partial x_2} - x_2\frac{\partial}{\partial x_3} = \sin\varphi\frac{\partial}{\partial \theta} + \cot\theta\cos\varphi\frac{\partial}{\partial \varphi},$$

$$\tau(Y_2) = x_1\frac{\partial}{\partial x_3} - x_3\frac{\partial}{\partial x_1} = -\cos\varphi\frac{\partial}{\partial \theta} + \cot\theta\sin\varphi\frac{\partial}{\partial \varphi},$$

$$\tau(Y_3) = x_2\frac{\partial}{\partial x_1} - x_1\frac{\partial}{\partial x_2} = -\frac{\partial}{\partial \varphi},$$

and

$$\tau(H_0) = 2i\tau(X_3) = -2i\frac{\partial}{\partial \varphi},$$

$$\tau(E_0) = \tau(X_1) + i\tau(X_2) = e^{i\varphi}\left(-i\frac{\partial}{\partial \theta} + \cot\theta\frac{\partial}{\partial \varphi}\right),$$

$$\tau(F_0) = -\tau(X_1) + i\tau(X_2) = -e^{-i\varphi}\left(i\frac{\partial}{\partial \theta} + \cot\theta\frac{\partial}{\partial \varphi}\right).$$

These operators do not involve the variable r. This is because $\tau(Y)$ commutes with the dilations. In the following we will consider the restrictions of the polynomials in $\mathcal{H}_\ell$ to the unit sphere in $\mathbb{R}^3$ defined by $r = 1$.

Put

$$f_\ell(x_1, x_2, x_3) = (x_1 + ix_2)^\ell = (\sin\theta)^\ell e^{i\ell\varphi}.$$

This is a polynomial in $\mathcal{H}_\ell$, and

$$\tau(H_0)f_\ell = 2\ell f_\ell,$$
$$\tau(E_0)f_\ell = 0.$$

From our study of the irreducible representations of $\mathfrak{sl}(2, \mathbb{C})$ (see Section 7.5) we know that we get a basis of $\mathcal{H}_\ell$ by letting the successive powers of $\tau(F_0)$ act on f_ℓ. Put

$$f_k = \tau(F_0)^{\ell-k}f_\ell, \quad -\ell \leq k \leq \ell.$$

We know that $\tau(H_0)f_k = 2kf_k$, which can be written

$$\frac{\partial f_k}{\partial \varphi} = ikf_k.$$

Hence f_k can be written

$$f_k(\theta, \varphi) = e^{ik\varphi} F_k(\theta).$$

Note that $F_\ell(\theta) = (\sin\theta)^\ell$. The relation $f_{k-1} = \tau(F_0)f_k$ can be written

$$e^{i(k-1)\varphi} F_{k-1} = -e^{-i\varphi} \left(i\frac{\partial}{\partial\theta} + \cot\theta \frac{\partial}{\partial\varphi} \right) e^{ik\varphi} F_k$$

$$= e^{i(k-1)\varphi} \left(-i\frac{dF_k}{d\theta} - ik\cot\theta\, F_k \right),$$

or

$$\frac{dF_k}{d\theta} + k\cot\theta\, F_k = i F_{k-1}.$$

Put $\cos\theta = t$, $F_k(\theta) = i^{\ell-k} p_k(t)$, then, for $0 \le \theta \le \pi$,

$$\frac{dF_k}{d\theta} = -(-i)^{\ell-k}(1-t^2)^{1/2}\frac{dp_k}{dt},$$

therefore

$$(1-t^2)^{1/2}\left(\frac{dp_k}{dt} - k\frac{t}{1-t^2}p_k \right) = p_{k-1}.$$

Now put

$$u_k(t) = (1-t^2)^{k/2} p_k(t).$$

We get

$$\frac{du_k}{dt} = u_{k-1},$$

and, since $u_\ell(t) = (1-t^2)^\ell$,

$$u_k(t) = \left(\frac{d}{dt} \right)^{\ell-k} (1-t^2)^\ell,$$

$$p_k(t) = (1-t^2)^{-k/2}\left(\frac{d}{dt} \right)^{\ell-k} (1-t^2)^\ell.$$

Finally,

$$f_k(\theta, \varphi) = i^{\ell-k} e^{ik\varphi} p_k(\cos\theta).$$

For $k = 0$, p_0 is proportional to the Legendre polynomial P_ℓ which is given by

$$P_\ell(t) = \frac{1}{\ell!2^\ell}\left(\frac{d}{dt} \right)^\ell (t^2 - 1)^\ell.$$

For $k = -\ell$,

$$p_{-\ell}(t) = (1 - t^2)^{\ell/2} \left(\frac{d}{dt}\right)^{2\ell} (1 - t^2)^{\ell} = C(1 - t^2)^{\ell/2},$$

and

$$f_{-\ell}(x_1, x_2, x_3) = C'(x_1 - ix_2)^{\ell}.$$

7.7 Exercises

1. Let S^2 be the unit sphere in $\mathbb{R}^3$, whose equation is

$$x_1^2 + x_2^2 + x_3^2 = 1.$$

Let ϕ be the *stereographic projection* $\phi : S^2 \setminus \{-e_3\} \to \mathbb{C}$.

(a) Show that ϕ is given by the formula

$$z = \frac{x_1 + ix_2}{1 + x_3},$$

and that its inverse ϕ^{-1} is given by

$$x_1 + ix_2 = \frac{2z}{1 + |z|^2}, \quad x_3 = \frac{1 - |z|^2}{1 + |z|^2}.$$

(b) For

$$g = \begin{pmatrix} \alpha & \beta \\ -\bar{\beta} & \bar{\alpha} \end{pmatrix} \in SU(2),$$

let $T(g)$ be the fractional linear transformation defined by

$$T(g)(z) = \frac{\alpha z + \beta}{-\bar{\beta} z + \bar{\alpha}}.$$

The aim of this exercise is to show that $R(g) = \phi^{-1} \circ T(g) \circ \phi$ is the restriction to S^2 of a rotation in $SO(3)$.

(c) Let $\mathcal{P}_2$ denote the space of polynomials in two variables with complex coefficients, homogeneous of degree 2, and π_2 the representation of $G = SU(2)$ on $\mathcal{P}_2$ defined by

$$\left(\pi_2(g)f\right)(u, v) = f(\alpha u - \bar{\beta}v, \beta u + \bar{\alpha}v),$$

for

$$g = \begin{pmatrix} \alpha & \beta \\ -\bar{\beta} & \bar{\alpha} \end{pmatrix}, \quad |\alpha|^2 + |\beta|^2 = 1.$$

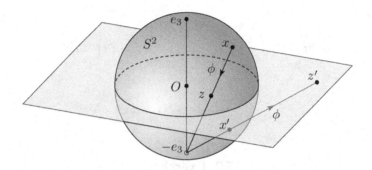

Figure 4

One puts, for $z \in \mathbb{C}$,

$$f_z(u, v) = \frac{z}{1 + |z|^2} u^2 + \frac{1 - |z|^2}{1 + |z|^2} uv - \frac{\bar{z}}{1 + |z|^2} v^2.$$

Show that

$$\pi_2(g) f_0 = f_z,$$

with $z = T(g)(0)$. Then prove that

$$\pi_2(g) f_z = f_{T(g)(z)}.$$

(d) Show that

$$\pi_2(g) f_{\phi(x)} = f_{\phi\left(R(g)x\right)},$$

and then that $R(g)$ is the restriction to S^2 of a transformation in $SO(3)$. Show that R is a surjective morphism from $SU(2)$ onto $SO(3)$.

2. Let m be a positive integer, and let $\mathcal{V}$ be the space of polynomials in one variable of degree $\leq m$ with coefficients in $\mathbb{K} = \mathbb{R}$ or $\mathbb{C}$, and let A, B, C be the endomorphism of $\mathcal{V}$ defined by

$$Af = -mf + 2x\frac{df}{dx}, \quad Bf = mxf - x^2\frac{df}{dx}, \quad Cf = \frac{df}{dx}.$$

(a) Show that A, B, and C generate a Lie algebra isomorphic to $\mathfrak{g} = \mathfrak{sl}(2, \mathbb{K})$. Hence one gets a representation ρ of $\mathfrak{g}$ on $\mathcal{V}$.

(b) Let $G = SL(2, \mathbb{K})$. For

$$g = \begin{pmatrix} a & b \\ c & d \end{pmatrix},$$

one puts, for $f \in \mathcal{V}$,

$$\big(\pi(g)f\big)(x) = (bx+d)^m f\left(\frac{ax+c}{bx+d}\right).$$

Show that π is a representation of G, and ρ is the derived representation of π.

(c) Show that the representation π is equivalent to the representation π_m defined in Section 7.2.

3. Let G be a compact group. For $\lambda \in \hat{G}$ let π_λ be an irreducible representation in the class λ. The representation $\pi_\lambda \otimes \pi_\mu$ can be decomposed as a sum of irreducible representations:

$$\pi_\lambda \otimes \pi_\mu = \bigoplus_{\nu \in E(\lambda,\mu)} c(\lambda, \mu;, \nu)\pi_\nu$$

$(E(\lambda, \mu) \subset \hat{G}, \; c(\lambda, \mu; \nu) \in \mathbb{N}^*)$. The numbers $c(\lambda, \mu; \nu)$ are called Clebsch–Gordan coefficients.

(a) Show that

$$\chi_\lambda \cdot \chi_\mu = \sum_{\nu \in E(\lambda,\mu)} c(\lambda, \mu; \nu)\chi_\nu.$$

(b) In the case of $G = SU(2)$, the set $\hat{G}$ can be identified with $\mathbb{N}$. Show that

$$\chi_p \cdot \chi_q = \sum_{k=0}^{\min(p,q)} \chi_{p+q-2k}.$$

Then show that

$$\pi_p \otimes \pi_q = \bigoplus_{k=0}^{\min(p,q)} \pi_{p+q-2k}.$$

That is

$$E(p, q) = \{m = |p-q| + 2j \mid j = 0, \ldots, \min(p, q)\},$$

and, for every $m \in E(p, q)$, $c(p, q; m) = 1$.

4. (a) Consider the three following operators acting on the space $\mathcal{C}^\infty(\mathbb{R})$:

$$Af = -\frac{i}{2}x^2 f, \quad Bf = -\frac{i}{2}\frac{d^2}{dx^2}f, \quad Cf = x\frac{d}{dx}f + \frac{1}{2}f.$$

Show that the vector space generated by A, B, C is a Lie algebra isomorphic to $\mathfrak{sl}(2, \mathbb{R})$, and hence one gets a representation ρ of $\mathfrak{sl}(2, \mathbb{R})$ on $\mathcal{C}^\infty(\mathbb{R})$.

(b) Show that

$$f_0(x) = e^{-x^2/2}$$

is an eigenfunction of $D = A - B$. Put $U = C + i(A + B)$. Compute $[D, U]$, and show that $f_k = U^k f_0$ is an eigenfunction of D. Notice that:

$$U = \frac{1}{2}\left(x + \frac{d}{dx}\right)^2.$$

5. On the space $\mathcal{P}_m$ of polynomials in two variables with complex coefficients, homogeneous of degree m, consider the Hermitian inner product given by

$$(p|q) = \frac{1}{\pi^2}\int_{\mathbb{C}^2} p(u, v)\overline{q(u, v)}e^{-(|u|^2 + |v|^2)}\lambda(du)\lambda(dv),$$

where λ denotes the Lebesgue measure on $\mathbb{C} \simeq \mathbb{R}^2$.

(a) Show that, for this inner product, the representation π_m of $SU(2)$ is unitary.

(b) Compute the norm of f_j. Show that

$$\|f_j\|^2 = j!(m - j)!.$$

6. On the space $\mathcal{P}_m$ of polynomials in two variables with complex coefficients, homogeneous of degree m, consider the Hermitian inner product given by

$$(p|q) = \int_{SU(2)} p(\alpha, \beta)\overline{q(\alpha, \beta)}\mu(dg),$$

where

$$g = \begin{pmatrix} \alpha & \beta \\ -\bar{\beta} & \bar{\alpha} \end{pmatrix},$$

and μ is the normalised Haar measure of $SU(2)$.

(a) Show that, for this inner product, the representation π_m of $SU(2)$ is unitary.

(b) Show that this inner product is proportional to the inner product considered in the preceding exercise, and compute the factor of proportionality.

7. (a) Show that, for every φ, the polynomial

$$\left(x_3 + i(x_1 \cos\varphi + x_2 \sin\varphi)\right)^\ell$$

belongs in $\mathcal{H}_\ell$. Put

$$f(x_1, x_2, x_3) = \frac{1}{2\pi}\int_0^{2\pi} \left(x_3 + i(x_1 \cos\varphi + x_2 \sin\varphi)\right)^\ell e^{-ik\varphi}d\varphi.$$

Show that f is a polynomial which belongs to $\mathcal{H}_\ell$, and f is an eigen-function of the operator $\tau(H_0)$ (which was introduced in Section 7.6).

(b) Then establish the following integral representation of the function F_k (introduced at the end of Section 7.6):

$$F_k(\theta) = c(k, \ell) \frac{1}{2\pi} \int_0^{2\pi} (\cos\theta + i\sin\theta\cos\varphi)^\ell e^{-ik\varphi}d\varphi,$$

where $c(k, \ell)$ is a constant.

8. The representation π_m of the group $G = SU(2)$ can be realised on the space $\mathcal{P}_m$ of polynomials in one variable with complex coefficients of degree $\leq m$ as follows

$$\left(\pi_m(g)f\right)(z) = (\beta z + \bar\alpha)^m f\left(\frac{\alpha z - \bar\beta}{\beta z + \bar\alpha}\right),$$

for

$$g = \begin{pmatrix} \alpha & \beta \\ -\bar\beta & \bar\alpha \end{pmatrix}.$$

(a) Consider on $\mathcal{P}_m$ the Hermitian inner product defined by

$$(f_1|f_2) = \frac{m+1}{\pi} \int_{\mathbb{C}} f_1(z)\overline{f_2(z)}(1 + |z|^2)^{-m-2}d\lambda(z),$$

where λ is the Lebesgue measure on $\mathbb{C} \simeq \mathbb{R}^2$. Show that the representation π_m is unitary.

Hint. Let F be a non-negative measurable function on $\mathbb{C}$. Show, for

$$g = \begin{pmatrix} \alpha & \beta \\ -\bar\beta & \bar\alpha \end{pmatrix} \in G,$$

that

$$\int_{\mathbb{C}} F\left(\frac{\alpha z - \bar\beta}{\beta z + \bar\alpha}\right)(1 + |z|^2)^{-2}d\lambda(z) = \int_{\mathbb{C}} F(w)(1 + |w|^2)^{-2}d\lambda(w),$$

and, for

$$w = \frac{\alpha z - \bar\beta}{\beta z + \bar\alpha},$$

that

$$1 + |w|^2 = \frac{1}{|\beta z + \bar\alpha|^2}(1 + |z|^2).$$

(b) Put $f_j(z) = z^j$. Compute $\|f_0\|$. Show that, for every $f \in \mathcal{P}_m$,

$$(f|f_0) = f(0).$$

Show that

$$\|f_j\|^2 = \frac{j!(m-j)!}{m!}.$$

Hint. By using the relation $d\pi(E)^* = d\pi(F)$, where

$$E = \begin{pmatrix} 0 & 1 \\ 0 & 0 \end{pmatrix}, \quad F = \begin{pmatrix} 0 & 0 \\ 1 & 0 \end{pmatrix},$$

show that

$$\|f_j\|^2 = \frac{j}{m-j+1}\|f_{j-1}\|^2.$$

(c) If $\{\varphi_j\}$ is an orthonormal basis of $\mathcal{P}_m$, put, for $z, w \in \mathbb{C}$,

$$K(z, w) = \sum_{j=0}^{m} \varphi_j(z)\overline{\varphi_j(w)}.$$

Show that the kernel K does not depend on the choice of basis. It is called the *reproducing kernel* of the Hilbert space $\mathcal{P}_m$. Show that

$$K(z, w) = (1 + z\bar{w})^m.$$

Show that, for every $f \in \mathcal{P}_m$, and every $w \in \mathbb{C}$,

$$f(w) = (f|K_w), \text{ if } K_w(z) = K(z, w).$$

9. The aim of this exercise is to construct an equivariant isomorphism from the space $\mathcal{P}_{2\ell}$ onto the space $\mathcal{H}_\ell$ of harmonic polynomials in three variables, and homogeneous of degree ℓ.

(a) To every $x = (x_1, x_2, x_3) \in \mathbb{R}^3$ one associates the symmetric complex matrix

$$M(x) = \begin{pmatrix} x_1 + ix_2 & x_3 \\ x_3 & -x_1 + ix_2 \end{pmatrix}.$$

Show that the image of the map $M : \mathbb{R}^3 \to Sym(2, \mathbb{C})$ is the space $\{X \in Sym(2, \mathbb{C}) \mid \bar{X} = JXJ\}$, where

$$J = \begin{pmatrix} 0 & -1 \\ 1 & 0 \end{pmatrix}.$$

Let $g \in G = SU(2)$. Show that there is an orthogonal transformation $\gamma \in O(3)$ such that

$$\forall x \in \mathbb{R}^3, \quad g^t M(x)g = M(x\gamma).$$

Hint. Observe that $JgJ = -\bar{g}$, and consider the determinant of $M(x)$.

Show that the map $\tau : G \to O(3)$, $g \mapsto \gamma$ is a group morphism.
Determine its image and its kernel.

(b) For $x \in \mathbb{R}^3$, $z \in \mathbb{C}$, define

$$H(x, z) = (x_1 + ix_2)z^2 + 2zx_3 + (-x_1 + ix_2).$$

Show that, for $\ell \in \mathbb{N}$, z fixed, the function $x \mapsto H(x, z)^\ell$ is a harmonic
polynomial.

Hint. Show that, if $a, b, c \in \mathbb{C}$ satisfy $a^2 + b^2 + c^2 = 0$, then the polynomial F,

$$F(x) = (ax_1 + bx_2 + cx_3)^\ell,$$

is harmonic.

(c) Let

$$g = \begin{pmatrix} \alpha & \beta \\ -\bar\beta & \bar\alpha \end{pmatrix} \in G, \quad \text{and} \quad \gamma = \tau(g).$$

Show that

$$H(x\gamma, z) = (-\bar\beta z + \bar\alpha)^2 H\left(x, \frac{\alpha z + \beta}{-\bar\beta z + \bar\alpha}\right).$$

Hint. Observe that

$$H(x, z) = (z \quad 1) M(x) \begin{pmatrix} z \\ 1 \end{pmatrix}.$$

(d) Let $\mathcal{H}_\ell$ be the space of harmonic polynomials in three real variables
x_1, x_2, x_3, homogeneous of degree ℓ. Let T_ℓ be the representation of
$SO(3)$ on $\mathcal{H}_\ell$ defined by

$$\big(T_\ell(\gamma)F\big)(x) = F(x\gamma) \quad \big(\gamma \in SO(3)\big).$$

For $f \in \mathcal{P}_{2\ell}$ put

$$(A_\ell f)(x) = \frac{2\ell + 1}{\pi} \int_{\mathbb{C}} f(z)\overline{H(x, z)}^\ell (1 + |z|^2)^{-2\ell - 2} d\lambda(z).$$

Show that $A_\ell f \in \mathcal{H}_\ell$, and that

$$(A_\ell f_0)(x) = (-x_1 + ix_2)^\ell, \quad (A_\ell f_{2\ell})(x) = (x_1 + ix_2)^\ell.$$

Show that, for $g \in G$ and $\gamma = \tau(\bar{g})$,

$$A_\ell \circ \pi_{2\ell}(g) = T_\ell(\gamma) \circ A_\ell,$$

and that A_ℓ is an isomorphism from $\mathcal{P}_{2\ell}$ onto $\mathcal{H}_\ell$.

10. The aim of this exercise is to show that $SU(4)$ is a covering of order two
of $SO(6)$.

(a) Consider the representation π of $SL(4, \mathbb{C})$ on the space $\Lambda^2(\mathbb{C}^4)$ such that, if $\alpha = u \wedge v \in \Lambda^2(\mathbb{C}^4)$, with $u, v \in \mathbb{C}^4$, then

$$\pi(g)\alpha = (gu) \wedge (gv).$$

Show that the kernel of π is $\{\pm I\}$.

Hint. For g in the kernel of π, consider a basis $\{u_1, u_2, u_3, u_4\}$ of $\mathbb{C}^4$ with respect to which the matrix of g is upper triangular.

Let ω be the element in $\Lambda^4(\mathbb{C}^4)$ defined by

$$\omega = e_1 \wedge e_2 \wedge e_3 \wedge e_4,$$

where $\{e_1, e_2, e_3, e_4\}$ is the canonical basis of $\mathbb{C}^4$. Show that there is a bilinear form B on $\Lambda^2(\mathbb{C}^4)$ such that, for $\alpha, \beta \in \Lambda^2(\mathbb{C}^4)$,

$$\alpha \wedge \beta = B(\alpha, \beta)\omega.$$

Show that the bilinear form B is symmetric, and non-degenerate, and that, for $g \in SL(4, \mathbb{C})$,

$$B\big(\pi(g)\alpha, \pi(g)\beta\big) = B(\alpha, \beta).$$

Show that $d\pi$ is an isomorphism from $\mathfrak{sl}(4, \mathbb{C})$ onto $\mathfrak{so}(B, \mathbb{C})$, and that the image of π is equal to $SO(B, \mathbb{C})$.

Hint. Compare the dimensions of $SL(4, \mathbb{C})$ and $SO(B, \mathbb{C})$, and use the fact that $SO(B, \mathbb{C})$ is connected (see Exercise 5 of Chapter 2).

(b) To a 4×4 skewsymmetric matrix $A = (a_{ij})$ one associates $\alpha \in \Lambda^2(\mathbb{C}^4)$ defined by

$$\alpha = \sum_{i,j=1}^{4} a_{ij} e_i \wedge e_j,$$

and the space $\Lambda^2(\mathbb{C}^4)$ is endowed with the Hermitian inner product defined by

$$(\alpha|\beta) = \text{tr}(AB^*) = -\text{tr}(A\bar{B}).$$

Let π_0 denote the restriction of the representation π to the subgroup $SU(4)$. Show that π_0 is a unitary representation of $SU(4)$ on $\Lambda^2(\mathbb{C}^4)$. Consider the basis of $\Lambda^2(\mathbb{C}^4)$ consisting of the following matrices:

$$\alpha_1 = e_1 \wedge e_2 + e_3 \wedge e_4, \quad \alpha_2 = i(e_1 \wedge e_2 - e_3 \wedge e_4),$$
$$\alpha_3 = e_1 \wedge e_3 + e_2 \wedge e_4, \quad \alpha_4 = i(e_1 \wedge e_3 - e_2 \wedge e_4),$$
$$\alpha_5 = e_1 \wedge e_4 + e_2 \wedge e_3, \quad \alpha_6 = i(e_1 \wedge e_4 - e_2 \wedge e_3).$$

Show that, if

$$\alpha = \sum_{j=1}^{6} z_j \alpha_j,$$

then

$$B(\alpha, \alpha) = 2 \sum_{j=1} z_j^2,$$

$$(\alpha | \alpha) = 4 \sum_{j=1}^{6} |z_j|^2.$$

Show that, for $g \in SU(4)$, the matrix of $\pi(g)$ with respect to the basis $\{\alpha_j\}$ belongs to $SO(6)$, and the image of π_0 is equal to $SO(6)$.

Conclude that $SL(4, \mathbb{C})$ is a covering of order two of $SO(6, \mathbb{C})$, and $SU(4)$ is a covering of order two of $SO(6)$.

8

Analysis on the group $SU(2)$

After we have determined the irreducible representations of the group $SU(2)$, we can make explicit the Peter–Weyl and Plancherel Theorems we saw in Chapter 6. We will see how the properties of classical Fourier series extend in this setting. Finally we will show how Fourier analysis can be used to solve the Cauchy problem for the heat equation on the group $SU(2)$.

8.1 Fourier series on $SO(2)$

Let us recall first some properties of classical Fourier series expansions for functions defined on the group $G = SO(2) \simeq \mathbb{R}/2\pi\mathbb{Z}$. In this case $\hat{G} \simeq \mathbb{Z}$, and the *Fourier coefficient* $\hat{f}(m)$ of an integrable function f on G is defined by

$$\hat{f}(m) = \frac{1}{2\pi} \int_0^{2\pi} f(x)e^{-imx} dx.$$

If f is square integrable, then the *Fourier series* of f converges to f in the mean, that is in the sense of $L^2(G)$,

$$f(x) = \sum_{m \in \mathbb{Z}} \hat{f}(m)e^{imx}.$$

The Plancherel formula can be written as

$$\frac{1}{2\pi} \int_0^{2\pi} |f(x)|^2 dx = \sum_{m \in \mathbb{Z}} |\hat{f}(m)|^2.$$

If f is continuous and if

$$\sum_{m \in \mathbb{Z}} |\hat{f}(m)| < \infty,$$

then the Fourier series of f converges absolutely and uniformly. This is the case if f is a C^1 function. In fact, by integrating by parts, one establishes that the Fourier coefficients of the derivative f' are related to those of f as follows:

$$\widehat{f'}(m) = im\hat{f}(m),$$

and, by the Schwarz inequality,

$$\sum_{m\neq 0} |\hat{f}(m)| = \sum_{m\neq 0} \frac{1}{|m|} |\widehat{f'}(m)|$$

$$\leq \left(\sum_{m\neq 0} \frac{1}{m^2}\right)^{1/2} \left(\sum_{m\neq 0} |\widehat{f'}(m)|^2\right)^{1/2} < \infty,$$

since

$$\sum_{m\in\mathbb{Z}} |\widehat{f'}(m)|^2 = \frac{1}{2\pi} \int_0^{2\pi} |f'(x)|^2 dx.$$

More generally, if f is C^k, then

$$\widehat{f^{(k)}}(m) = (im)^k \hat{f}(m),$$

and hence

$$|m|^k |\hat{f}(m)| \leq \frac{1}{2\pi} \int_0^{2\pi} |f^{(k)}(x)| dx.$$

Therefore, if f is C^∞ then, for all $k \in \mathbb{N}$,

$$\sup_{m\in\mathbb{Z}} |m|^k |\hat{f}(m)| < \infty.$$

Conversely, if this condition holds, then the Fourier series of f can be differentiated termwise

$$f^{(k)}(x) = \sum_{m\in\mathbb{Z}} (im)^k \hat{f}(m) e^{imx}.$$

This shows that f is C^∞. Hence the Fourier transform which, to a function f, associates the sequence of its Fourier coefficients,

$$f \mapsto \left(\hat{f}(m)\right),$$

is an isomorphism from $C^\infty(\mathbb{R}/2\pi\mathbb{Z})$ onto the space $\mathcal{S}(\mathbb{Z})$ of rapidly decreasing sequences.

In this chapter we will establish analogous properties for the Fourier series on the group $SU(2)$

8.2 Functions of class C^k

We make precise the definition of a function of class C^k on a linear Lie group G, and give some of the properties of C^k functions. Let $U \subset G$ be an open set. We will say that a complex valued function f defined on U is of class C^1 if

(i) for every $g \in U$, and $X \in \mathfrak{g} = Lie(G)$, the function

$$t \mapsto f(g \exp tX)$$

is differentiable at $t = 0$, and then one puts

$$(\rho(X)f)(g) = \frac{d}{dt} f(g \exp tX)\big|_{t=0},$$

(ii) the map

$$\mathfrak{g} \times U \to \mathbb{C}, \quad (X, g) \mapsto (\rho(g)f)(g),$$

is continuous.

Let $C^1(U)$ denote the space of C^1 functions on U. One can show that, if f is a C^1 function on U, then, for every $g \in U$, the function $X \mapsto f(g \exp X)$ is C^1 on a neighbourhood V of 0 in $\mathfrak{g}$. In particular the map $X \mapsto \rho(X)f$ is linear.

Let us recall the notation

$$(L(g)f)(x) = f(g^{-1}x), \quad (R(g)f)(x) = f(xg).$$

(a) If $f \in C^1(U)$ then $L(g)f \in C^1(gU)$, and

$$\rho(X)L(g)f = L(g)\rho(X)f.$$

(b) If $f \in C^1(U)$ then $R(g)f \in C^1(Ug^{-1})$, and

$$\rho(X)R(g)f = R(g)\rho(\mathrm{Ad}(g^{-1})X)f.$$

In fact,

$$(R(g)f)(x \exp tX) = f(x \exp tXg) = f(xg \exp(t\,\mathrm{Ad}(g^{-1})X)).$$

This can also be written

$$R(g)\rho(X)R(g^{-1}) = \rho(\mathrm{Ad}(g)X).$$

One defines the space $C^k(U)$ of C^k functions on U recursively with respect to k: a function f is C^k on U if $f \in C^1$, and if, for every $X \in \mathfrak{g}$, the function $\rho(X)f$ is C^{k-1}. A function f is C^∞ if it is C^k for every k.

Let V be a neighbourhood of 0 in $\mathfrak{g}$ and W a neighbourhood of g in G. Assume that the map $X \mapsto g \exp X$ is a diffeomorphism from V onto W. Let f be a function defined on W. One can show that f is C^k on W ($1 \leq k \leq \infty$) if and only if the map $X \mapsto f(g \exp X)$ is C^k on V.

·

(c) If $f \in C^2(U)$, $X, Y \in \mathfrak{g}$,

$$[\rho(X), \rho(Y)]f = \rho([X, Y])f.$$

In fact, it follows from (b) that

$$\big(\rho(Y)f\big)(x \exp tX) = \Big(\rho\big(\mathrm{Ad}(\exp tX)Y\big)f\Big)\big(R(\exp tX)f\big)(x).$$

Taking the derivatives with respect to t at $t = 0$ we get

$$\rho(X)\rho(Y)f = \rho([X, Y])f + \rho(Y)\rho(X)f.$$

These relations say that the right regular representation R, acting on the space $C^\infty(G)$, is differentiable, and that its differential is equal to the representation ρ.

(d) If $f \in C^2(U)$, $X, Y \in \mathfrak{g}$,

$$\frac{d}{dt} f(g \exp tX \exp tY)\big|_{t=0} = \big(\rho(X + Y)f\big)(g),$$

$$\frac{d^2}{dt^2} f(g \exp tX \exp tY)\big|_{t=0} = \big(\rho(X + Y)^2 f\big)(g) + \rho([X, Y])f(g).$$

In fact,

$$\frac{d}{dt} f(g \exp tX \exp tY)$$

$$= \big(\rho(X)R(\exp tY)f\big)(g \exp tX) + \big(\rho(Y)f\big)(g \exp tX \exp tY),$$

and

$$\frac{d}{dt^2} f(g \exp tX \exp tY)\big|_{t=0}$$

$$= \Big(\rho(X)\big(\rho(X) + \rho(Y)\big)f\Big)(g) + \Big(\big(\rho(X) + \rho(Y)\big)\rho(Y)f\Big)(g).$$

The statement follows by noticing that

$$\rho(X)^2 + 2\rho(X)\rho(Y) + \rho(Y)^2 = \big(\rho(X) + \rho(Y)\big)^2 + [\rho(X), \rho(Y)].$$

Let $(\pi, \mathcal{V})$ be a finite dimensional representation of G. Recall that $\mathcal{M}_\pi$ denotes the subspace of $C(G)$ generated by the coefficients of π. A function f in $\mathcal{M}_\pi$ can be written

$$f(g) = \mathrm{tr}\big(A\pi(g)\big) \quad \big(A \in \mathcal{L}(\mathcal{V})\big).$$

Then $f \in C^\infty(G)$ and, for $X \in \mathfrak{g}$,

$$\big(\rho(X)f\big)(g) = \mathrm{tr}\big(d\pi(X)A\pi(g)\big).$$

Let G be a linear Lie group, and consider on its Lie algebra $\mathfrak{g} = Lie(G)$ a Euclidean inner product. For an orthonormal basis $\{X_1, \ldots, X_n\}$ put

$$\Delta f(x) = \sum_{i=1}^{n} \frac{d^2}{dt^2} f(x \exp t X_i)\Big|_{t=0},$$

where f is a $\mathcal{C}^2$ function on G, that is

$$\Delta = \sum_{i=1}^{n} \rho(X_i)^2.$$

The differential operator Δ which is defined by this formula does not depend on the chosen orthonormal basis, and is left invariant:

$$\Delta\big(f \circ L(g)\big) = (\Delta f) \circ L(g).$$

Furthermore, if the Euclidean inner product on $\mathfrak{g}$ is invariant under the adjoint representation, then the operator Δ is right invariant as well:

$$\Delta\big(f \circ R(g)\big) = (\Delta f) \circ R(g).$$

In that case Δ is called a *Laplace operator*.

Assume G to be compact. There exists on $\mathfrak{g}$ a Euclidean inner product which is invariant under the adjoint representation. Let us denote by μ the normalised Haar measure on G. For $f, \varphi \in \mathcal{C}^1(G)$,

$$\big(\rho(X)f \mid \varphi\big) = -(f \mid \rho(X)\varphi)$$

with respect to the inner product on $L^2(G)$. In fact,

$$\int_G \frac{d}{dt}\Big|_{t=0} f(g \exp t X)\overline{\varphi(g)}\mu(dg) = \frac{d}{dt}\Big|_{t=0} \int_G f(g \exp t X)\overline{\varphi(g)}\mu(dg)$$

$$= \frac{d}{dt}\Big|_{t=0} \int_G f(g)\overline{\varphi(g \exp -t X)}\mu(dg)$$

$$= \int_g f(g)\frac{d}{dt}\Big|_{t=0} \overline{\varphi(g \exp -t X)}\mu(dg).$$

Therefore, the Laplace operator Δ is symmetric: if f and $\varphi \in \mathcal{C}^2(G)$,

$$(\Delta f \mid \varphi) = (f \mid \Delta\varphi),$$

and $-\Delta$ is positive since

$$-(\Delta f \mid f) = \int_G \sum_{i=1}^{n} |\rho(X_i)f(g)|^2 \mu(dg).$$

Assume also that G is connected. Let π be an irreducible representation of G on a complex vector space $\mathcal{V}$. There exists a number κ_π such that $\Omega_\pi = -\kappa_\pi$ (Corollary 6.7.2).

Proposition 8.2.1 *A function $f \in \mathcal{M}_\pi$ is an eigenfunction of the Laplace operator Δ,*

$$\Delta f = -\kappa_\pi f.$$

Proof. A function $f \in \mathcal{M}_\pi$ can be written

$$f(x) = \operatorname{tr}\big(A\pi(g)\big),$$

where A is an endomorphism of $\mathcal{V}$, and

$$\Delta f(g) = \operatorname{tr}\big(\Omega_\pi A\pi(g)\big) = -\kappa_\pi f(g).$$ □

8.3 Laplace operator on the group $SU(2)$

Let us consider on the Lie algebra $\mathfrak{su}(2)$ of the group $SU(2)$ the Euclidean inner product defined by

$$(X|Y) = \tfrac{1}{2}\operatorname{tr}(XY^*) = -\tfrac{1}{2}\operatorname{tr}(XY).$$

This Euclidean inner product is invariant under the adjoint representation and the basis $\{X_1, X_2, X_3\}$ is orthonormal, with

$$X_1 = \begin{pmatrix} i & 0 \\ 0 & -i \end{pmatrix}, \quad X_2 = \begin{pmatrix} 0 & 1 \\ -1 & 0 \end{pmatrix}, \quad X_3 = \begin{pmatrix} 0 & i \\ i & 0 \end{pmatrix}.$$

Let ρ be a representation of $\mathfrak{su}(2)$ on a finite dimensional complex vector space $\mathcal{V}$. The representation ρ extends as a $\mathbb{C}$-linear representation of $\mathfrak{sl}(2, \mathbb{C})$.

Proposition 8.3.1

$$-\Omega_\rho = \rho(H)^2 + 2\rho(H) + 4\rho(F)\rho(E).$$

Proof. Recall the notation:

$$H = \begin{pmatrix} 1 & 0 \\ 0 & -1 \end{pmatrix}, \quad E = \begin{pmatrix} 0 & 1 \\ 0 & 0 \end{pmatrix}, \quad F = \begin{pmatrix} 0 & 0 \\ 1 & 0 \end{pmatrix},$$

and observe that

$$X_1 = iH, \quad X_2 = E - F, \quad X_3 = i(E + F).$$

Hence

$$\rho(X_1)^2 = -\rho(H)^2,$$
$$\rho(X_2)^2 = \rho(E)^2 + \rho(F)^2 - \rho(E)\rho(F) - \rho(F)\rho(E),$$
$$\rho(X_3)^2 = -\rho(E)^2 - \rho(F)^2 - \rho(E)\rho(F) - \rho(F)\rho(E),$$

therefore

$$\Omega_\rho = -\rho(H)^2 - 2\rho(E)\rho(F) - 2\rho(F)\rho(E).$$

Using the relation $[E, F] = H$ we get

$$-\Omega_\rho = \rho(H)^2 + 2\rho(H) + 4\rho(F)\rho(E). \qquad \square$$

Let us consider the representation $\rho = \rho_m$ of $\mathfrak{sl}(2, \mathbb{R})$ on the space $\mathcal{P}_m$ of polynomials in two variables which are homogeneous of degree m. This representation was introduced in Section 7.5. Let $\Omega_m = \Omega_{\rho_m}$ denote the associated Casimir operator. There exists a number κ_m such that

$$\Omega_m = -\kappa_m I$$

(Corollary 6.7.2).

Proposition 8.3.2

$$\kappa_m = m(m + 2).$$

Proof. For every $f \in \mathcal{P}_m$, $\Omega_m f = -\kappa_m f$. Consider the monomial f_m:

$$f_m(u, v) = u^m.$$

We saw that

$$\rho_m(H) f_m = m f_m,$$
$$\rho_m(E) f_m = 0.$$

Therefore, by Proposition 8.3.1,

$$-\Omega_m f_m = (m^2 + 2m) f_m. \qquad \square$$

A function f in $\mathcal{M}_m$, the subspace of $\mathcal{C}\big(SU(2)\big)$ generated by the coefficients of the representation π_m, is an eigenfunction of the Laplace operator,

$$\Delta f = -m(m + 2) f$$

(Proposition 8.2.1). Hence the decomposition of $L^2(G)$,

$$L^2(G) = \widehat{\bigoplus_{m \in \mathbb{N}}} \mathcal{M}_m,$$

can be seen as the decomposition of $L^2(G)$ as orthogonal sum of the eigenspaces of the Laplace operator.

If $f \in C^2(G)$ is a central function then the function Δf is central as well. A central function is determined by its restriction to the diagonal matrices

$$a(\theta) = \begin{pmatrix} e^{i\theta} & 0 \\ 0 & e^{-i\theta} \end{pmatrix}.$$

Put

$$f_0(\theta) = f\big(a(\theta)\big).$$

Proposition 8.3.3 *If f is a central function*

$$(\Delta f)_0 = Lf_0,$$

with

$$
\begin{aligned}
Lf_0 &= \frac{d^2 f_0}{d\theta^2} + 2\cot\theta \frac{df_0}{d\theta} \\
&= \frac{1}{\sin^2\theta} \frac{d}{d\theta}\left(\sin^2\theta f_0\right) \\
&= \frac{1}{\sin\theta}\left(\frac{d^2}{d\theta^2} + 1\right)\sin\theta f_0.
\end{aligned}
$$

Observe that, if $f = \chi_m$ is the character of the representation π_m, then, by Proposition 7.5.4,

$$f_0(\theta) = \frac{\sin(m+1)\theta}{\sin\theta},$$

and

$$Lf_0 = -m(m+2)f_0,$$

and this agrees with Proposition 8.3.2.

We saw that the Laplace operator Δ is symmetric: for $f, \varphi \in C^2(G)$,

$$\int_G \Delta f(x)\varphi(x)\mu(dx) = \int_G f(x)\Delta\varphi(x)\mu(dx),$$

and one can check that, if f and φ are central,

$$\int_0^\pi Lf_0(\theta)\varphi_0(\theta)\sin^2\theta \, d\theta = \int_0^\pi f_0(\theta)L\varphi_0(\theta)\sin^2\theta \, d\theta.$$

We will give two proofs of this proposition. The first is simpler, but we will see in Chapter 10 how the second proof can be generalised.

First proof. The function f, being radial, only depends on the trace:

$$f(x) = F\left(\tfrac{1}{2}\operatorname{tr} x\right),$$

where the function F is defined on $[-1, 1]$. We get

$$\left(\rho(X_1)^2 f\right)\left(a(\theta)\right) = \frac{d^2}{d\theta^2} F(\cos\theta)$$
$$= F''(\cos\theta)\sin^2\theta - F'(\cos\theta)\cos\theta,$$

$$\left(\rho(X_2)^2 f\right)\left(a(\theta)\right) = \frac{d^2}{dt^2}\bigg|_{t=0} F(\cos\theta\cos t)$$
$$= -F'(\cos\theta)\cos\theta,$$

$$\left(\rho(X_3)^2 f\right)\left(a(\theta)\right) = \frac{d^2}{dt^2}\bigg|_{t=0} F(\cos\theta\cos t)$$
$$= -F'(\cos\theta)\cos\theta.$$

Furthermore

$$f_0'(\theta) = -F'(\cos\theta)\sin\theta,$$
$$f_0''(\theta) = F''(\cos\theta)\sin^2\theta - F'(\cos\theta)\cos\theta.$$

The stated formulae follow easily.   □

Second proof. If f is central, then

$$f(\exp sT\, g \exp -sT) = f(g).$$

By taking the second derivative at $s = 0$ we obtain relations leading to the computation of the radial part L. This equation can also be written

$$f\left(g\exp\left(s\,\mathrm{Ad}(g^{-1})T\right)\exp(-sT)\right) = f(g).$$

We will take

$$g = a(\theta) = \begin{pmatrix} e^{i\theta} & 0 \\ 0 & e^{-i\theta} \end{pmatrix},$$

$$T = T(z) = \begin{pmatrix} 0 & z \\ -\bar{z} & 0 \end{pmatrix} \quad (z \in \mathbb{C}).$$

Observe that $T(1) = X_2$, $T(i) = X_3$. We will use the relations

$$\mathrm{Ad}\left(a(\theta)\right)T(z) = T(e^{2i\theta}z),$$
$$[T(z), T(w)] = -2\operatorname{Im}(z\bar{w})X_1.$$

Let us apply (d) in Section 8.2 with

$$X = \mathrm{Ad}\left(a(-\theta)\right)T(z) = T(e^{-2i\theta}z), \quad Y = -T(z).$$

Then

$$X + Y = T((e^{-2i\theta} - 1)z),$$
$$[X, Y] = -2 \sin 2\theta |z|^2 X_1.$$

If $z = i e^{i\theta}$, then

$$X + Y = 2 \sin \theta X_2,$$
$$[X, Y] = -2 \sin 2\theta X_1,$$

and, for a radial function f,

$$4 \sin^2 \theta \big(\rho(X_2)^2 f\big)\big(a(\theta)\big) - 2 \sin 2\theta \big(\rho(X_1)f\big)\big(a(\theta)\big) = 0,$$

or

$$\big(\rho(X_2)^2 f\big)\big(a(\theta)\big) = \cot \theta \big(\rho(X_1)f\big)\big(a(\theta)\big).$$

Similarly, if $z = e^{i\theta}$, we get

$$\big(\rho(X_3)^2 f\big)\big(a(\theta)\big) = \cot \theta \big(\rho(X_1)f\big)\big(a(\theta)\big).$$

Finally,

$$\begin{aligned} \Delta f\big(a(\theta)\big) &= \rho(X_1)^2 f\big(a(\theta)\big) + \rho(X_2)^2 f\big(a(\theta)\big) + \rho(X_3)^2 f\big(a(\theta)\big) \\ &= \frac{d^2 f_0}{d\theta^2} + 2 \cot \theta \frac{d f_0}{d\theta}. \end{aligned}$$

$\square$

8.4 Uniform convergence of Fourier series on the group $SU(2)$

In Section 7.5 we considered the representation $(\pi_m, \mathcal{P}_m)$ of the group $G = SU(2)$ on the space $\mathcal{P}_m$ of polynomials in two variables with complex coefficients which are homogeneous of degree m. The space $\mathcal{P}_m$ can be equipped with a Euclidean inner product for which the representation π_m is unitary. This representation is irreducible, and every irreducible representation of G is equivalent to one of the representations π_m. Hence $\hat{G} \simeq \mathbb{N}$. The *Fourier coefficient* $\hat{f}(m)$ ($m \in \mathbb{N}$) of an integrable function f on G is the endomorphism $\hat{f}(m)$ of $\mathcal{P}_m$ defined by

$$\hat{f}(m) = \int_G f(x) \pi_m(x^{-1}) \mu(dx).$$

The *Fourier series* of f can be written as

$$\sum_{m=0}^{\infty} (m+1) \operatorname{tr}\big(\hat{f}(m)\pi_m(x)\big).$$

If $f \in L^2(G)$

$$\sum_{m=0}^{\infty} (m+1) \operatorname{tr}\big(\hat{f}(m)\pi_m(x)\big) = f(x),$$

the Fourier series converges in $L^2(G)$, and the Plancherel formula can be written as

$$\int_G |f(x)|^2 \mu(dx) = \sum_{m=0}^{\infty} (m+1) \||\hat{f}(m)\||^2$$

(Theorem 6.4.2).

We will see, using the Laplace operator Δ, that the Fourier series of a $\mathcal{C}^2$ function on the group $SU(2)$ converges uniformly.

Proposition 8.4.1 *If $f \in \mathcal{C}^2(G)$, then*

$$\widehat{\Delta f}(m) = -m(m+2)\hat{f}(m).$$

Proof. Let $\pi = \pi_m$, $u, v \in \mathcal{H}_\pi$, then

$$(\hat{f}(m)u|v)_{\mathcal{H}_\pi} = (f|\varphi)_{L^2(G)}$$

with $\varphi(x) = (\pi(x)v|u)$, and

$$\begin{aligned}
(\widehat{\Delta f}(m)u|v)_{\mathcal{H}_\pi} &= (\Delta f|\varphi)_{L^2(G)} \\
&= (f|\Delta\varphi)_{L^2(G)} \\
&= -m(m+2)(f|\varphi)_{L^2(G)} \\
&= -m(m+2)\big(\hat{f}(m)u|v\big)_{\mathcal{H}_\pi}. \qquad \square
\end{aligned}$$

Theorem 8.4.2 *If $f \in \mathcal{C}^2(G)$, then*

$$\sum_{m=0}^{\infty} (m+1)^{3/2} \||\hat{f}(m)\|| < \infty,$$

and

$$f(g) = \sum_{m=0}^{\infty} (m+1) \operatorname{tr}\big(\hat{f}(m)\pi_m(g)\big),$$

the series converging uniformly and absolutely.

Proof. By Proposition 8.4.1,

$$\hat{f}(m) = -\frac{1}{m(m+2)}\widehat{\Delta f}(m) \qquad (m > 0),$$

and using the Schwarz inequality we get

$$\sum_{m=1}^{\infty}(m+1)^{3/2}|||\hat{f}(m)||| = \sum_{m=1}^{\infty}\frac{(m+1)^{3/2}}{m(m+2)}|||\widehat{\Delta f}(m)|||$$

$$\leq \left(\sum_{m=1}^{\infty}\frac{(m+1)^2}{m^2(m+2)^2}\right)^{1/2}\left(\sum_{m=1}^{\infty}(m+1)|||\widehat{\Delta f}(m)|||^2\right)^{1/2} < \infty$$

because, by Theorem 6.4.2,

$$\sum_{m=0}^{\infty}(m+1)|||\widehat{\Delta f}(m)|||^2 = \int_G |\Delta f(x)|^2 \mu(dx).$$

The statement of the theorem follows (Proposition 6.6.1). $\square$

Theorem 8.4.3 *Let f be a continuous function on G. The function f belongs to $\mathcal{C}^{\infty}(G)$ if and only if*

$(*)$ $\forall k > 0, \; \sup_{m \in \mathbb{N}} m^k |||\hat{f}(m)||| < \infty.$

Proof. (a) Assume that $f \in \mathcal{C}^{\infty}(G)$. Then

$$\widehat{\Delta^k f}(m) = \left(-m(m+2)\right)^k \hat{f}(m),$$

and

$$m^k(m+2)^k|||\hat{f}(m)||| \leq \sqrt{m+1}\int_G |\Delta^k f(x)|\mu(dx).$$

This shows that condition $(*)$ holds for the sequence $\left(\hat{f}(m)\right)$.

(b) Conversely, assume that condition $(*)$ holds for $\left(\hat{f}(m)\right)$. Then

$$\sum_{m=0}^{\infty}(m+1)^{3/2}|||\hat{f}(m)||| < \infty,$$

and

$$f(g) = \sum_{m=0}^{\infty}(m+1)\,\mathrm{tr}\left(\hat{f}(m)\pi_m(g)\right),$$

the convergence being uniform. For $X \in \mathfrak{g}$ let us consider the series

$$f(g\exp tX) = \sum_{m=0}^{\infty}(m+1)\,\mathrm{tr}\left(\hat{f}(m)\pi_m(g\exp tX)\right).$$

We will show that it can be diferentiated termwise. Put

$$\varphi_m(t) = (m+1)\,\mathrm{tr}\big(\hat{f}(m)\pi_m(g\exp tX)\big)$$
$$= (m+1)\,\mathrm{tr}\big(\pi_m(\exp tX)\hat{f}(m)\pi_m(g)\big).$$

The derivative of the function φ_m is given by

$$\varphi'_m(t) = (m+1)\,\mathrm{tr}\big(d\pi_m(X)\hat{f}(m)\pi_m(g\exp tX)\big). \qquad \square$$

Lemma 8.4.4 *The derived representation $d\pi_m$ satisfies the following estimate:*

$$\|\!|d\pi_m(X)|\!\| \le m\sqrt{m+1}\,\|X\|,$$

where

$$\|X\|^2 = \tfrac{1}{2}\,\mathrm{tr}(XX^*).$$

Proof. The eigenvalues of the operator $d\pi(X_1)$, where

$$X_1 = \begin{pmatrix} i & 0 \\ 0 & -i \end{pmatrix},$$

are the numbers $i(m-2j)\ (0 \le j \le m)$ and $|i(m-2j)| \le m$. Therefore,

$$\|\!|d\pi_m(X_1)|\!\| \le m\sqrt{m+1}.$$

A matrix $X \in \mathfrak{g} = \mathfrak{su}(2)$ can be diagonalised in an orthogonal basis, and its eigenvalues are pure imaginary and opposite. It follows that there exists $g \in G = SU(2)$ and $\lambda \ge 0$ such that $X = \mathrm{Ad}(g)\lambda X_1$, and $\|X\| = \lambda\|X_1\| = \lambda$. Since

$$d\pi_m(X) = \lambda\pi_m(g)d\pi_m(X_1)\pi_m(g^{-1}),$$

it follows that

$$\|\!|d\pi_m(X)|\!\| = \lambda\|\!|d\pi_m(X_1)|\!\| \le m\sqrt{m+1}\,\|X\|. \qquad \square$$

From this lemma one gets the following inequalities

$$|\varphi'_m(t)| \le (m+1)\|\!|d\pi_m(X)|\!\|\,\|\!|\hat{f}(m)|\!\|$$
$$\le m(m+1)^{3/2}\|X\|\,\|\!|\hat{f}(m)|\!\|.$$

Furthermore, by assumption,

$$\sum_{m=0}^{\infty} m(m+1)^{3/2}\|\!|\hat{f}(m)|\!\| < \infty.$$

Hence it is posssible to differentiate termwise:

$$\rho(X) = \sum_{m=0}^{\infty}(m+1)\,\mathrm{tr}\big(d\pi_m(X)\hat{f}(m)\pi_m(g)\big).$$

Therefore one can show recursively with respect to k that $f \in C^k$ for every k.   □

The Fourier series of a central function f can be written

$$f(g) = \sum_{m=0}^{\infty}(m + 1)a(m)\chi_m(g),$$

with

$$(m + 1)a(m) = \int_{SU(2)} f(x)\chi_m(x)\mu(dx),$$

and, also, by using the integration formula in Corollary 7.2.2,

$$(m + 1)a(m) = \frac{2}{\pi}\int_0^\pi f\big(a(\theta)\big)\frac{\sin(m + 1)\theta}{\sin \theta}\sin^2 \theta \, d\theta$$
$$= \frac{2}{\pi}\int_0^\pi f\big(a(\theta)\big)\sin(m + 1)\theta \sin \theta \, d\theta.$$

Let us consider the Fourier expansion on the group $SO(2) \simeq \mathbb{R}/2\pi\mathbb{Z}$ of a C^2 even function f:

$$f(\theta) = \sum_{m=0}^{\infty} a_m \cos m\theta.$$

Differentiating termwise we get

$$-\frac{1}{\sin \theta} f'(\theta) = \sum_{m=0}^{\infty} a_{m+1} \frac{\sin(m + 1)\theta}{\sin \theta},$$

and we can see this series as the Fourier expansion of a central function on the group $SU(2)$.

For instance, from the classical expansion

$$\frac{1 - r^2}{1 - 2r\cos\theta + r^1} = 1 + 2\sum_{m=1}^{\infty} r^m \cos m\theta \quad (|r| < 1),$$

we get

$$\frac{1 - r^2}{(1 - 2r\cos\theta + r^2)^2} = \sum_{m=0}^{\infty}(m + 1)r^m \frac{\sin(m + 1)\theta}{\sin \theta},$$

and it can be seen that

$$\det(I - rg)^{-2} = \frac{1}{1 - r^2}\sum_{m=0}^{\infty}(m + 1)r^m \chi_m(g) \quad \big(g \in SU(2)\big).$$

8.5 Heat equation on $SO(2)$

Let us first recall some classical results for the *heat equation* on $G = SO(2) \simeq \mathbb{R}/2\pi\mathbb{Z}$. This equation can be written

$$\frac{\partial u}{\partial t} = \frac{\partial^2 u}{\partial x^2},$$

where f is a C^2 function on $]\alpha, \beta[\times G$. The Cauchy problem can be stated as follows. Given a continuous function f on G, determine a continuous function u on $[0, \infty[\times G$, which is C^2 on $]0, \infty[\times G$, such that

$$\frac{\partial u}{\partial t} = \frac{\partial^2 u}{\partial x^2},$$

$$u(0, x) = f(x).$$

In order to show that the solution, if it exists, is unique, one can observe that the energy

$$E(t) = \int_0^{2\pi} u(x, t)^2 dx$$

is decreasing. In fact, for $t > 0$,

$$E'(t) = 2 \int_0^{2\pi} u(t, x) \frac{\partial u}{\partial t}(t, x) \, dx$$

$$= 2 \int_0^{2\pi} u(t, x) \frac{\partial^2 u}{\partial x^2}(t, x) \, dx$$

$$= -2 \int_0^{2\pi} \left(\frac{\partial u}{\partial x}(t, x) \right)^2 dx \leq 0.$$

One can show that the solution exists by the Fourier method. Observe that the functions

$$e^{-m^2 t} e^{imx} \quad (m \in \mathbb{Z})$$

are solutions of the heat equation. Assume first that the function f is C^1. Then

$$\sum_{m \in \mathbb{Z}} |\hat{f}(m)| < \infty,$$

and the Fourier series of f converges uniformly to f:

$$f(x) = \sum_{m \in \mathbb{Z}} \hat{f}(m) e^{imx}.$$

Put

$$u(t, x) = \sum_{m \in \mathbb{Z}} \hat{f}(m) e^{-m^2 t} e^{imx}.$$

For $t > 0$ one can differentiate termwise and check that u is solution of the heat equation. Therefore u is solution of the Cauchy problem.

By writing

$$\hat{f}(m) = \frac{1}{2\pi} \int_0^{2\pi} e^{-imy} f(y) dy,$$

and permuting integral and sum, one gets

$$u(t, x) = \frac{1}{2\pi} \int_0^{2\pi} h(t, x - y) f(y) dy,$$

where h is the *heat kernel* defined on $]0, \infty[\times G$ by

$$h(t, x) = \sum_{m \in \mathbb{Z}} e^{-m^2 t} e^{imx}.$$

The heat kernel h has the following properties

(1) $$h(t, x) \geq 0,$$

(2) $$\tfrac{1}{2\pi} \int_0^{2\pi} h(t, x) dx = 1,$$

(3) $$\forall \eta, \ 0 < \eta < \pi, \ \lim_{t \to 0} \tfrac{1}{2\pi} \int_{-\eta}^{\eta} h(t, x) dx = 1.$$

Now let f be a continuous function on G, and let u be the function defined on $]0, \infty[\times G$ by

$$u(t, x) = \frac{1}{2\pi} \int_0^{2\pi} h(t, x - y) f(y) dy.$$

By integrating termwise one obtains

$$u(t, x) = \sum_{m \in \mathbb{Z}} \hat{f}(m) e^{-m^2 t} e^{imx}.$$

This termwise integration is justified since

$$|\hat{f}(m)| \leq \frac{1}{2\pi} \int_0^{2\pi} |f(x)| dx.$$

For $t > 0$ this series can be differentiated termwise and one can check here also that u is solution of the heat equation.

The function u can be written

$$u(t, x) = \frac{1}{2\pi} \int_{-\pi}^{\pi} h(t, y) f(x - y) dy.$$

Because it is continuous, the function f is uniformly continuous. Hence, for every $\varepsilon > 0$, there exists $\eta > 0$ such that, if $|y| \leq \eta$, then $|f(x - y) - f(x)| \leq \varepsilon$.

Using properties (1), (2) and (3) one gets

$$|u(t, x) - f(x)| = \frac{1}{2\pi} \left| \int_{-\pi}^{\pi} h(t, y)\big(f(x - y) - f(x)\big)dy \right|$$

$$\leq \frac{\varepsilon}{2\pi} \int_{-\eta}^{\eta} h(t, y)dy + 2 \sup |f| \frac{1}{2\pi} \int_{|x| \geq \eta} h(t, y)dy$$

$$\leq \varepsilon + 2 \sup |f| \left(1 - \frac{1}{2\pi} \int_{-\eta}^{\eta} h(t, y)dy \right).$$

It follows that

$$\lim_{t \to 0} u(t, x) = f(x),$$

and the limit is uniform in x.

The heat kernel h can also be written

$$h(t, x) = \sqrt{\frac{\pi}{t}} \sum_{k=-\infty}^{\infty} e^{-(x - 2k\pi)^2/4t}.$$

This can be established using the Poisson summation formula.

Poisson summation formula *Let the function f belong to the Schwartz space $\mathcal{S}(\mathbb{R})$. Put*

$$\check{f}(x) = \int_{-\infty}^{\infty} e^{ix\xi} f(\xi)d\xi.$$

Then

$$\sum_{m=-\infty}^{\infty} f(m)e^{imx} = \sum_{k=-\infty}^{\infty} \check{f}(x - 2k\pi).$$

Proof. The function φ, which is defined on $\mathbb{R}$ by

$$\varphi(x) = \sum_{k=-\infty}^{\infty} \check{f}(x - 2k\pi),$$

is continuous and 2π-periodic. One obtains its Fourier coefficients

$$\hat{\varphi}(m) = \frac{1}{2\pi} \int_{0}^{2\pi} \varphi(x)e^{-mx}dx$$

by integrating termwise the series:

$$\hat{\varphi}(m) = \frac{1}{2\pi} \sum_{k=-\infty}^{\infty} \check{f}(x - 2k\pi) e^{-imx} dx$$

$$= \frac{1}{2\pi} \int_{-\infty}^{\infty} \check{f}(x) e^{-imx} dx$$

$$= f(m).$$

The function φ is $\mathcal{C}^1$ (and even $\mathcal{C}^\infty$), hence its Fourier series converges to φ:

$$\varphi(x) = \sum_{m=-\infty}^{\infty} f(m) e^{imx}. \qquad \square$$

Take

$$f(\xi) = e^{-t\xi^2} \quad (t > 0),$$

then

$$\check{f}(x) := \int_{-\infty}^{\infty} e^{-t\xi^2} e^{ix\xi} d\xi = \sqrt{\frac{\pi}{t}} e^{-x^2/4t},$$

and we get

$$h(t, x) = \sum_{m=-\infty}^{\infty} e^{-tm^2} e^{imx}$$

$$= \sqrt{\frac{\pi}{t}} \sum_{k=-\infty}^{\infty} e^{-(x-2k\pi)^2/4t}.$$

It follows that, for every $c, 0 < c < \pi^2/4$,

$$h(t, x) = \sqrt{\frac{\pi}{t}} e^{-x^2/4t} + \mathcal{O}\big(e^{-c/t}\big),$$

if $|x| \le \pi, 0 < t \le 1$. One obtains this estimate from

$$\sum_{k=1}^{\infty} e^{-\frac{(x-2k\pi)^2}{4t}} \le \sum_{k=1}^{\infty} e^{-(2k-1)^2\pi^2/4t}$$

$$\le e^{-\frac{\pi^2}{4t}} \sum_{k=1}^{\infty} e^{-k(k-1)\pi^2}.$$

8.6 Heat equation on $SU(2)$

The *heat equation* for the group $G = SU(2)$ can be written

$$\frac{\partial u}{\partial t} = \Delta u,$$

where u is a C^2 function on $]\alpha, \beta[\times G$. We will study the following Cauchy problem. Given a continuous function f on G, determine a continuous function u on $[0, \infty[\times G$, which is C^2 on $]0, \infty[\times G$ such that

$$\frac{\partial u}{\partial t} = \Delta u \text{ for } t > 0,$$

$$u(0, x) = f(x).$$

We will first show that the solution, if it exists, is unique. For that we will present two methods. The first uses the maximum principle for the heat equation.

Proposition 8.6.1 *Let u be a continuous function on $[0, T] \times G$, which is C^2 on $]0, T[\times G$ such that*

$$\frac{\partial u}{\partial t} = \Delta u \quad (0 < t < T).$$

Then, for $(t, x) \in [0, T] \times G$,

$$\min_{x \in G} u(0, x) \le u(t, x) \le \max_{x \in G} u(0, x).$$

Proof. Fix $0 < T_0 < T$ and $\varepsilon > 0$, and put

$$u_\varepsilon(t, x) = u(t, x) + \varepsilon t.$$

Let $(t_0, x_0) \in [0, T_0] \times G$ be such that

$$u_\varepsilon(t_0, x_0) = \min\{u_\varepsilon(t, x) \mid (t, x) \in [0, T_0] \times G\}.$$

We will show that $t_0 = 0$. For that assume the converse, that $t_0 > 0$. At (t_0, x_0)

$$\frac{\partial u_\varepsilon}{\partial t}(t_0, x_0) \le 0 \quad (= 0 \text{ if } t_0 < T_0),$$

$$\Delta u_\varepsilon(t_0, x_0) \ge 0,$$

and this is impossible since

$$\frac{\partial u_\varepsilon}{\partial t} - \Delta u_\varepsilon = \varepsilon > 0.$$

Therefore, since $u_\varepsilon(0, x) = u(0, x)$,

$$u_\varepsilon(t, x) \ge \min_{x \in G} u(0, x),$$

or

$$u(t, x) \geq \min_{x \in G} u(0, x) - \varepsilon t.$$

This inequality holds for all $\varepsilon > 0$, hence

$$u(t, x) \geq \min_{x \in G} u(0, x).$$

One obtains the inequality

$$u(t, x) \leq \max_{x \in G} u(0, x)$$

from the preceding inequality by replacing u by $-u$.   □

The second method uses the decrease in energy.

Proposition 8.6.2 *Let u be a solution of the heat equation on* $]0, T[\times G$. *The energy, which is defined by*

$$E(t) = \int_G u(t, x)^2 \mu(dx),$$

is decreasing.

Proof. In fact,

$$E'(t) = 2 \int_G \frac{\partial u}{\partial t}(t, x) u(tx,) \mu(dx)$$

$$= 2 \int_G \Delta u(t, x) u(t, x) \mu(dx) \leq 0,$$

since $-\Delta$ is a positive operator.   □

We will establish the existence of the solution of the Cauchy problem using the Fourier method. One observes that a function of the form

$$u(t, x) = e^{-m(m+2)t} v(x),$$

where $v \in \mathcal{M}_m$, is a solution of the heat equation since

$$\Delta v = -m(m + 2)v.$$

The Fourier method consists in seeking for a solution of the Cauchy problem as a sum:

$$u(t, x) = \sum_m e^{-m(m+2)t} v_m(x),$$

where $v_m \in \mathcal{M}_m$. The initial condition can be written

$$\sum_m v_m(x) = f(x).$$

Assume first that the function f is C^2. We know that its Fourier coefficients $\hat{f}(m)$ satisfy

$$\sum_{m=0}^{\infty} (m+1)^{3/2} \, \|\|\hat{f}(m)\|\| < \infty,$$

and the Fourier series of f

$$f(x) = \sum_{m=0}^{\infty} (m+1) \operatorname{tr}\big(\hat{f}(m)\pi_m(x)\big)$$

converges absolutely and uniformly. Put

$$u(t,x) = \sum_{m=0}^{\infty} (m+1)e^{-m(m+2)t} \operatorname{tr}\big(\hat{f}(m)\pi_m(x)\big).$$

This series converges uniformly and absolutely on $[0, \infty[\times G$. For $t > 0$ the function u is C^∞ and is a solution of the heat equation. It is the solution of the Cauchy problem.

Let us define the *heat kernel H* by

$$H(t,x) = \sum_{m=0}^{\infty} (m+1)e^{-m(m+2)t} \chi_m(x) \quad (t > 0, x \in G).$$

For $t \geq t_0 > 0$, this series converges uniformly and absolutely since

$$|\chi_m(x)| \leq m+1.$$

The solution $u(t,x)$ can be written, for $t > 0$,

$$u(t,x) = \int_G H(t, xy^{-1})f(y)\mu(dy).$$

In fact,

$$\operatorname{tr}\big(\pi_m(x)\hat{f}(m)\big) = \operatorname{tr}\left(\int_G \pi_m(x)\pi_m(y^{-1})f(y)\mu(dy)\right)$$

$$= \int_G \chi_m(xy^{-1})f(y)\mu(dy),$$

and, by the uniform convergence of the series

$$H(t, xy^{-1})f(y) = \sum_{m=0}^{\infty} (m+1)e^{-m(m+2)t} \chi_m(xy^{-1})f(y),$$

it is possible to integrate termwise.

Proposition 8.6.3 *The heat kernel H has the following properties:*

(i) $H(t, x) \geq 0$,

(ii) $\int_G H(t, x)\mu(dx) = 1$,

(iii) *for every neighbourhood V of e,*

$$\lim_{t \to 0} \int_V H(t, x)\mu(dx) = 1.$$

Proof. (i) Let the function $f \geq 0$ be C^2 on G. Then the function u defined for $t > 0$ by

$$u(t, x) = \int_G H(t, xy^{-1})f(y)\mu(dy),$$

is the solution of the Cauchy problem with the initial data f. By the maximum principle (Proposition 8.6.1), $u(t, x) \geq 0$. Hence, for every C^2 function $f \geq 0$,

$$\int_G H(t, y)f(y)\mu(dy) \geq 0.$$

It follows that $H(t, y) \geq 0$.

(ii) For $m \geq 1$

$$\int_G \chi_m(x)\mu(dx) = 0,$$

and $\chi_0(x) = 1$, hence

$$\int_G H(t, x)\mu(dx) = \int_G \chi_0(x)\mu(dx) = 1.$$

(iii) Let V be a neighbourhood of e. There exists a C^2 function f on G, such that

$$0 \leq f(x) \leq 1, \quad f(e) = 1, \quad f(x) = 0 \text{ on } V^c.$$

We know that

$$\lim_{t \to 0} \int_G H(t, x)f(x)\mu(dx) = f(e) = 1.$$

And,

$$\int_G H(t, x)f(x)\mu(dx) = \int_V H(t, x)f(x)\mu(dx) \leq \int_V H(t, x)\mu(dx) \leq 1.$$

Therefore

$$\lim_{t \to 0} \int_V H(t, x)\mu(dx) = 1. \qquad \square$$

Proposition 8.6.4 *Let f be a continuous function on G. The solution u of the Cauchy problem is given, for t > 0, by*

$$u(t, x) = \int_G H(t, xy^{-1}) f(y) \mu(dy).$$

This integral can also be written

$$u(t, x) = \int_G H(t, y) f(xy^{-1}) \mu(dy).$$

Proof. One checks first that, for $t > 0$, the function u as given above is a solution of the heat equation. We will show that

$$\lim_{t \to 0} u(t, x) = f(x)$$

uniformly on G. Since the function f is continuous and the group G is compact, f is uniformly continuous. Let $\varepsilon > 0$. There exists a neighbourhood V of e such that, if $y \in V$, then, for every $x \in G$,

$$|f(xy^{-1}) - f(x)| \le \varepsilon.$$

Hence

$$|u(t, x) - f(x)| = \left| \int_G H(t, y) \big(f(xy^{-1}) - f(x) \big) \mu(dx) \right|$$

$$\le \int_G H(t, y) |f(xy^{-1}) - f(x)| \mu(dx)$$

$$\le \varepsilon \int_V H(t, y) \mu(dy) + 2 \sup |f| \int_{V^c} H(t, y) \mu(dy)$$

$$\le \varepsilon + 2 \sup |f| \int_{V^c} H(t, y) \mu(dy),$$

and, by Proposition 8.6.3,

$$\lim_{t \to 0} \int_{V^c} H(t, y) \mu(dy) = 0.$$

The statement follows.   □

Recall that the heat kernel $h(t, \theta)$ of the group $SO(2) \simeq \mathbb{R}/2\pi\mathbb{Z}$ is given by

$$h(t, \theta) = 1 + 2 \sum_{m=1}^{\infty} e^{-m^2 t} \cos m\theta.$$

Let $H_0(t, \theta)$ denote the restriction of the heat kernel of the group $SU(2)$ to the subgroup of diagonal matrices:

$$H_0(t, \theta) = H\big(t, a(\theta)\big), \quad a(\theta) = \begin{pmatrix} e^{i\theta} & 0 \\ 0 & e^{-i\theta} \end{pmatrix}.$$

Proposition 8.6.5

$$H_0(t, \theta) = -\frac{e^t}{2\sin\theta} \frac{\partial}{\partial\theta} h(t, \theta).$$

Proof. In fact,

$$H_0(t, \theta) = \sum_{m=0}^{\infty} (m+1) e^{-m(m+2)t} \chi_m\big(a(\theta)\big)$$

$$= \sum_{m=0}^{\infty} (m+1) e^{-m(m+2)t} \frac{\sin(m+1)\theta}{\sin\theta},$$

and on the other hand

$$\frac{\partial}{\partial\theta} h(t, \theta) = -2 \sum_{m=1}^{\infty} m e^{-m^2 t} \sin m\theta$$

$$= -2 \sum_{m=0}^{\infty} (m+1) e^{-(m+1)^2 t} \sin(m+1)\theta. \qquad \square$$

We saw in the preceding section that

$$h(t, \theta) = \sqrt{\frac{\pi}{t}} \sum_{k=-\infty}^{\infty} e^{-(\theta - 2k\pi)^2/4t}.$$

From Proposition 8.6.5 one can deduce the following formula.

Proposition 8.6.6

$$H_0(t, \theta) = \frac{\sqrt{\pi}}{4} \frac{e^t}{t\sqrt{t}} \sum_{k=-\infty}^{\infty} \frac{\theta - 2k\pi}{\sin\theta} e^{-(\theta - 2k\pi)^2/4t}.$$

For θ close to 0 the dominant term of this series is the one which corresponds to $k = 0$:

$$\frac{\sqrt{\pi}}{4} \frac{e^t}{t\sqrt{t}} \frac{\theta}{\sin\theta} e^{-\theta^2/4t}.$$

The group $SU(2)$ can be identified with the unit sphere S^3 in $\mathbb{R}^4$. One gets the geodesic distance to the identity element e for the usual Riemannian metric of S^3 as follows: if $x = ga(\theta)g^{-1}$ with $|\theta| \le \pi$, then $r = d(e, x) = |\theta|$. The Riemannian measure m does not agree with the normalised Haar measure but is proportional to it:

$$m = 2\pi^2 \mu.$$

The factor $2\pi^2$ is the volume of $SU(2)$ for this Riemannian measure. By dividing by this factor one obtains the classical estimate in Riemannian geometry for the heat kernel:

$$\frac{1}{2\pi^2}H(t,x) \simeq \frac{1}{(2\sqrt{\pi t})^3}e^{-r^2/4t} \quad (t \to 0).$$

By Proposition 8.6.6 one has

$$H(t,x) = \frac{\sqrt{\pi}}{4}\frac{e^t}{t\sqrt{t}}\frac{r}{\sin r}e^{-r^2/4t} + R(t,x),$$

with a remainder term $R(t,x)$ satisfying

$$\int_G |R(t,x)|\mu(dx) = O\big(e^{-c/t}\big),$$

where $c > 0$ is a constant.

8.7 Exercises

1. Let τ be the representation of $\mathfrak{so}(3)$ on $C^\infty(\mathbb{R}^3)$ defined by

$$\big(\tau(Y)f\big)(x) = \frac{d}{dt}f(x\exp tY)\big|_{t=0}.$$

The Casimir operator Ω is defined by

$$\Omega = \tau(Y_1)^2 + \tau(Y_2)^2 + \tau(Y_3)^2$$

(using the notation of Section 7.6).
(a) Show that

$$\Omega = \|x\|^2\Delta - E(E+I),$$

where E is the Euler operator,

$$E = x_1\frac{\partial}{\partial x_1} + x_2\frac{\partial}{\partial x_2} + x_3\frac{\partial}{\partial x_3},$$

and Δ the Laplace operator,

$$\Delta = \frac{\partial^2}{\partial x_1^2} + \frac{\partial^2}{\partial x_2^2} + \frac{\partial^2}{\partial x_3^2}.$$

(b) Show that, if $f \in \mathcal{H}_\ell$, then

$$\Omega f = -\ell(\ell+1)f.$$

(c) Show that, in terms of spherical coordinates, the Casimir operator can be written

$$\Omega = \frac{1}{\sin\theta}\frac{\partial}{\partial\theta}\left(\sin\theta\frac{\partial}{\partial\theta}\right) + \frac{1}{\sin^2\theta}\frac{\partial^2}{\partial\varphi^2},$$

and the Laplace operator

$$\Delta = \frac{\partial^2}{\partial r^2} + \frac{2}{r}\frac{\partial}{\partial r} + \frac{1}{\sin\theta}\frac{\partial}{\partial\theta}\left(\sin\theta\frac{\partial}{\partial\theta}\right) + \frac{1}{\sin^2\theta}\frac{\partial^2}{\partial\varphi^2}.$$

2. *Abel convergence for Fourier series.* For $0 < r < 1$, let P_r be the function defined on $G = SU(2)$ by

$$P_r(x) = \sum_{m=0}^{\infty}(m+1)r^m\chi_m(x).$$

(a) Show that, if

$$x = a(\theta) = \begin{pmatrix} e^{i\theta} & 0 \\ 0 & e^{-i\theta} \end{pmatrix},$$

then

$$P_r(x) = \frac{1-r^2}{(1-2r\cos\theta+r^2)^2}.$$

Hint. Differentiate with respect to θ the relation

$$1 + 2\sum_{m=1}^{\infty}r^m\cos m\theta = \frac{1-r^2}{1-2r\cos\theta+r^2}.$$

(b) Establish the following properties:
   (i) $P_r(x) \geq 0$,
   (ii) $\int_G P_r(dx)\mu(dx) = 1$,
   (iii) for every neighbourhood V of e,

$$\lim_{r\to 1}\int_V P_r(x)\mu(dx) = 1.$$

(c) Let f be a continuous function on G. For $0 < r < 1$, put

$$f_r(x) = \sum_{m=0}^{\infty}(m+1)r^m\operatorname{tr}\left(\hat{f}(m)\pi_m(x)\right).$$

Show that the convergence of this series is uniform on G and that

$$f_r(x) = \int_G P_r(xy^{-1})f(y)\mu(dy).$$

(d) Show that

$$\lim_{r\to 1}f_r(x) = f(x),$$

uniformly on G.

3. *Elementary solution of* $-\Delta + \lambda$.

(a) For $\alpha \in \mathbb{C}$, let q_α be the central function defined on $SU(2) \setminus \{I\}$ by, for $0 < \theta \leq \pi$,

$$q_\alpha(a(\theta)) = \frac{\sin \alpha(\pi - \theta)}{\sin \theta},$$

where

$$a(\theta) = \begin{pmatrix} e^{i\theta} & 0 \\ 0 & e^{-i\theta} \end{pmatrix}.$$

Show that q_α is integrable. Define

$$c(m) = \frac{1}{m+1} \int_{SU(2)} q_\alpha(x) \chi_m(x) \mu(dx),$$

where μ is the normalised Haar measure on $SU(2)$. Show that

$$c(m) = \frac{2}{\pi} \frac{\sin \alpha \pi}{(m+1)^2 - \alpha^2}.$$

(b) Let λ be a complex number, and f a continuous function on $SU(2)$. We propose to solve the equation

$$-\Delta u + \lambda u = f,$$

where u is a $\mathcal{C}^2$ function on $SU(2)$. Show that, if u is a solution, then

$$\bigl(m(m+2) + \lambda\bigr)\hat{u}(m) = \hat{f}(m).$$

Deduce that, if $\lambda \neq -m(m+2)$, for every $m \in \mathbb{N}$ then, if there is one solution, it is unique and is given by

$$u(x) = \sum_{m=0}^{\infty} (m+1) \frac{1}{m(m+2) + \lambda} \operatorname{tr}\bigl(\hat{f}(m)\pi_m(x)\bigr).$$

To prove that the series converges uniformly, show, using the Schwarz inequality, that

$$\sum_{m=0}^{\infty} \frac{(m+1)^{3/2}}{|m(m+2) + \lambda|} \||\hat{f}(m)\|| \leq C \left(\sum_{m=0}^{\infty} (m+1) \||\hat{f}(m)\||^2 \right)^{1/2},$$

where C is a constant which depends only on λ.

(c) Show that

$$u(x) = \frac{\pi}{2} \frac{1}{\sin \alpha \pi} \int_{SU(2)} q_\alpha(xy^{-1}) f(y) \mu(dy),$$

where α is a complex number such that $\lambda = 1 - \alpha^2$.

Hint. Show that both sides of the equation have the same Fourier coefficients.

4. *Elementary solution of* $-\Delta$.
   (a) Let q be the central function defined on $SU(2) \setminus \{I\}$ by, for $0 < \theta \leq \pi$,

   $$q(a(\theta)) = \frac{(\pi - \theta)\cos\theta}{2\sin\theta},$$

   where

   $$a(\theta) = \begin{pmatrix} e^{i\theta} & 0 \\ 0 & e^{-i\theta} \end{pmatrix}.$$

   Show that the function q is integrable. Define

   $$c(m) = \frac{1}{m+1} \int_{SU(2)} q(x)\chi_m(x)\mu(dx),$$

   where μ is the normalised Haar measure of $SU(2)$. Show that

   $$c(m) = \begin{cases} \dfrac{1}{m(m+2)} & \text{if } m \geq 1, \\ \dfrac{1}{4} & \text{if } m = 0. \end{cases}$$

   (b) Let f be a continuous function on $SU(2)$. Consider the equation

   $$-\Delta u = f,$$

   where u is a C^2 function on $SU(2)$. Show that, if u is a solution, then

   $$m(m+2)\hat{u}(m) = \hat{f}(m),$$

   and that, if there is a solution, then necessarily

   $$\int_{SU(2)} f(x)\mu(dx) = 0.$$

   (c) Assume that the equation admits a solution. Show that every solution can be written:

   $$u(x) = C + \sum_{m=1}^{\infty} (m+1)\frac{1}{m(m+2)} \operatorname{tr}\big(\hat{f}(m)\pi_m(x)\big).$$

   (d) Show that every solution can also be written:

   $$u(x) = C + \int_{SU(2)} q(xy^{-1})f(y)\mu(dy).$$

9

Analysis on the sphere and the Euclidean space

The special orthogonal group $SO(n)$ acts transitively on the unit sphere $S = S^{n-1}$ in $\mathbb{R}^n$ and then on the space $\mathcal{C}(S)$ of continuous functions on S, and also on the space $L^2(S)$ of square integrable functions with respect to the uniform measure. For $n \geq 3$, the space $\mathcal{Y}_\ell$ of spherical harmonics of degree ℓ is an invariant irreducible subspace. The properties of the representations of a compact group we studied in Chapter 7 lead to remarkable applications for analysis on the sphere $S = S^{n-1}$. The Laplace operator Δ_S of the sphere commutes with the action of $SO(n)$ and plays a role of primary importance in this analysis.

9.1 Integration formulae

As in Section 7.2, we consider the differential form ω of degree $n-1$ on $\mathbb{R}^n$ defined by

$$\omega = \sum_{i=1}^{n} (-1)^{i-1} x_i dx_1 \wedge \cdots \wedge \widehat{dx_i} \wedge \cdots \wedge dx_n.$$

Its restriction to the unit sphere $S = S^{n-1}$ in $\mathbb{R}^n$,

$$S = \left\{ x \in \mathbb{R}^n \mid x_1^2 + \cdots + x_n^2 = 1 \right\},$$

defines a measure on S which is invariant under $G = SO(n)$. Let σ denote the corresponding normalised measure:

$$\int_S f(x)\sigma(dx) = \frac{1}{\varpi_n} \int_S f\omega,$$

186

with

$$\varpi_n = \int_S \omega.$$

We will see below that

$$\varpi_n = 2\frac{\pi^{n/2}}{\Gamma\left(\frac{n}{2}\right)}.$$

Recall that, for $n = 4$, the sphere S^3 can be identified with the group $SU(2)$; the measure σ is then the normalised Haar measure of $SU(2)$, and the action of $SO(4)$ on S^3 is nothing but the action of the group $SU(2) \times SU(2)$ on $SU(2)$ given by

$$x \mapsto g_1 x g_2^{-1} \quad \left(g_1, g_2 \in SU(2)\right).$$

In fact we saw that $SO(4) \simeq SU(2) \times SU(2)/\{\pm I\}$ (Proposition 7.1.2).

We will first establish the integration formula corresponding to the polar decomposition. The map

$$\varphi :]0, \infty[\times S \to \mathbb{R}^n \setminus \{0\}, \quad (r, u) \mapsto ru,$$

is a diffeomorphism. Let λ denote the Lebesgue measure on $\mathbb{R}^n$, normalised in such a way that the unit hypercube built on the vectors of the canonical basis has measure one.

Proposition 9.1.1 *Let f be an integrable function on $\mathbb{R}^n$:*

$$\int_{\mathbb{R}^n} f(x)\lambda(dx) = \varpi_n \int_0^\infty \left(\int_S f(ru)\sigma(du)\right) r^{n-1} dr.$$

In particular, if f is radial, $f(x) = F(\|x\|)$, where F is a function defined on $]0, \infty[$, that is if f is $O(n)$-invariant,

$$\int_{\mathbb{R}^n} f(x)\lambda(dx) = \varpi_n \int_0^\infty F(r) r^{n-1} dr.$$

The constant ϖ_n can be evaluated by applying the above formula to the function

$$f(x) = e^{-\|x\|^2}.$$

In fact

$$\int_{\mathbb{R}^n} f(x)\lambda(dx) = \prod_{i=1}^n \int_{\mathbb{R}} e^{-x_i^2} dx_i = (\sqrt{\pi})^n,$$

$$\int_0^\infty e^{-r^2} r^{n-1} dr = \frac{1}{2}\Gamma\left(\frac{n}{2}\right),$$

hence

$$\varpi_n = 2\frac{\pi^{n/2}}{\Gamma\left(\frac{n}{2}\right)}.$$

In particular $\varpi_2 = 2\pi$, $\varpi_3 = 4\pi$, $\varpi_4 = 2\pi^2$.

Proof. Let α be the following differential form on $\mathbb{R}^n$, to which the Lebesgure measure λ is associated:

$$\alpha = dx_1 \wedge \cdots \wedge dx_n.$$

For n vectors $X_1, \ldots, X_n$ in $\mathbb{R}^n$,

$$\alpha(X_1, \ldots, X_n) = \det(X_1, \ldots, X_n),$$

and

$$\omega_x(X_1, \ldots, X_{n-1}) = \alpha(x, X_1, \ldots, X_{n-1}).$$

We will show that

$$\varphi^*\alpha = r^{n-1}dr \otimes \omega$$

(here ω is seen as a differential form on S). Let $X \in \mathbb{R}$, and Y a tangent vector to the sphere S at u, which is orthogonal to u:

$$(D\varphi)_{(r,u)}(X, Y) = Xu + rY.$$

Hence, if $Y_1, \ldots, Y_{n-1}$ are $n - 1$ tangent vectors to S at u,

$$
\begin{aligned}
(\varphi^*\alpha)_{(r,u)}&(X, Y_1, \ldots, Y_{n-1}) \\
&= \alpha\big((D\varphi)_{(r,u)}X, (D\varphi)_{(r,u)}Y_1, \ldots, (D\varphi)_{(r,u)}Y_{n-1}\big) \\
&= \alpha(Xu, rY_1, , \ldots, rY_{n-1}) \\
&= Xr^{n-1}\alpha(u, Y_1, \ldots, Y_{n-1}) \\
&= Xr^{n-1}\omega_u(Y_1, \ldots, Y_{n-1}).
\end{aligned}
$$

Therefore

$$\varphi^*\alpha = r^{n-1}dr \otimes \omega.$$

The integration formula follows. $\qquad\square$

We will also need the integration formula which gives the integral of a zonal function. We will say that a function, which is defined on the sphere S, is *zonal* if it is constant on every 'parallel' $x_n = c$. This is a function which is invariant under the isotropy subgroup $K \equiv SO(n - 1)$ of the 'north pole' e_n. Such a function f can be written

$$f(x) = F(x_n),$$

where F is a function defined on $[-1, 1]$.

Let $S_0 = S^{n-2}$ denote the unit sphere in $\mathbb{R}^{n-1}$, identified with the hyperplane with equation $x_n = 0$, and σ_0 the normalised uniform measure on S_0. (S_0 is the 'equator'.) The map

$$\varphi :]0, \pi[\times S_0 \to S \setminus \{\pm e_n\}, \quad (\theta, u) \mapsto \sin\theta u + \cos\theta e_n,$$

is a diffeomorphism.

Proposition 9.1.2 *Let f be an integrable function on S. Then*

$$\int_S f(x)\sigma(dx)$$

$$= \frac{\Gamma\left(\frac{n}{2}\right)}{\sqrt{\pi}\,\Gamma\left(\frac{n-1}{2}\right)} \int_0^\pi \left(\int_{S_0} f(\sin\theta u + \cos\theta e_n)\sigma_0(du) \right) \sin^{n-2}\theta \, d\theta.$$

In particular, if the function f is zonal, $f(x) = F(x_n)$, where F is a function defined on $[-1, 1]$, then

$$\int_S f(x)\sigma(dx) = \frac{\Gamma\left(\frac{n}{2}\right)}{\sqrt{\pi}\,\Gamma\left(\frac{n-1}{2}\right)} \int_0^\pi F(\cos\theta)\sin^{n-2}\theta \, d\theta$$

$$= \frac{\Gamma\left(\frac{n}{2}\right)}{\sqrt{\pi}\,\Gamma\left(\frac{n-1}{2}\right)} \int_{-1}^1 F(t)(1 - t^2)^{(n-3)/2} dt.$$

For $n = 4$, $S^3 \simeq SU(2)$, and this proposition corresponds to Corollary 7.2.2.

Proof. Let ω_0 be the differential form on S_0 with degree $n - 2$ given by

$$\omega_0 = \sum_{i=1}^{n-1} (-1)^{i-1} u_i du_1 \wedge \cdots \wedge \widehat{du_i} \wedge \cdots \wedge du_{n-1}.$$

We will show that

$$\varphi^* \omega = \sin^{n-2}\theta \, d\theta \otimes \omega_0.$$

Let $X \in \mathbb{R}$, and Y a tangent vector to S_0 at u. Then

$$(D\varphi)_{(\theta, u)}(X, Y) = (\cos\theta u - \sin\theta e_n)X + \sin\theta Y.$$

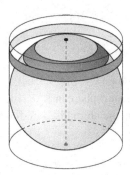

Figure 5

Hence, if $Y_1, \ldots, Y_{n-1}$ are $n - 2$ tangent vectors at u, then

$$
\begin{aligned}
(\varphi^*\omega)_{(\theta,u)}&(X, Y_1, \ldots, Y_{n-2}) \\
&= \omega_{\varphi(\theta,u)}\big((D\varphi)_{(\theta,u)}X, (D\varphi)_{(\theta,u)}Y_1, \ldots, (D\varphi)_{(\theta,u)}Y_{n-2}\big) \\
&= \det\big(\sin\theta u + \cos\theta e_n, (\cos\theta u - \sin\theta e_n)X, \sin\theta Y_1, \ldots, \sin\theta Y_{n-2}\big) \\
&= (-1)^{n-1} X \sin^{n-2}\theta \; (\omega_0)_u(Y_1, \ldots, Y_{n-2}).
\end{aligned}
$$

Therefore

$$
\varphi^*\omega = (-1)^n \sin^{n-2}\theta \, d\theta \otimes \omega_0,
$$

and the statement follows since

$$
\frac{\varpi_{n-1}}{\varpi_n} = \frac{\Gamma\left(\frac{n}{2}\right)}{\sqrt{\pi}\,\Gamma\left(\frac{n-1}{2}\right)}.
$$

$\square$

From this proposition, the projection on a diameter of the measure σ is a measure on $[-1, 1]$ with density

$$
\frac{\Gamma\left(\frac{n}{2}\right)}{\sqrt{\pi}\,\Gamma\left(\frac{n-1}{2}\right)}(1 - t^2)^{(n-3)/2}.
$$

In particular, for $n = 3$, this density is constant, equal to $\frac{1}{2}$. This property can be stated as follows: the axial projection from S onto the cylinder tangent to the sphere along the equator preserves the measure. This property was observed by Archimedes.

9.2 Laplace operator

The Laplace operator Δ of the Euclidean space $\mathbb{R}^n$ is defined by

$$\Delta f = \sum_{i=1}^{n} \frac{\partial^2 f}{\partial x_i^2},$$

where f is a C^2 function which is defined on a domain Ω in $\mathbb{R}^n$. The operator Δ is invariant under the orthogonal group $G = O(n)$:

$$\Delta(f \circ g) = (\Delta f) \circ g \quad (g \in G).$$

Assume Ω to be G-invariant, for instance $\Omega = B(0, R)$, the open ball with centre 0 and radius R, and f to be G-invariant as well:

$$\forall g \in G = O(n), \quad f(g \cdot x) = f(x).$$

Then one can write

$$f(x) = F(r), \ r = \|x\|,$$

where the function F is defined on an interval in $[0, \infty[$.

Proposition 9.2.1

$$(\Delta f)(x) = (LF)(r),$$

where the operator L is given by

$$LF = \frac{d^2 F}{dr^2} + \frac{n-1}{r} \frac{dF}{dr},$$
$$= \frac{1}{r^{n-1}} \frac{d}{dr} \left(r^{n-1} \frac{dF}{dr} \right).$$

The operator L is called the *radial part* of the Laplace operator Δ.

Lemma 9.2.2 *Let f be a C^2 function which is defined on an open set Ω in a finite dimensional vector space V. Let U be an endomorphism of V, and $a \in V$. Let $\varepsilon > 0$ such that, for $|t| < \varepsilon$, $\exp tU \cdot a \in \Omega$. Assume that, for $|t| < \varepsilon$,*

$$f(\exp tU \cdot a) = f(a).$$

Then

$$(Df)_a(U \cdot a) = 0,$$
$$(D^2 f)_a)(U \cdot a, U \cdot a) + (Df)_a(U^2 \cdot a) = 0.$$

Proof. This simply amounts to computing the first and second derivatives of the map $t \mapsto f(\exp tU \cdot a)$ at $t = 0$. $\square$

Proof of Proposition 9.2.1. If X is a skewsymmetric matrix, $X \in Skew(n, \mathbb{R})$, then, for every $t \in \mathbb{R}$, $\exp tX$ is an orthogonal matrix, $\exp tX \in O(n)$. If the function f is $O(n)$-invariant, then

$$f(\exp tX \cdot a) = f(a) \quad (a \in \Omega).$$

Let $a = re_1, r > 0$. For $2 \le i \le n$, consider the skewsymmetric matrix

$$X = E_{i1} - E_{1i}.$$

Then

$$X \cdot a = re_i, \quad X^2 \cdot a = -re_1 \quad (r = \|a\|).$$

By Lemma 9.2.2, for $2 \le i \le n$,

$$\frac{\partial f}{\partial x_i}(a) = 0, \quad \text{and} \quad r^2 \frac{\partial^2 f}{\partial x_i^2}(a) - r \frac{\partial f}{\partial x_1}(a) = 0,$$

or

$$\frac{\partial^2 f}{\partial x_i^2}(a) = \frac{1}{r} \frac{dF}{dr}.$$

Since

$$\frac{\partial^2 f}{\partial x_1^2}(a) = \frac{d^2 F}{dr^2},$$

the statement follows. $\qquad\qquad\square$

Consider an open set $\Omega \subset \mathbb{R}^n$ which is invariant under $K = SO(n - 1)$. We put

$$x = r(\sin\theta u + \cos\theta e_n),$$

with $r \ge 0, 0 \le \theta \le \pi, u \in S_0 = S^{n-2}$.

Proposition 9.2.3 *Let f be a C^2 function which is defined on Ω and K-invariant. Such a function can be written $f(x) = F(r, \theta)$. Then*

$$\Delta f = \frac{\partial F}{\partial r^2} + \frac{n-1}{r}\frac{\partial F}{\partial r} + \frac{1}{r^2}\left(\frac{\partial^2 F}{\partial \theta^2} + (n-2)\cot\theta\frac{\partial F}{\partial \theta}\right).$$

Proof. Let us write $x = (x^0, x_n)$ with $x^0 = (x_1, \ldots, x_{n-1}) \in \mathbb{R}^{n-1}$, and consider the polar decomposition of x^0:

$$x^0 = \rho u \quad (\rho \ge 0, \ u \in S_0).$$

By Proposition 9.2.1,

$$\Delta = \frac{\partial^2}{\partial \rho^2} + \frac{n-2}{\rho} \frac{\partial}{\partial \rho} + \frac{\partial^2}{\partial x_n^2}.$$

Observe that $\rho = r \sin \theta$, $x_n = r \cos \theta$. Hence

$$\frac{\partial^2}{\partial \rho^2} + \frac{\partial^2}{\partial x_n^2} = \frac{\partial^2}{\partial r^2} + \frac{1}{r} \frac{\partial}{\partial r} + \frac{1}{r^2} \frac{\partial^2}{\partial \theta^2}.$$

On the other hand,

$$\frac{\partial}{\partial \rho} = \frac{\partial r}{\partial \rho} \frac{\partial}{\partial r} + \frac{\partial \theta}{\partial \rho} \frac{\partial}{\partial \theta}.$$

From the relation $r^2 = \rho^2 + x_n^2$ it follows that

$$\frac{\partial r}{\partial \rho} = \frac{\rho}{r},$$

and, from $\tan \theta = \rho / x_n$, that

$$\frac{\partial \theta}{\partial \rho} = \frac{\cos^2 \theta}{x_n} = \frac{\cos \theta}{r}.$$

One gets

$$\frac{1}{\rho} \frac{\partial}{\partial \rho} = \frac{1}{r} \frac{\partial}{\partial r} + \frac{1}{r^2} \cot \theta \frac{\partial}{\partial \theta},$$

and finally

$$\Delta = \frac{\partial^2}{\partial r^2} + \frac{n-1}{r} \frac{\partial}{\partial r} + \frac{1}{r^2} \left(\frac{\partial^2}{\partial \theta^2} + (n-2) \cot \theta \frac{\partial}{\partial \theta} \right). \qquad \square$$

We will denote by Δ_S the Laplace operator of the sphere S. It can be defined as follows. Let f be a C^2 function on S. The function f extends to $\mathbb{R}^n \setminus \{0\}$ as a function $\tilde{f}$ which is homogeneous of degree 0:

$$\tilde{f}(x) = f \left(\frac{x}{\|x\|} \right).$$

Then the Laplace operator $\Delta_S f$ applied to f is the restriction to S of $\Delta \tilde{f}$:

$$\Delta_S f = (\Delta \tilde{f})\big|_S.$$

From the preceding proposition one obtains the following result.

Corollary 9.2.4 *Let f be a C^2 function on S which is zonal. It can be written*

$$f(x) = F(\theta) \quad \text{if } x = \sin \theta u + \cos \theta e_n \ (u \in S_0),$$

with a function F defined on $[0, \pi]$. *Then* $\Delta_S f = LF$, *where*

$$
L = \frac{d^2}{d\theta^2} + (n-2)\cot\theta \frac{d}{d\theta}
$$
$$
= \frac{1}{\sin^{n-2}\theta} \frac{d}{d\theta}\left(\sin^{n-2}\theta \frac{d}{d\theta}\right).
$$

For $n = 4$, then $S^3 \simeq SU(2)$, and the Laplace operator Δ_S is the Laplace operator of the group $SU(2)$ we considered in Section 8.3. The above corollary corresponds to Proposition 8.3.3.

One can show that, if f is a C^2 function on an open set in $\mathbb{R}^n$, then, for the polar decomposition $x = ru$ ($r \geq 0$, $u \in S$), the Laplace operator can be written

$$
\Delta f = \frac{\partial^2 f}{\partial r^2} + \frac{n-1}{r}\frac{\partial f}{\partial r} + \frac{1}{r^2}\Delta_S f.
$$

For $n = 3$, then $S = S^2$, and the equator S_0 is a circle. Consider on S^2 spherical coordinates:

$$
x_1 = \sin\theta\cos\varphi,
$$
$$
x_2 = \sin\theta\sin\varphi,
$$
$$
x_3 = \cos\theta
$$

$(0 \leq \theta \leq \pi, 0 \leq \varphi \leq 2\pi.)$ In terms of these coordinates the measure σ is given by

$$
\sigma(dx) = \frac{1}{4\pi}\sin\theta\, d\theta\, d\varphi,
$$

and the Laplace operator on S^2 by

$$
\Delta_S = \frac{\partial^2}{\partial\theta^2} + \cot\theta\frac{\partial}{\partial\theta} + \frac{1}{\sin^2\theta}\frac{\partial^2}{\partial\varphi^2}.
$$

9.3 Spherical harmonics

Let $\mathcal{P}$ denote the space of polynomials in n variables, with complex coefficients, and $\mathcal{P}_m$ the subspace of those polynomials which are homogeneous of degree m. A basis of $\mathcal{P}_m$ consists of monomials of degree m:

$$
x^\alpha = x_1^{\alpha_1}\dots x_n^{\alpha_n},
$$

with $\alpha_1, \ldots, \alpha_n \in \mathbb{N}$, $\alpha_1 + \cdots + \alpha_n = m$. The dimension of $\mathcal{P}_m$ is

$$\delta_m = \binom{m + n - 1}{n - 1}.$$

This evaluation can be obtained by observing that δ_m is the coefficient of t^m in the power expansion of $(1 - t)^{-n}$:

$$(1 - t)^{-n} = \sum_{m=0}^{\infty} \delta_m t^m \quad (|t| < 1).$$

In fact

$$(1 - t)^{-n} = (1 + t + \cdots + t^m + \cdots)^n$$
$$= \sum_{\alpha \in \mathbb{N}^n} t^{\alpha_1} t^{\alpha_2} \ldots t^{\alpha_n}$$
$$= \sum_{m=0}^{\infty} \#\{\alpha \in \mathbb{N}^n \mid \alpha_1 + \cdots + \alpha_n = m\} t^m.$$

One equips the space $\mathcal{P}$ with the Hermitian inner product defined by

$$\langle p, q \rangle = \left(p\left(\frac{\partial}{\partial x}\right) \bar{q} \right)(0).$$

If

$$p(x) = \sum_{\alpha} a_{\alpha} x^{\alpha}, \quad q(x) = \sum_{\alpha} b_{\alpha} x^{\alpha},$$

then

$$\langle p, q \rangle = \sum_{\alpha} \alpha! a_{\alpha} \overline{b_{\alpha}},$$

with $\alpha! = \alpha_1! \ldots \alpha_n!$. Hence $\mathcal{P}$ is a preHilbert space, and the subspaces $\mathcal{P}_m$ are pairwise orthogonal.

Let

$$Q(x) = x_1^2 + \cdots + x_n^2.$$

The Laplace operator is the constant coefficient differential operator associated to the polynomial Q:

$$\Delta = Q\left(\frac{\partial}{\partial x}\right) = \frac{\partial^2}{\partial x_1^2} + \cdots + \frac{\partial^2}{\partial x_n^2}.$$

A $\mathcal{C}^2$ function f defined on an open set in $\mathbb{R}^n$ is said to be *harmonic* if it is a solution of the *Laplace equation*:

$$\Delta f = 0.$$

We will denote by $\mathcal{H}_m$ the space of harmonic polynomials which are homogeneous of degree m.

Theorem 9.3.1 *A polynomial p which is homogeneous of degree m decomposes uniquely as*

$$p = \sum_{k=0}^{[n/2]} Q^k h_k,$$

with $h_k \in \mathcal{H}_{m-2k}$.

Proof. (a) The Laplace operator Δ is a surjective map from $\mathcal{P}_m$ onto $\mathcal{P}_{m-2}$. To see this we will show that the orthogonal of the image reduces to $\{0\}$. Let $r \in \mathcal{P}_{m-2}$ be such that, for every $p \in \mathcal{P}_m$,

$$\langle r, \Delta p \rangle = 0.$$

Take $p = rQ$. Then

$$\langle rQ, rQ \rangle = 0,$$

hence $rQ = 0$, and, since $Q \neq 0$, then $r = 0$. It follows that

$$d_m := \dim \mathcal{H}_m = \delta_m - \delta_{m-2} = (2m + n - 2)\frac{(m+n-3)!}{(n-2)!m!}.$$

(b) Let us show that $p \in \mathcal{P}_m$ decomposes uniquely as

$$p = Qp_1 + h,$$

where $p_1 \in \mathcal{P}_{m-2}$ and $h \in \mathcal{H}_m$. For that we will show that $\mathcal{M} = Q\mathcal{P}_{m-2}$ and $\mathcal{N} = \mathcal{H}_m$ are two complementary subspaces in $\mathcal{P}_m$. Their intersection reduces to $\{0\}$, since, if

$$\Delta(Qp) = 0,$$

then

$$p\left(\frac{\partial}{\partial x}\right) \Delta(Qp) = 0 \quad \text{or} \quad \langle Qp, Qp \rangle = 0,$$

hence $p = 0$. On the other hand, by (a)

$$\dim \mathcal{M} + \dim \mathcal{N} = \dim \mathcal{P}_m.$$

(c) One continues, with p_1 instead of p, until one obtains a polynomial p_i of degree ≤ 1. $\square$

Recall that $S = S^{n-1}$ denotes the unit sphere in $\mathbb{R}^n$. The restriction to S of a polynomial in $\mathcal{H}$ is called a *spherical harmonic* of degree m. Let $\mathcal{Y}_m$ be the space of spherical harmonics of degree m, that is the space of restrictions to S of polynomials in $\mathcal{H}_m$. The restriction map $\mathcal{H}_m \to \mathcal{Y}_m$ is an isomorphism.

Theorem 9.3.2 (i) *The subspaces $\mathcal{Y}_m$ are pairwise orthogonal in $L^2(S, \sigma)$.*
 (ii)

$$L^2(S, \sigma) = \widehat{\bigoplus_{m \geq 0}} \mathcal{Y}_m.$$

Proof. (a) Let us recall the Green formula in the case of the unit ball

$$B = \{x \in \mathbb{R}^n \mid \|x\| \leq 1\}.$$

For $\mathcal{C}^2$ functions u and v on B,

$$\varpi_n \int_S \left(u \frac{\partial v}{\partial \nu} - v \frac{\partial u}{\partial \nu} \right) \sigma(dy) = \int_B (u \Delta v - v \Delta u) \lambda(dx),$$

where $\partial/\partial \nu$ denotes the outer normal derivative (see Corollary 9.7.2 below). For an ℓ-homogeneous harmonic polynomial p,

$$\frac{\partial p}{\partial \nu} = \ell p,$$

and an m-homogeneous harmonic polynomial q,

$$0 = \int_S \left(p \frac{\partial q}{\partial \nu} - q \frac{\partial p}{\partial \nu} \right) \sigma(dy) = (m - \ell) \int_S pq\sigma(dy).$$

 (b) From Theorem 9.3.1 it follows that

$$\sum_{m=0}^{\infty} \mathcal{Y}_m$$

is the space of restrictions to S of all polynomials. This space is an algebra which separates points, contains constants and if f belongs to it, then $\bar{f}$ does as well. By the Stone–Weierstrass Theorem (recalled above: Theorem 6.4.3) this space is dense in the space $\mathcal{C}(S)$ of complex valued continuous functions on S for the uniform convergence topology. The statement follows. $\square$

Let T be the representation of the group $G = SO(n)$ on the space $\mathcal{C}(\mathbb{R}^n)$ defined by

$$(T(g)f)(x) = f(xg).$$

If f is $\mathcal{C}^2$, then

$$\Delta(f \circ T(g)) = (\Delta f) \circ T(g).$$

The subspace $\mathcal{H}_m$ is invariant under the representation T, and the restriction T_m of T to $\mathcal{H}_m$ is a finite dimensional representation of G. We saw in Chapter 7 that, for $n = 3$, the representation T_m is irreducible (Theorem 7.6.5). We will show that this holds for $n \geq 3$ as well. The method will be different. From now on we assume that $n \geq 3$.

Let K be the isotropy subgroup of $e_n = (0, \ldots, 0, 1)$. We saw that K is isomorphic to $SO(n - 1)$ (Section 1.5). Let $\mathcal{H}_m^K$ denote the subspace of K-invariant polynomials in $\mathcal{H}_m$:

$$\mathcal{H}_m^K = \{p \in \mathcal{H}_m \mid \forall k \in K, \ T_m(k)p = p\}.$$

One defines similarly the subspace $\mathcal{Y}_m^K$ in $\mathcal{Y}_m$. A function on S is said to be zonal if it is K-invariant. Such a function f can be written

$$f(x) = F(x_n) \quad (x = (x_1, \ldots, x_n) \in S),$$

with a function F defined on $[-1, 1]$.

Theorem 9.3.3

$$\dim \mathcal{Y}_m^K = \dim \mathcal{H}_m^K = 1.$$

Proof. Let us orthogonalise the sequence of the functions f_m:

$$f_m(x) = (x_n)^m \quad (x = (x_1, \ldots, x_n) \in S)$$

with respect to the inner product in $L^2(S, \sigma)$. We get a sequence of functions of the form

$$\varphi_m(x) = p_m(x_n),$$

where p_m is a polynomial of degree m. We will show that every function in $\mathcal{Y}_m^K$ is proportional to φ_m. Let $f \in \mathcal{H}_m^K$. We can write

$$f(x) = \sum_{k=0}^{m} x_n^k q_k(x_1, \ldots, x_{n-1}),$$

where q_k is a polynomial in $n - 1$ variables which is homogeneous of degree $m - k$. Since f is K-invaraint, $q_k = 0$ if $m - k$ is odd and, if $m - k$ is even, $m - k = 2j$, then

$$q_{m-2j}(x) = c_j \left(x_1^2 + \cdots + x_{n-1}^2 \right)^j.$$

Hence

$$f(x) = \sum_{k+2j=m} c_j x_n^k \left(1 - x_n^2 \right)^j \quad (x \in S).$$

Therefore the restriction $\tilde{f}$ of f to S is a linear combination of the functions f_k ($k \leq m$). Furthermore $\tilde{f}$ is orthogonal to the space

$$\mathcal{Y}_0 \oplus \cdots \oplus \mathcal{Y}_{m-1}$$

which contains the functions $f_0, \ldots, f_{m-1}$; hence $\tilde{f}$ is proportional to φ_m.

It remains to prove that $\mathcal{Y}_m^K$ does not reduce to $\{0\}$. If $a_1, \ldots, a_n$ are n complex numbers such that $a_1^2 + \cdots + a_n^2 = 0$, then the polynomial

$$f(x) = (a_1 x_1 + \cdots + a_n x_n)^m$$

is harmonic and homogeneous of degree m: $f \in \mathcal{H}_m$. In particular, the polynomial

$$q(x) = (x_n + i x_1)^m$$

belongs to $\mathcal{H}_m$, and $q(e_n) = 1$. Put

$$p(x) = \int_K q(xk)\mu_0(dk),$$

where μ_0 is the normalised Haar measure of K. The polynomial q belongs to $\mathcal{H}_m^K$ and is not equal to 0; in fact $p(e_n) = 1$. □

Theorem 9.3.4 *The representation* $(T_m, \mathcal{H}_m)$ *is irreducible.*

Proof. Let $\mathcal{Y} \neq \{0\}$ be a G-invariant subspace of $\mathcal{Y}_m$, and $f_1 \neq 0$ a function in $\mathcal{Y}$. There exists $x \in S$ such that $f_1(x) \neq 0$. Since G acts transitively on S, there exists $g \in G$ such that $x = e_n \cdot g$. The function $f_2 = T(g)f_1$ belongs to $\mathcal{Y}$ and

$$f_2(e_n) = f_1(e_n \cdot g) \neq 0.$$

Put

$$f_0(x) = \int_K f_2(x \cdot k)\mu_0(dk).$$

Then $f_0 \in \mathcal{Y}_m^K$, and $f_0(e_n) = f_2(e_n) \neq 0$. Since $\dim \mathcal{Y}_m^K = 1$, it follows that $\mathcal{Y}_m^K \subset \mathcal{Y}$. Let us prove that the orthogonal $\mathcal{Y}^\perp$ of $\mathcal{Y}$ in $\mathcal{Y}_m$ reduces to $\{0\}$. In fact, if $\mathcal{Y}^\perp$ did not reduce to $\{0\}$, one would show as above that $\mathcal{Y}_m^K \subset \mathcal{Y}^\perp$, a contradiction. □

An important property of the spaces $\mathcal{Y}_m$ is that they provide a spectral decomposition of the Laplace operator Δ_S.

Proposition 9.3.5 *The space* $\mathcal{Y}_m$ *is an eigenspace of the Laplace operator* Δ_S: *if* $f \in \mathcal{Y}_m$, *then*

$$\Delta_S f = -m(m + n - 2)f.$$

Proof. Recall that $\Delta_S f$ is the restriction to S of $\Delta \tilde{f}$, $\tilde{f}$ being the function which extends f to $\mathbb{R}^n \setminus \{0\}$ and which is homogeneous of degree 0. A function $f \in \mathcal{Y}_m$ is the restriction to S of a polynomial $p \in \mathcal{H}_m$, hence

$$\tilde{f}(x) = f\left(\frac{x}{\|x\|}\right) = \frac{1}{\|x\|^m} p(x) \quad (x \neq 0).$$

Recall also that, for $\mathcal{C}^2$ functions u and v on an open set in $\mathbb{R}^n$,

$$\Delta(uv) = (\Delta u)v + 2(\nabla u | \nabla v) + u \Delta v,$$

where ∇u denotes the gradient of u, that is the vector valued function whose components are the partial derivatives of u. We get

$$\Delta \tilde{f} = \Delta\left(\frac{1}{\|x\|^m}\right) p + 2\left(\nabla \frac{1}{\|x\|^m} | \nabla p\right) + \frac{1}{\|x\|^m} \Delta p.$$

On the one hand,

$$\nabla \frac{1}{\|x\|^m} = -m \frac{x}{\|x\|^{m+2}},$$

and, being homogeneous, the polynomial p satisfies the Euler equation

$$\sum_{i=1}^{n} x_i \frac{\partial p}{\partial x_i} = mp, \quad \text{or} \quad (x|\nabla p) = mp,$$

therefore

$$\left(\nabla \frac{1}{\|x\|^m} | \nabla p\right) = -m^2 \frac{1}{\|x\|^{m+2}} p.$$

On the other hand, using Proposition 9.2.1, one obtains

$$\Delta\left(\frac{1}{\|x\|^m}\right) = m(m - n + 2)\frac{1}{\|x\|^{m+2}}.$$

Finally, since $\Delta p = 0$,

$$\Delta \tilde{f} = -m(m + n - 2)\frac{1}{\|x\|^{m+2}} p(x),$$

and therefore

$$\Delta_S f = -m(m + n - 2)f. \qquad \square$$

9.4 Spherical polynomials

In this section we will study the sequence of polynomials p_m which appear in the proof of Theorem 9.3.3. We call them *spherical polynomials*. For $\nu > -\frac{1}{2}$

we consider on the space $\mathbb{C}[T]$ of polynomials in one variable with complex coefficients the following inner product:

$$(p|q) = \frac{\Gamma(v+1)}{\sqrt{\pi}\,\Gamma\left(v+\frac{1}{2}\right)} \int_{-1}^{1} p(t)\overline{q(t)}(1-t^2)^{v-1/2}dt.$$

Observe that, for $v = \frac{n-2}{2}$, by Proposition 9.1.2,

$$(p|q) = \int_S p(x_n)\overline{q(x_n)}\sigma(dx),$$

where S is the unit sphere in $\mathbb{R}^n$, and σ the normalised uniform measure on S. The polynomials p_m are obtained by orthogonalising the sequence of monomials $1, t, \ldots, t^m, \ldots$. The polynomial p_m will be normalised by the condition $p_m(1) = 1$.

Proposition 9.4.1 (Rodrigues' formula)

$$(1-t^2)^{v-1/2} p_m(t) = (-1)^m 2^{-m} \frac{\Gamma\left(v+\frac{1}{2}\right)}{\Gamma\left(v+m+\frac{1}{2}\right)} \left(\frac{d}{dt}\right)^m (1-t^2)^{v+m-1/2}.$$

Proof. Put

$$q_m(t) = (1-t^2)^{-v+1/2} \left(\frac{d}{dt}\right)^m (1-t^2)^{v+m-1/2}.$$

We will show that q_m is orthogonal to $f_\ell(t) = t^\ell$ if $0 \le \ell < m$. As a result the polynomials p_m and q_m will be proven to be proportional. By performing ℓ integrations by part we get

$$(q_m|f_\ell) = \gamma(v) \int_{-1}^{1} \left(\frac{d}{dt}\right)^m (1-t^2)^{v+m-1/2} t^\ell dt$$

$$= (-1)^\ell \gamma(v)\ell! \int_{-1}^{1} \left(\frac{d}{dt}\right)^{m-\ell} (1-t^2)^{v+m-1/2} dt = 0.$$

Letting $t = 1 - u$, we obtain

$$q_m(1-u) = (-1)^m (2u-u^2)^{-v-1/2} \left(\frac{d}{du}\right)^m (2u-u^2)^{v+m-1/2},$$

and

$$q_m(1) = (-1)^m 2^m \left(v+m-\frac{1}{2}\right)\left(v+m-1-\frac{1}{2}\right)\ldots\left(v+1-\frac{1}{2}\right). \qquad \square$$

The properties of the polynomials p_m can be obtained from the Rodrigues formula. But we prefer to establish them using the link to spherical harmonics.

Consider an orthonormal basis $\{\psi_j\}$ of the space $\mathcal{Y}_m$ ($1 \le j \le d_m$), and put

$$\mathcal{K}_m(x, y) = \sum_{j=1}^{d_m} \psi_j(x)\overline{\psi_j(y)} \quad (x, y \in S).$$

One can check that this definition does not depend on the choice of basis, and that the kernel $\mathcal{K}_m$ is invariant under the group $G = SO(n)$ in the following sense:

$$\mathcal{K}_m(xg, yg) = \mathcal{K}_m(x, y) \quad (g \in G, x, y \in S).$$

The kernel $\mathcal{K}_m$ is called the *reproducing kernel* of the space $\mathcal{Y}_m$; this means that, for every function $f \in \mathcal{Y}_m$,

$$\int_S \mathcal{K}_m(x, y) f(y)\sigma(dy) = f(x).$$

It is also the kernel of the orthogonal projection P_m from $L^2(S)$ onto $\mathcal{Y}_m$.

Proposition 9.4.2

$$\mathcal{K}_m(x, y) = d_m\, p_m\big((x|y)\big).$$

Proof. Since the kernel $\mathcal{K}_m$ is G-invariant, and the group G acts transitively on S, the value $\mathcal{K}_m(x, x)$ does not depend on x: $\mathcal{K}_m(x, x) = C_m$. On the other hand,

$$\mathcal{K}_m(x, x) = \sum_{j=1}^{d_m} |\psi_j(x)|^2.$$

By integrating over S one gets

$$C_m = \sum_{j=1}^{d_m} \int_S |\psi_j(x)|^2 \sigma(dx) = d_m.$$

Fix $y = e_n$; the function $x \mapsto \mathcal{K}_m(x, e_n)$ belongs to $\mathcal{Y}_m$ and is K-invariant. Hence it is proportional to $\varphi_m(x) = p_m(x_n)$ by Theorem 9.3.3, and, since $p_m(1) = 1$,

$$\mathcal{K}_m(x, e_n) = d_m\, p_m(x_n).$$

By the invariance of $\mathcal{K}_m$, it follows that

$$\mathcal{K}_m(x, y) = d_m\, p_m\big((x|y)\big). \qquad \square$$

Corollary 9.4.3

$$\frac{\Gamma\left(\frac{n}{2}\right)}{\sqrt{\pi}\,\Gamma\left(\frac{n-1}{2}\right)} \int_{-1}^{1} p_m(t)^2 (1 - t^2)^{(n-3)/2} dt = \frac{1}{d_m}.$$

Proof. Since P_m is a projection, $P_m^2 = P_m$, and this gives

$$\int_S \mathcal{K}_m(x, z)\mathcal{K}_m(z, y)\sigma(dz) = \mathcal{K}_m(x, y).$$

Taking $x = y = e_n$ we get $\mathcal{K}_m(x, z) = \mathcal{K}_m(z, y) = d_m p_m(z_n)$, hence

$$d_m^2 \int_S p_m(z_n)^2 \sigma(dz) = d_m.$$

Using the formula giving the integral of a zonal function (Proposition 9.1.2), the statement follows. □

As we saw at the end of the proof of Theorem 9.3.3, the polynomial

$$q(x) = (x_n + ix_1)^m$$

belongs to $\mathcal{H}_m$, and

$$\int_K q(xk)\mu_0(dk) = \varphi_m(x),$$

and this can also be written

$$\int_{S_0} (x_n + i\sqrt{1 - x^2}u)^m \sigma_0(du) = p_m(x_n).$$

By Proposition 9.1.2 one has the following.

Proposition 9.4.4

(i) $\quad p_m(\cos\theta) = \dfrac{\Gamma\left(\frac{n-1}{2}\right)}{\sqrt{\pi}\,\Gamma\left(\frac{n-2}{2}\right)} \displaystyle\int_0^\pi (\cos\theta + i\sin\theta\cos\varphi)^m \sin^{n-3}\varphi\,d\varphi.$

(ii) *For* $-1 \le t \le 1$, $|p_m(t)| \le 1$.

The function φ_m, since it belongs to the space $\mathcal{Y}_m$, is an eigenfunction of the Laplace operator Δ_S (Proposition 9.3.5):

$$\Delta_S\varphi_m = -m(m + n - 2)\varphi_m.$$

Using Corollary 9.2.4 it follows that p_m satisfies the following differential equation:

$$\left(\frac{d^2}{d\theta^2} + (n-2)\cot\theta\frac{d}{d\theta}\right)p_m(\cos\theta) = -m(m+n-2)p_m(\cos\theta),$$

and, by putting $\cos\theta = t$, we obtain the following result.

Proposition 9.4.5

$$\left((1 - t^2)\frac{d^2}{dt^2} - (n-1)t\frac{d}{dt}\right)p_m = -m(m + n - 2)p_m.$$

For $n = 4$, $S^3 \simeq SU(2)$, and the function φ_m is proportional to the character χ_m of the representation we considered in Section 7.5:

$$\varphi_m(x) = \frac{1}{m+1} \chi_m(x).$$

In this case p_m is, up to a factor, the Chebyshev polynomial usually denoted by U_m:

$$p_m(\cos\theta) = \frac{1}{m+1} U_m(\cos\theta) = \frac{1}{m+1} \frac{\sin(m+1)\theta}{\sin\theta}.$$

For $n = 3$, p_m is the Legendre polynomial of degree m.

9.5 Funk–Hecke Theorem

In this section we will analyse the operators acting on the space $\mathcal{C}(S)$ of the form:

$$Af(x) = \int_S a\big((x|y)\big) f(y)\sigma(dy),$$

where a is a continuous complex valued function on $[-1, 1]$. We denote by $\mathcal{A}$ the set of operators of this form. Such an operator is invariant. This means that it commutes with the representation T of $G = SO(n)$ on the space $\mathcal{C}(S)$ defined by $\big(T(g)f\big)(x) = f(xg)$: for every $g \in G$, $AT(g) = T(g)A$. In fact, using the invariance of the measure σ, one gets

$$\big(AT(g)f\big)(x) = \int_S a\big((x|y)\big) f(yg)\sigma(dy)$$

$$= \int_S a\big((x|y'g^{-1})\big) f(y')\sigma(dy')$$

$$= \int_S a\big((xg|y')\big) f(y')\sigma(dy') = \big(T(g)Af\big)(x).$$

Proposition 9.5.1 *The set $\mathcal{A}$ is a commutative algebra.*

Clearly $\mathcal{A}$ is a vector space. In order to show that the product of two operators in $\mathcal{A}$ belongs to $\mathcal{A}$ we will use the following lemma.

Lemma 9.5.2 *Let H be a continuous invariant kernel on S: H is a continuous function on $S \times S$ such that*

$$H(xg, yg) = H(x, y) \quad (g \in G).$$

Then there exists a continuous function h on $[-1, 1]$ *such that*

$$H(x, y) = h\big((x|y)\big).$$

Proof. The function $x \mapsto H(x, e_n)$ is K-invariant, hence zonal. Therefore it can be written

$$H(x, e_n) = h(x_n),$$

where h is a continuous function on $[-1, 1]$. Both kernels $H(x, y)$ and $H_1(x, y) = h\big((x|y)\big)$ are invariant, and

$$H(x, e_n) = H_1(x, e_n).$$

Since G acts transitively on S, it follows that they are equal. $\qquad\square$

Proof of Proposition 9.5.1. Consider the product AB of two operators A and B in $\mathcal{A}$:

$$(ABf)(x) = \int_S a\big((a|z)\big) \int_S b\big((z|y)\big) f(y)\sigma(dy)$$
$$= \int_S H(x, y) f(y)\sigma(dy),$$

with

$$H(x, y) = \int_S a\big((x|z)\big) b\big((z|y)\big) \sigma(dz).$$

The kernel H is continuous and invariant. The invariance is easily obtained from the invariance of the measure σ. Hence by Lemma 9.5.2 there exists a continuous function h on $[-1, 1]$ such that

$$H(x, y) = h\big((x|y)\big).$$

This shows that $\mathcal{A}$ is an algebra. It also follows that the kernel H is symmetric: $H(y, x) = H(x, y)$ and, as a result, $AB = BA$. Hence $\mathcal{A}$ is a commutative algebra. $\qquad\square$

The projection P_m onto the space $\mathcal{Y}_m$ belongs to the algebra $\mathcal{A}$:

$$(P_m f)(x) = d_m \int_S P_m\big((x|y)\big) f(y)\sigma(dy).$$

Theorem 9.5.3 (Funk–Hecke Theorem) *Let A be an operator in the algebra $\mathcal{A}$. Then $\mathcal{Y}_m$ is an eigenspace of A for the eigenvalue*

$$\widehat{a}(m) = \frac{\Gamma\big(\frac{n}{2}\big)}{\sqrt{\pi}\,\Gamma\big(\frac{n-1}{2}\big)} \int_{-1}^{1} a(t) p_m(t)(1 - t^2)^{(n-3)/2} dt.$$

Proof. By Proposition 9.5.1,

$$AP_m = P_m A.$$

Therefore $A(\mathcal{Y}_m) \subset \mathcal{Y}_m$. Since the opeator A is invariant, for $g \in G$, then

$$T_m(g)A = AT_m(g).$$

Because the representation T_m is irreducible, by Schur's Lemma (Theorem 6.1.3), it follows that $\mathcal{Y}_m$ is an eigenspace of A: there exists $\lambda_m \in \mathbb{C}$ such that, for $f \in \mathcal{Y}_m$,

$$Af = \lambda_m f.$$

By taking $f(x) = p(x_n)$ one gets

$$\lambda_m = Af(e_n) = \int_S a(y_n)p_m(y_n)\sigma(dy)$$

$$= \frac{\Gamma\left(\frac{n}{2}\right)}{\sqrt{\pi}\,\Gamma\left(\frac{n-1}{2}\right)} \int_{-1}^{1} a(t)p_m(t)(1-t^2)^{n-3/2}dt = \widehat{a}(m). \qquad \square$$

Proposition 9.5.4 *Let a be a continuous function on $[-1, 1]$.*

(i) *The Plancherel formula can be written*

$$\sum_{m=0}^{\infty} d_m |\widehat{a}(m)|^2 = \frac{\Gamma\left(\frac{n}{2}\right)}{\sqrt{\pi}\,\Gamma\left(\frac{n-1}{2}\right)} \int_{-1}^{1} |a(t)|^2 (1-t^2)^{(n-3)/2}dt.$$

(ii) *If*

$$\sum_{m=0}^{\infty} d_m |\widehat{a}(m)| < \infty,$$

then

$$a(t) = \sum_{m=0}^{\infty} d_m \widehat{a}(m) p_m(t),$$

and the series converges uniformly on $[-1, 1]$.

(iii) *If a is C^{2k} with $2k > \frac{n-1}{2}$ (that is if $4k \geq n$), then*

$$\sum_{m=0}^{\infty} d_m |\widehat{a}(m)| < \infty.$$

Proof. (a) The Plancherel formula (i) follows from the fact that the polynomials $\sqrt{d_m}\,p_m$ form a Hilbert basis of

$$L^2\left([-1, 1]; \frac{\Gamma\left(\frac{n}{2}\right)}{\sqrt{\pi}\,\Gamma\left(\frac{n-1}{2}\right)}(1-t^2)^{(n-3)/2}dt\right).$$

(b) Put

$$a_0(t) = \sum_{m=0}^{\infty} d_m \widehat{a}(m) p_m(t).$$

Since $|p_m(t)| \leq 1$ on $[-1, 1]$, the series converges uniformly on $[-1, 1]$, and hence a_0 is continuous on $[-1, 1]$. For every m,

$$\int_{-1}^{1} \big(a(t) - a_0(t)\big) p_m(t)(1 - t^2)^{(n-3)/2} dt = 0.$$

By the Weierstrass Theorem, it follows that, for every continuous function f on $[-1, 1]$,

$$\int_{-1}^{1} \big(a(t) - a_0(t)\big) f(t)(1 - t^2)^{(n-3)/2} dt = 0,$$

and, taking $f(t) = \overline{a(t) - a_0(t)}$, one gets

$$a_0(t) = a(t).$$

(c) Let L denote the differential operator

$$L = (1 - t^2)\frac{d^2}{dt^2} - (n-1)t\frac{d}{dt}.$$

Observe that

$$(1 - t^2)^{(n-3)/2} L = \frac{d}{dt}\left((1 - t^2)^{(n-1)/2}\frac{d}{dt}\right).$$

Therefore, if u and v are two C^2 functions on $[-1, 1]$, then

$$\int_{-1}^{1} Lu(t)v(t)(1 - t^2)^{(n-3)/2} dt = \int_{-1}^{1} u'(t)v'(t)(1 - t^2)^{(n-1)/2} dt$$

$$= \int_{-1}^{1} u(t)Lv(t)(1 - t^2)^{(n-3)/3} dt.$$

Since the spherical polynomial p_m is an eigenfunction of L:

$$Lp_m = -(m(m + n - 2)p_m$$

(Proposition 9.4.5), it follows that, if a is C^2,

$$\widehat{La}(m) = -m(m + n - 2)\widehat{a}(m).$$

If a is C^{2k}, by the Plancherel formula,

$$\frac{\Gamma\left(\frac{n}{2}\right)}{\sqrt{\pi}\,\Gamma\left(\frac{n-1}{2}\right)} \int_{-1}^{1} |L^k a(t)|^2 (1 - t^2)^{(n-3)/2} dt = \sum_{m=0}^{\infty} d_m \big(m(m + n - 2)\big)^{2k} |\widehat{a}(m)|^2.$$

Using the Schwarz inequality one has

$$
\left(\sum_{m=1}^{\infty} d_m |\hat{a}(m)| \right)^2
$$

$$
\leq \sum_{m=1}^{\infty} \frac{d_m}{\left(m(m+n-2)\right)^{2k}} \sum_{m=1}^{\infty} d_m \left(m(m+n-2)\right)^{2k} |\hat{a}(m)|^2.
$$

One observes that d_m is a polynomial of degree $n-2$ in m. Therefore there exists a constant C such that

$$
d_m \leq C(1+m)^{n-2}.
$$

Hence the series

$$
\sum_{m=1}^{\infty} \frac{d_m}{\left(m(m+n-2)\right)^{2k}}
$$

converges for $2k > \frac{n-1}{2}$ and, as a result,

$$
\sum_{m=0}^{\infty} d_m |\widehat{a}(m)| < \infty.
$$

$\square$

9.6 Fourier transform and Bochner–Hecke relations

The Fourier transform $\hat{f} = \mathcal{F}f$ of an integrable function f on $\mathbb{R}^n$ is defined by

$$
\hat{f}(\xi) = \int_{\mathbb{R}^n} e^{-i(\xi|x)} f(x) \lambda(dx).
$$

For $g \in GL(n, \mathbb{R})$ put

$$
\left(T(g)f\right)(x) = f(xg).
$$

Then

$$
\mathcal{F} \circ T(g) = |\det(g)|^{-1} T(g^{T-1}) \circ \mathcal{F}.
$$

Hence, if $g \in O(n)$, then $g^{T-1} = g$, and

$$
\mathcal{F} \circ T(g) = T(g) \circ \mathcal{F}.
$$

If f is $O(n)$-invariant, that is if f is radial, the same holds for its Fourier transform $\hat{f}$.

Let us consider the Fourier transform of the measure σ, the normalised uniform measure on the unit sphere S, seen as a measure on $\mathbb{R}^n$:

$$\hat{\sigma}(\xi) = \int_S e^{-i(\xi|u)}\sigma(du).$$

Since the measure σ is $O(n)$-invariant, its Fourier transform $\hat{\sigma}$ is radial. Put $\xi = \rho v$ ($\rho \geq 0$, $v \in S$). Then

$$\hat{\sigma}(\xi) = \int_S e^{-i\rho(v|u)}\sigma(du).$$

Using Proposition 9.1.2 we get

$$\hat{\sigma}(\xi) = \frac{\Gamma\left(\frac{n}{2}\right)}{\sqrt{\pi}\Gamma\left(\frac{n-1}{2}\right)} \int_{-1}^1 e^{-i\rho t}(1 - t^2)^{(n-3)/2}dt,$$

with $\rho = \|\xi\|$.

We define the *Bessel function* $\mathcal{J}_v$, for $v > -\frac{1}{2}$, by

$$\mathcal{J}_v(\tau) = \frac{\Gamma(v + 1)}{\sqrt{\pi}\Gamma(v + \frac{1}{2})} \int_{-1}^1 e^{-i\tau t}(1 - t^2)^{v-1/2}dt.$$

The function $\mathcal{J}_v$ is sometimes called the reduced Bessel function. It is related to the usual Bessel function J_v by

$$J_v(\tau) = \frac{1}{\Gamma(v + 1)} \left(\frac{\tau}{2}\right)^v \mathcal{J}_v(\tau).$$

By expanding in power series the exponential function $e^{-i\tau t}$ and by integrating termwise one gets the power series expansion of the function $\mathcal{J}_v$:

$$\mathcal{J}_v(\tau) = \sum_{k=0}^{\infty}(-1)^k \frac{\Gamma(v + 1)}{\Gamma(k + v + 1)} \frac{1}{k!} \left(\frac{\tau}{2}\right)^{2k},$$

which converges for every $\tau \in \mathbb{C}$. This follows from the evaluation

$$\int_{-1}^1 t^{2k}(1 - t^2)^{v-1/2}dt = \frac{\Gamma\left(k + \frac{1}{2}\right)\Gamma\left(v + \frac{1}{2}\right)}{\Gamma\left(k + v + 1\right)},$$

and the relation

$$\Gamma\left(k + \frac{1}{2}\right) = \frac{1 \cdot 3 \cdot 5 \cdots (2n - 1)}{2^n}\sqrt{\pi}.$$

Observe that, for $v = \frac{1}{2}$,

$$\mathcal{J}_{1/2}(\tau) = \frac{\sin \tau}{\tau}.$$

The Fourier transform $\hat{\sigma}$ of the measure σ can be written:

$$\hat{\sigma}(\xi) = \mathcal{J}_{(n-2)/2}(\|\xi\|).$$

The *Hankel transform* $\mathcal{H}_\nu$ is defined by

$$(\mathcal{H}_\nu F)(\rho) = \frac{2^{-\nu}}{\Gamma(\nu+1)} \int_0^\infty \mathcal{J}_\nu(\rho r) F(r) r^{2\nu+1} dr \quad (\rho \geq 0),$$

where F is a measurable function on $[0, \infty[$ such that

$$\int_0^\infty |F(r)| r^{2\nu+1} dr < \infty.$$

Proposition 9.6.1 *Let f be an integrable function on $\mathbb{R}^n$ which is radial. It can be written*

$$f(x) = F(\|x\|),$$

with a measurable function F on $[0, \infty[$ such that

$$\int_0^\infty |F(r)| r^{n-1} dr < \infty.$$

The Fourier transform $\hat{f}$ of f is radial:

$$\hat{f}(\xi) = \tilde{F}(\|\xi\|),$$

with

$$\tilde{F} = (2\pi)^{n/2} \mathcal{H}_{(n-2)/2} F.$$

Proof. Using the integration formula of Proposition 9.1.1 we get, if $\xi = \rho v$ $(\rho \geq 0, v \in S)$,

$$\hat{f}(\xi) = \varpi_n \int_0^\infty \left(\int_S e^{-i\rho r(v|u)} \sigma(du) \right) F(r) r^{n-1} dr$$

$$= \varpi_n \int_0^\infty \mathcal{J}_{(n-2)/2}(\rho r) F(r) r^{n-1} dr$$

$$= (2\pi)^{n/2} \mathcal{H}_{(n-2)/2} F(\rho). \qquad \square$$

Consider now a function f of the following form

$$f(x) = F(\|x\|) h(x),$$

where F is a measurable function on $[0, \infty[$, and h a harmonic polynomial which is homogeneous of degree m. We assume that

$$\int_0^\infty |F(r)| r^{m+n-1} dr < \infty.$$

The function f is then integrable on $\mathbb{R}^n$. We will see that its Fourier transform has the same form:

$$\hat{f}(\xi) = \tilde{f}(\|\xi\|)h(\xi).$$

Let us compute the Fourier transform of f by using the integration formula of Proposition 9.1.1, by putting $\xi = \rho v$ with $\rho \geq 0$, $v \in S$:

$$\hat{f}(\xi) = \varpi_n \int_0^\infty \left(\int_S e^{-i\rho r(v|u)} h(u) d\sigma(u) \right) F(r) r^{m+n-1} dr.$$

Let $a(t; \tau) = e^{-i\tau t}$ and

$$\hat{a}(m; \tau) = \frac{\Gamma\left(\frac{n}{2}\right)}{\sqrt{\pi}\Gamma\left(\frac{n-1}{2}\right)} \int_{-1}^1 e^{-i\tau t} p_m(t)(1 - t^2)^{(n-3)/2} dt.$$

By the Funk–Hecke Theorem (Theorem 9.5.3),

$$\int_S e^{-i\rho r(u|v)} h(u) d\sigma(u) = \hat{a}(m; \rho r) h(v).$$

We will express $\hat{a}(m; \tau)$ in terms of the Bessel function $\mathcal{J}_\nu$.

Proposition 9.6.2

$$\hat{a}(m; \tau) = \frac{\Gamma\left(\frac{n}{2}\right)}{\Gamma\left(\frac{n}{2} + m\right)} \left(-i\frac{\tau}{2}\right)^m \mathcal{J}_{(n-2)/2+m}(\tau).$$

Proof. By the Rodrigues formula (Proposition 9.4.1),

$$p_m(t)(1 - t^2)^{(n-3)/2} = \left(-\frac{1}{2}\right)^m \frac{\Gamma\left(\frac{n-1}{2}\right)}{\Gamma\left(\frac{n-3}{2} + m\right)} \left(\frac{d}{dt}\right)^m (1 - t^2)^{(n-1)/2+m}.$$

By carrying out m integrations by parts one gets

$$
\begin{aligned}
\hat{a}(m; \tau) &= \frac{\Gamma\left(\frac{n}{2}\right)}{\sqrt{\pi}\Gamma\left(\frac{n-1}{2} + m\right)} \left(-\frac{1}{2}\right)^m \int_{-1}^1 e^{-i\tau t} \left(\frac{d}{dt}\right)^m (1 - t^2)^{(n-3)/2+m} dt \\
&= \frac{\Gamma\left(\frac{n}{2}\right)}{\sqrt{\pi}\Gamma\left(\frac{n-1}{2} + m\right)} \left(-i\frac{\tau}{2}\right)^m \int_{-1}^1 e^{-i\tau t} (1 - t^2)^{(n-3)/2+m} dt \\
&= \frac{\Gamma\left(\frac{n}{2}\right)}{\Gamma\left(\frac{n}{2} + m\right)} \left(-i\frac{\tau}{2}\right)^m \mathcal{J}_{(n-2)/2+m}(\tau). \qquad \square
\end{aligned}
$$

Theorem 9.6.3 (Bochner–Hecke relations) *The Fourier transform of the function f on $\mathbb{R}^n$,*

$$f(x) = F(\|x\|)h(x),$$

where h is a harmonic polynomial, homogeneous of degree m, and F a measurable function on $[0, \infty[$ such that

$$\int_0^\infty |F(r)| r^{m+n-1} dr < \infty,$$

is equal to

$$\hat{f}(\xi) = (-i)^m \tilde{F}(\|\xi\|) h(\xi),$$

with

$$\tilde{F}(\rho) = (2\pi)^{n/2} \mathcal{H}_{(n-2)/2+m} F.$$

Furthermore, if F is an even function in the Schwarz space $S(\mathbb{R})$, then this holds for $\tilde{F}$ as well.

Let us consider the important example of a Gaussian function. The Fourier transform of the function f_0 defined on $\mathbb{R}^n$ by

$$f_0(x) = e^{-\|x\|^2/2}$$

is equal to

$$\hat{f}_0(\xi) = (2\pi)^{n/2} e^{-\|\xi\|^2/2}.$$

Hence, if $F(r) = e^{-r^2/2}$, then, for every n,

$$\mathcal{H}_{(n-2)/2} F = F.$$

It follows that the Fourier transform of

$$f(x) = e^{-\|x\|^2/2} h(x),$$

where h is a harmonic polynomial, homogeneous of degree m, is equal to

$$\hat{f}(\xi) = (2\pi)^{n/2} (-i)^m e^{-\|\xi\|^2/2} h(\xi).$$

9.7 Dirichlet problem and Poisson kernel

Let Ω be an open set in $\mathbb{R}^n$. A function F defined on Ω is said to be *harmonic* if it is C^2 and a solution of the *Laplace equation*

$$\Delta F = 0.$$

We will recall basic properties of harmonic functions with a hint of how they can be established.

We recall first the following theorem of Gauss. If ξ is a C^1 vector field which is defined on an open set in $\mathbb{R}^n$, its *divergence* is the function defined by

$$\operatorname{diver} \xi = \sum_{i=1}^{n} \frac{\partial \xi_i}{\partial x_i}.$$

Let Ω be an open set whose boundary $\partial\Omega$ is C^1. If $y \in \partial\Omega$ let $\nu(y)$ denote the outer unit normal vector at y. Consider the differential form α which is defined on $\partial\Omega$ by

$$\alpha_y(X_1, \ldots, X_{n-1}) = \det(\nu(y), X_1, \ldots, X_{n-1}),$$

and let Σ denote the associated positive measure on Ω: $\Sigma = |\alpha|$. We will say, for an open set Ω, that a function is C^k on $\overline{\Omega}$ if it is C^k in Ω and extends as a continuous function on $\overline{\Omega}$ with its derivatives of order $\leq k$.

Theorem 9.7.1 (Gauss' Theorem) *Let Ω be an open set in $\mathbb{R}^n$ with a C^1 boundary. Let ξ be a C^1 vector field on $\overline{\Omega}$. Then*

$$\int_{\partial\Omega} \big(\xi(y)|\nu(y)\big)\Sigma(dy) = \int_{\Omega} \operatorname{diver}\xi(x)\lambda(dx),$$

where λ is the Lebesgue measure for which $\lambda([0,1]^n) = 1$.

(The left-hand side is the flux of the vector field ξ through the boundary of Ω.)

This theorem is a special case of Stokes' Theorem. In fact, to the vector field ξ, one can associate the differential form ω of degree $n - 1$ which is defined by

$$\omega_x(X_1, \ldots, X_{n-1}) = \det(\xi(x), X_1, \ldots, X_{n-1}),$$

that is

$$\omega = \sum_{i=1}^{n}(-1)^{i-1}\xi_i(x)dx_1 \wedge \cdots \wedge \widehat{dx_i} \wedge \cdots \wedge dx_n.$$

Then

$$d\omega = \operatorname{diver}\xi \, dx_1 \wedge \cdots \wedge dx_n.$$

On the other hand, if $y \in \partial\Omega$, and if $X_1, \ldots, X_{n-1} \in T_y(\partial\Omega)$, then

$$\omega_y(X_1, \ldots, X_{n-1}) = \big(\xi(y)|\nu(y)\big)\alpha_y(X_1, \ldots, X_{n-1}).$$

By Stokes' Theorem,

$$\int_{\partial\Omega} \omega = \int_{\Omega} d\omega.$$

The theorem of Gauss follows.

Corollary 9.7.2 (Green's formula) *For C^2 functions u and v on $\overline{\Omega}$,*

$$\int_{\Omega} (u\Delta v - v\Delta u)\lambda(dx) = \int_{\partial\Omega} \left(u\frac{\partial v}{\partial \nu} - v\frac{\partial u}{\partial \nu} \right) \Sigma(dy),$$

where the outer normal derivative $\partial u / \partial \nu$ is defined at $y \in \partial\Omega$ by

$$\frac{\partial u}{\partial \nu}(y) = \big(\nabla u(y) \mid \nu(dy)\big).$$

One applies Gauss' Theorem to the vector field

$$\xi = u\nabla v - v\nabla u.$$

Proposition 9.7.3 *For $R > 0$ let v_0 be the function defined by*

$$v_0(x) = \frac{1}{(n-2)\varpi_n} \left(\frac{1}{\|x\|^{n-2}} - \frac{1}{R^{n-2}} \right) \quad (x \neq 0).$$

For a C^2 function u on the ball $\overline{B(0,R)}$,

$$\int_S u(Rz)\sigma(dz) = u(0) + \int_{B(0,R)} \Delta u(x)v_0(x)\lambda(dx).$$

Proof. Let us apply Green's formula for the open set

$$\Omega_{\varepsilon,R} = \{x \in \mathbb{R}^n \mid \varepsilon < \|x\| < R\} \quad (0 < \varepsilon < R).$$

Since the function v_0 is harmonic in $\Omega_{\varepsilon,R}$ we get

$$-\int_{\Omega_{\varepsilon,R}} \Delta u(x)v_0(x)\lambda(dx) = \int_{\partial\Omega_{\varepsilon,R}} \left(u\frac{\partial v_0}{\partial \nu} - v_0\frac{\partial u}{\partial \nu} \right) \Sigma(dy).$$

The boundary $\partial\Omega_{\varepsilon,R}$ is the union of the spheres S_ε and S_R with radius ε and R. On the one hand,

$$\int_{S_\varepsilon} u\frac{\partial v_0}{\partial \nu}\Sigma(dy) = \frac{1}{\varpi_n \varepsilon^{n-1}} \int_{S_\varepsilon} u(y)\Sigma(dy)$$

$$= \int_S u(\varepsilon z)\sigma(dz),$$

and

$$\lim_{\varepsilon \to 0} \int_S u(\varepsilon z)\sigma(dz) = u(0).$$

Furthermore

$$\int_{S_\varepsilon} v_0\frac{\partial u}{\partial \nu}\Sigma(dy) = O(\varepsilon) \quad (\varepsilon \to 0).$$

On the other hand,

$$\int_{S_R} u \frac{\partial v_0}{\partial \nu} \Sigma(dy) = -\frac{1}{\varpi_n R^{n-1}} \int_{S_R} u(y) \Sigma(dy)$$

$$= -\int_S u(Rz)\sigma(dz).$$

One obtains the statement as $\varepsilon \to 0$, by observing that the function v_0 is integrable. $\qquad\square$

As a result one can see that a harmonic function has the *mean value property*. Let the function u be harmonic on an open set Ω in $\mathbb{R}^n$. If the closed ball $\overline{B(x_0, r)}$ is contained in Ω, then

$$\int_S u(x_0 + rz)\sigma(dz) = u(x_0).$$

From the mean value property one can deduce the maximum principle. Let Ω be a bounded open set with boundary $\partial\Omega$, and let F be a continuous function on $\overline{\Omega}$, which is harmonic in Ω. The maximum principle says that, for $x \in \Omega$,

$$\max_{x \in \overline{\Omega}} F(x) = \max_{y \in \partial\Omega} F(y).$$

Furthermore, if Ω is connected, and if the maximum of F on $\overline{\Omega}$ is reached at a point in Ω, then F is constant. Let Ω be a bounded open set in $\mathbb{R}^n$. A *Green's kernel* for Ω is a function on

$$\{(x, y) \in \Omega \times \Omega \mid x \neq y\}$$

of the following form

$$G(x, y) = \frac{1}{(n-2)\varpi_n} \frac{1}{\|x - y\|^{n-2}} - H(x, y).$$

For x fixed in Ω, the function $H_x(y) = H(x, y)$ is harmonic in Ω; the function $G_x(y) = G(x, y)$ extends as a continuous function on $\overline{\Omega} \setminus \{x\}$ and vanishes on the boundary $\partial\Omega$. If it exists, the Green kernel is unique. We will assume that the boundary Ω is $\mathcal{C}^1$, that the function H is $\mathcal{C}^2$ on $\overline{\Omega} \times \overline{\Omega}$. Let Ω be a bounded open set in $\mathbb{R}^n$ with a $\mathcal{C}^1$ boundary which admits a Green's kernel G. If u is $\mathcal{C}^2$ on $\overline{\Omega}$, then

$$-\int_{\partial\Omega} \frac{\partial}{\partial \nu_z} G(x, z) u(z) \Sigma(dz) = u(x) + \int_\Omega G(x, y) \Delta u(y) \lambda(dy).$$

The proof of this relation is similar to that of Proposition 9.7.3.

The *Dirichlet problem* is as follows: given a continuous function f on the boundary $\partial\Omega$ of the bounded open set Ω, determine a continuous function F

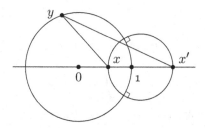

Figure 6

on $\overline{\Omega}$ which is harmonic in Ω and agrees with f on $\partial\Omega$. The solution, if it exists, is unique. This follows from the maximum principle. Under the above assumptions the solution of the Dirichlet problem admits the following integral representation:

$$F(x) = \int_{\partial\Omega} P(x, y)f(y)\Sigma(dy).$$

The kernel P, which is defined on $\Omega \times \partial\Omega$, is called the *Poisson kernel* of the open set Ω. It is given by

$$P(x, y) = -\frac{\partial}{\partial \nu_y} G(x, y).$$

Observe that, for x fixed in Ω, the function $H_x(y) = H(x, y)$ is the solution of the Dirichlet problem for the boundary data

$$f_x(y) = \frac{1}{(n - 2)\varpi_n}\frac{1}{\|x - y\|^{n-2}}.$$

Consider the case of Ω being the unit ball $B(0, 1)$. Then $\partial\Omega$ is the unit sphere S. In order to determine the Green kernel of Ω one uses the following geometric property. Let $x \in \Omega$ and x' its inverse by the inversion around 0:

$$x' = \frac{x}{\|x\|^2}.$$

Then, for every $y \in S$,

$$\frac{\|x - y\|}{\|x' - y\|} = \|x\|.$$

Therefore

$$H(x, y) = \frac{1}{(n - 2)\varpi_n}\frac{1}{\|x\|^{n-2}}\frac{1}{\|x' - y\|^{n-2}},$$

and the Green kernel of the ball is given by

$$G(x, y) = \frac{1}{(n-2)\varpi_n} \left(\frac{1}{\|x-y\|^{n-2}} - \frac{1}{\|x\|^{n-2}} \frac{1}{\|x'-y\|^{n-2}} \right) \quad (x, y \in \Omega).$$

If $x = 0$ one replaces the product $\|x\|\|x'-y\|$ in the above formulae by its limit as x goes to 0 which is equal to one. The Poisson kernel is equal to

$$P(x, y) = -\frac{d}{dt} G(x, ty)\Big|_{t=1} \quad (x \in \Omega, y \in S).$$

Hence one gets

$$P(x, y) = \frac{1}{\varpi_n} \frac{1-\|x\|^2}{\|x-y\|^n}.$$

If $x = ru$ is the polar decomposition of x, then

$$P(ru, v) = P_r\big((u|v)\big) \quad (0 \le r < 1, \ u, v \in S),$$

where

$$P_r(t) = \frac{1-r^2}{(1-2rt+r^2)^{n/2}}.$$

Let f be a continuous function on S. The solution F of the Dirichlet problem with the boundary data f is given by

$$F(ru) = \int_S P(ru, v) f(v) \sigma(dv) = \int_S P_r\big((u|v)\big) f(v) \sigma(dv).$$

If f is a spherical harmonic of degree m, $f \in \mathcal{Y}_m$, then F is a m-homogeneous harmonic polynomial whose restriction to S is equal to f:

$$F(ru) = r^m f(u).$$

By the Funk–Hecke Theorem (Theorem 9.5.3) it follows that

$$\widehat{P_r}(m) = r^m.$$

Proposition 9.7.4 *For* $0 < r < 1$,

(i) $\quad \dfrac{\Gamma\left(\frac{n}{2}\right)}{\sqrt{\pi}\,\Gamma\left(\frac{n-1}{2}\right)} \displaystyle\int_{-1}^{1} \dfrac{1-r^2}{(1-2rt+r^2)^{n/2}} P_m(t)(1-t^2)^{(n-3)/3} dt = r^m,$

(ii) $\quad P_r(t) = \dfrac{1-r^2}{(1-2rt+r^2)^{n/2}} = \displaystyle\sum_{m=0}^{\infty} d_m r^m p_m(t).$

Proof. (ii) follows from Proposition 9.5.4. $\qquad\qquad\qquad\qquad\qquad\qquad\square$

If f is a continuous function on S, for $0 \le r < 1$,

$$F(ru) = \int_S \left(\sum_{m=0}^{\infty} d_m r^m P_m\big((u|v)\big) \right) f(v)\sigma(dv).$$

The convergence is uniform in $v \in S$, and the series can be integrated termwise:

$$F(ru) = \sum_{m=0}^{\infty} d_m r^m \int_S P_m\big((u|v)\big) f(v)\sigma(dv).$$

This can be written, for $\|x\| < 1$,

$$F(x) = \sum_{m=0}^{\infty} F_m(x),$$

with

$$F_m(x) = d_m \int_S \|x\|^m P_m\left(\left(\frac{x}{\|x\|} \Big| v \right) \right) f(v)\sigma(dv).$$

By observing that, for v fixed in S, the function

$$x \mapsto \|x\|^m P_m\left(\left(\frac{x}{\|x\|} \Big| v \right) \right)$$

is a m-homogeneous harmonic polynomial, one can deduce from this integral representation that F_m is a m-homogeneous harmonic polynomial: $F_m \in \mathcal{H}_m$. The series

$$F(x) = \sum_{m=1}^{\infty} F_m(x)$$

converges uniformly on every ball $\overline{B(0,r)}$ with radius $r < 1$.

Theorem 9.7.5 *Let F be a harmonic function in the open ball $B(0, R)$. Then F admits an expansion as a series of homogeneous harmonic polynomials:*

$$F(x) = \sum_{m=0}^{\infty} F_m(x),$$

where $F_m \in \mathcal{H}_m$. The series converges uniformly on every ball with centre 0 and radius $r < R$. This expansion is unique.

Proof. (a) *Uniqueness.* Assume that such an expansion exists. Then, for $r < R$, $u \in S$,

$$F(ru) = \sum_{m=1}^{\infty} r^m F_m(u).$$

Uniqueness follows from the orthogonality of the subspaces $\mathcal{Y}_m$ in $L^2(S, \sigma)$.

(b) *Existence.* Fix $\rho < R$, and put

$$F^\rho(x) = F(\rho x).$$

The function F^ρ is continuous on the closed ball $\overline{B(0, 1)}$, and harmonic in the open ball $B(0, 1)$. By applying to F^ρ what was said above, one gets

$$F(x) = \sum_{m=0}^{\infty} F_m(x),$$

where $F_m \in \mathcal{H}_m$ is given by

$$F_m(x) = d_m \int_S \left\| \frac{x}{\rho} \right\|^m P_m \left(\left(\frac{x}{\|x\|}, v \right) \right) F(\rho v) \sigma(dv).$$

The series converges uniformly on every ball $\overline{B(0, r)}$ with radius $r < \rho$. □

For instance consider, for $a = Re_n$, the function

$$F(x) = \frac{1}{\|x - a\|^{n-2}}.$$

The function F is harmonic on $B(0, R)$, hence admits an expansion as a series of homogeneous harmonic polynomials:

$$F(x) = \sum_{m=0}^{\infty} F_m(x).$$

The function F is invariant under the group $K = \{g \in SO(n) \mid e_n g = e_n\}$. By uniqueness of the expansion, the polynomial F_m is K-invariant for every m. Therefore

$$F_m(x) = c_m \varphi_m(x) = c_m \|x\|^m P_m \left(\frac{x_n}{\|x\|} \right).$$

For $x = re_n$ ($r < R$) the expansion can be written

$$\frac{1}{(R - r)^{n-2}} = \sum_{m=0}^{\infty} c_m r^m.$$

From the power series expansion

$$(1 - z)^{-(n-2)} = \sum_{m=0}^{\infty} \binom{m + n - 3}{n - 3} z^m,$$

it follows that

$$c_m = R^{n-2} \binom{m + n - 3}{n - 3}.$$

Hence we get the following expansion.

Proposition 9.7.6

$$(1 - 2rt + t^2)^{-(n-2)/2} = \sum_{m=0}^{\infty} \binom{m+n-3}{n-3} r^m p_m(t).$$

The *Gegenbauer polynomials* C_m^ν ($\nu > \frac{1}{2}$) are usually defined by their generating function:

$$(1 - 2rt + r^2)^{-\nu} = \sum_{m=0}^{\infty} r^m C_m^\nu(t).$$

Hence

$$p_m(t) = \frac{(n-3)! m!}{(m+n-3)!} C_m^{(n-2)/2}(t).$$

9.8 An integral transform

In this section we consider the group $G = SU(2)$ and the subgroup K consisting of diagonal matrices:

$$K = \left\{ \begin{pmatrix} e^{i\varphi} & 0 \\ 0 & e^{-i\varphi} \end{pmatrix} \mid \varphi \in \mathbb{R} \right\}.$$

Let $\mathcal{C}^b(G)$ denote the space of K-biinvariant continuous functions h on G:

$$h(k_1 g k_2) = h(g) \quad (k_1, k_2 \in K).$$

From the computation of the product

$$
\begin{aligned}
k_1 g k_2 &= \begin{pmatrix} e^{i\varphi_1} & 0 \\ 0 & e^{-i\varphi_1} \end{pmatrix} \begin{pmatrix} \alpha & \beta \\ -\bar{\beta} & \bar{\alpha} \end{pmatrix} \begin{pmatrix} e^{i\varphi_2} & 0 \\ 0 & e^{-i\varphi_2} \end{pmatrix} \\
&= \begin{pmatrix} e^{i(\varphi_1+\varphi_2)}\alpha & e^{i(\varphi_1-\varphi_2)}\beta \\ -e^{-i(\varphi_1-\varphi_2)}\bar{\beta} & e^{-i(\varphi_1+\varphi_2)}\bar{\alpha} \end{pmatrix},
\end{aligned}
$$

it follows that a K-biinvariant function only depends on $|\alpha|$, and there exists a function h_0 on $[-1, 1]$ such that

$$h(g) = h_0(2|\alpha|^2 - 1), \quad \text{if } g = \begin{pmatrix} \alpha & \beta \\ -\bar{\beta} & \bar{\alpha} \end{pmatrix}.$$

In particular,

$$h(g) = h_0(\cos 2\theta), \quad \text{if } g = \begin{pmatrix} \cos\theta & \sin\theta \\ -\sin\theta & \cos\theta \end{pmatrix}.$$

Recall that the convolution product of two functions f_1 and f_2 on G is defined as

$$f_1 * f_2(g) = \int_G f_1(\gamma) f_2(\gamma^{-1}g) \mu(d\gamma).$$

The space $\mathcal{C}^\flat$ is a convolution algebra: if $h_1, h_2 \in \mathcal{C}(G)^\flat$, then $h_1 * h_2$ is as well. This convolution algebra is commutative. In fact for a function f on G one puts $\check{f}(g) = f(g^{-1})$. Then

$$(f_1 * f_2)^{\check{}} = \check{f}_2 * \check{f}_1.$$

On the other hand, if $h \in \mathcal{C}^\flat$, then $\check{h} = h$. The commutativity of the convolution algebra $\mathcal{C}(G)^\flat$ follows.

We saw that the adjoint representation $\tau = \mathrm{Ad}$ is a morphism from $SU(2)$ onto $SO(3)$, and is unitary if $\mathfrak{g} = \mathfrak{su}(2)$ is endowed with the Euclidean inner product

$$(X|Y) = \tfrac{1}{2}\,\mathrm{tr}(XY^*).$$

In this section we consider the following orthogonal basis of $\mathfrak{g} \simeq \mathbb{R}^3$:

$$X_1 = \begin{pmatrix} 0 & -1 \\ 1 & 0 \end{pmatrix}, \quad X_2 = \begin{pmatrix} 0 & i \\ i & 0 \end{pmatrix}, \quad X_3 = \begin{pmatrix} i & 0 \\ 0 & -i \end{pmatrix}.$$

Let $S = S^2$ denote the unit sphere in $\mathfrak{g}$. We choose on S the base point $x^0 = X_3$ ('north pole'). To a function f on S one associates the function $\tilde{f}$ which is defined on G by

$$\tilde{f}(g) = f\big(x^0 \tau(g)\big).$$

The function $\tilde{f}$ is left K-invariant, and this relation defines an isomorphism from $\mathcal{C}(S)$ onto $\mathcal{C}(K\backslash G)$, the space of continuous functions on G which are left K-invariant.

Proposition 9.8.1 *To a function $h \in \mathcal{C}(G)^\flat$ one associates the operator H on the space $\mathcal{C}(S)$ which is defined by*

$$\tilde{H}f = h * \tilde{f}.$$

Then the operator H belongs to the algebra $\mathcal{A}$, and

$$Hf(x) = \int_S h_0\big((x|y)\big) f(y)\sigma(dy).$$

It follows that the algebra $\mathcal{A}$, we introduced in Section 9.5, is isomorphic to the convolution algebra $\mathcal{C}(G)^\flat$.

Proof. Let A be the operator in the algebra $\mathcal{A}$ defined by

$$Af(x) = \int_S h_0\big((x|y)\big) f(y)\sigma(dy).$$

Since the operators H and A commute with the action of $SO(3)$, it is enough to show that, for every function $f \in C(S)$,

$$Hf(x^0) = Af(x^0).$$

Let us compute $Hf(x^0)$ using Euler angles (Proposition 7.4.1):

$$Hf(x^0) = (h * \tilde{f})(e) = \int_G h(g^{-1}) f\big(x^0\tau(g)\big)\mu(dg)$$

$$= \frac{1}{\pi}\int_0^{\pi/2} \sin 2\theta \, d\theta \int_0^\pi d\varphi \, h_0(\cos 2\theta) f(\sin 2\theta \cos 2\varphi, \sin 2\theta \sin 2\varphi, \cos 2\theta).$$

By putting $2\theta = \theta'$, $2\varphi = \varphi'$, and by using the integration formula on $S = S^2$ in terms of spherical coordinates one gets

$$Hf(x^0) = \int_S h(x_3) f(x)\sigma(dx). \qquad \square$$

Let f be a continuous function on G which is central. One associates to it the function $h = \mathcal{W}f$ which is defined on G by

$$h(g) = \int_K f(gk)\mu_0(dk).$$

The function h is continuous and K-biinvariant. The functions f and h can be written

$$f(g) = f_0(\mathrm{Re}\,\alpha), \quad h(g) = h_0(2|\alpha|^2 - 1), \quad \text{if } g = \begin{pmatrix} \alpha & \beta \\ -\bar{\beta} & \bar{\alpha} \end{pmatrix},$$

with functions f_0 and h_0 defined on $[-1,1]$.

Proposition 9.8.2 *With the preceding notation, the transform* $\mathcal{W}_0 : f_0 \mapsto h_0$ *can be written*

$$h_0(\cos 2\theta) = \frac{1}{\pi}\int_\theta^{\pi-\theta} f_0(\cos\psi)\frac{\sin\psi}{\sqrt{\cos^2\theta - \cos^2\psi}}d\psi.$$

Proof. Take

$$g = \begin{pmatrix} \cos\theta & \sin\theta \\ -\sin\theta & \cos\theta \end{pmatrix}, \quad k = \begin{pmatrix} e^{i\varphi} & 0 \\ 0 & e^{-i\varphi} \end{pmatrix}.$$

Then

$$gk = \begin{pmatrix} \cos\theta e^{i\varphi} & \sin\theta e^{-i\varphi} \\ -\sin\theta e^{i\varphi} & \cos\theta e^{-i\varphi} \end{pmatrix},$$
$$f(gk) = f_0(\cos\theta\cos\varphi),$$
$$h(g) = h_0(\cos 2\theta),$$

and we get

$$h_0(\cos 2\theta) = \frac{1}{2\pi}\int_0^{2\pi} f_0(\cos\theta\cos\varphi)d\varphi$$
$$= \frac{1}{\pi}\int_0^{\pi} f_0(\cos\theta\cos\varphi)d\varphi.$$

Observe that we may assume that $0 \le \theta \le \frac{\pi}{2}$. Put

$$\cos\theta\cos\varphi = \cos\psi,$$

then

$$d\varphi = \frac{\sin\psi}{\sqrt{\cos^2\theta - \cos^2\psi}}d\psi.$$

The formula follows. $\qquad\square$

The transform $\mathcal{W}$ is equivariant with respect to the Laplace operators of $G = SU(2)$ and of the sphere S^2.

Proposition 9.8.3 *If f is a C^2 central function on G, then*

$$\mathcal{W}(\Delta_{SU(2)}f) = 4\Delta_{S^2}\mathcal{W}f.$$

Proof. The function $h = \mathcal{W}f$ is given by

$$h_0(\cos 2\theta) = \frac{1}{\pi}\int_0^{\pi} f_0(\cos\theta\cos\varphi)d\varphi.$$

We will use the formulae giving the radial parts of $\Delta_{SU(2)}$ and Δ_{S^2}:

$$L_{SU(2)} = (1 - t^2)\frac{d^2}{dt^2} - 3t\frac{d}{dt},$$
$$L_{S^2} = (1 - t^2)\frac{d^2}{dt^2} - 2t\frac{d}{dt}.$$

On the one hand,

$$\mathcal{W}(\Delta_{SU(2)}f)(\cos 2\theta)$$
$$= \frac{1}{\pi}\int_0^\pi \Big(f_0''(\cos\theta\cos\varphi)(1-\cos^2\theta\sin^2\varphi)$$
$$- 3f_0'(\cos\theta\cos\varphi)\cos\theta\cos\varphi\Big)d\varphi.$$

On the other hand, by putting $t = \cos 2\theta$, we obtain

$$4(L_{S^2}h_0)(\cos 2\theta) = \left(\frac{d^2}{d\theta^2} + 2\cot 2\theta\,\frac{d}{d\theta}\right)h_0(\cos 2\theta),$$

hence

$$4(\Delta_{S^2}(\mathcal{W}f)_0)(\cos 2\theta)$$
$$= \frac{1}{\pi}\Big(f_0''(\cos\theta\cos\varphi)(\sin\theta\cos\varphi)^2$$
$$- f_0'(\cos\theta\cos\varphi)(\cos\theta + 2\cot 2\theta\sin\theta)\cos\varphi\Big)d\varphi.$$

Therefore,

$$\mathcal{W}_0\big(\Delta_{SU(2)}f(\cos 2\theta)\big) - 4\Delta_{S^2}\mathcal{W}_0 f(\cos 2\theta)$$
$$= \frac{1}{\pi}\int_0^\pi \Big(f_0''(\cos\theta\cos\varphi)\sin^2\varphi - f_0'(\cos\theta\cos\varphi)\frac{\cos\varphi}{\cos\theta}\Big)d\varphi$$
$$= -\frac{1}{\pi}\int_0^\pi \frac{d}{d\varphi}\left(f_0'(\cos\theta\cos\varphi)\frac{\sin\varphi}{\cos\theta}\right)d\varphi = 0. \qquad \square$$

Let us consider the transform of the character χ_m: $h = \mathcal{W}(\chi_m)$. If m is odd, then $\chi_m(-g) = -\chi_m(g)$, and therefore $\mathcal{W}(\chi_m) = 0$. If m is even, $m = 2\ell$, then $h = \mathcal{W}(\chi_{2\ell})$ is a function on S^2 which is K-invariant. It is an eigenfunction of Δ_{S^2}. In fact,

$$\Delta_{SU(2)}\chi_{2\ell} = -2\ell(2\ell + 2)\chi_{2\ell},$$

hence, by Proposition 9.8.3,

$$\Delta_{S^2}h = -\ell(\ell + 1)h.$$

The function h_0, given by

$$h_0(\cos 2\theta) = \frac{1}{\pi}\int_\theta^{\pi-\theta} \frac{\sin(2\ell + 1)\psi}{\sqrt{\cos^2\theta - \cos^2\psi}}d\psi,$$

is proportional to the spherical polynomial p_ℓ. Since

$$h_0(1) = \frac{1}{\pi}\int_0^\pi \frac{\sin(2\ell + 1)\psi}{\sin\psi}d\psi = 1,$$

we get $h_0 = p_\ell$. We have established the following formula.

Proposition 9.8.4

$$p_\ell(\cos 2\theta) = \frac{1}{\pi} \int_\theta^{\pi-\theta} \frac{\sin(2\ell+1)\psi}{\sqrt{\cos^2\theta - \cos^2\psi}} d\psi.$$

This is essentially the integral representation of the Legendre polynomials which is known as the *Dirichlet–Murphy formula*.

9.9 Heat equation

We will study the Cauchy problem for the *heat equation* on the sphere S,

$$\frac{\partial u}{\partial t} = \Delta_S u,$$

$$u(0, x) = f(x),$$

following a method similar to that of Section 8.6. We assume first that the initial data f is C^{2k} with $2k > \frac{n-1}{2}$. It is then possible to expand f as a series of spherical harmonics

$$f(x) = \sum_{m=0}^\infty f_m(x) \quad (f_m \in \mathcal{Y}_m),$$

which converges uniformly on S. The solution of the Cauchy problem is given, for $t \geq 0$, by

$$u(t, x) = \sum_{m=0}^\infty e^{-m(m+n-2)t} f_m(x).$$

One defines the *heat kernel* H, for $t > 0$, $x, y \in S$, by

$$H(t; x, y) = \sum_{m=0}^\infty d_m e^{-m(m+n-2)t} P_m\big((x|y)\big).$$

Then $u(t, x)$ is given by

$$u(t, x) = \int_S H(t; x, y) f(y)\sigma(dy).$$

Proposition 9.9.1 *The heat kernel H has the following properties:*

(i) $H(t; x, y) \geq 0$,

(ii) $\int_S H(t; x, y)\sigma(dy) = 1$,

(iii) *for every $x \in S$ and every neighbourhood V of x,*

$$\lim_{t \to 0} \int_V H(t; x, y)\sigma(dy) = 1.$$

The proof of this proposition is the same as the proof of Proposition 8.6.3 (one should write 'f is C^{2k} with $2k > \frac{n-1}{2}$' instead of 'f is C^2').

In the same way as Theorem 8.6.4 was established one can show the following.

Theorem 9.9.2 *Let f be a continuous function on S. The Cauchy problem admits a unique solution which is given, for $t > 0$, by*

$$u(t, x) = \int_S H(t; x, y) f(y) \sigma(dy).$$

Let H_t denote the operator which maps an initial data f onto the solution of the Cauchy problem at $t > 0$:

$$H_t f(x) = \int_S H(t; x, y) f(y) \sigma(dy).$$

This operator belongs to the algebra $\mathcal{A}$ we considered in Section 9.5. It is associated to the function

$$h^{(n-1)}(t; \tau) = \sum_{m=0}^{\infty} d_m e^{-m(m+n-2)t} p_m(\tau).$$

We saw in Section 8.5 that, for $n = 2$,

$$h^{(1)}(t; \cos\theta) = 1 + 2 \sum_{m=1}^{\infty} e^{-m^2 t} \cos m\theta$$

$$= \sqrt{\frac{\pi}{t}} \sum_{k=-\infty}^{\infty} e^{-(\theta - 2k\pi)^2/4t}.$$

For $n = 4$ the sphere S^3 can be identified with the group $SU(2)$ and we saw in Section 8.6 that

$$h^{(3)}(t; \cos\theta) = \sum_{m=0}^{\infty} (m + 1) e^{-m(m+2)t} \frac{\sin(m+1)\theta}{\sin\theta}$$

$$= \frac{\sqrt{\pi}}{4} \frac{e^t}{t\sqrt{t}} \sum_{k=-\infty}^{\infty} \frac{\theta - 2k\pi}{\sin\theta} e^{-(\theta - 2k\pi)^2/4t}.$$

For $n = 3$ one can deduce the function $h^{(2)}(t; \tau)$ from the function $h^{(3)}(t; \tau)$ by using the integral transform we considered in Section 9.8. In fact

$$p_\ell^{(2)}(\cos 2\theta) = \frac{1}{\pi} \int_\theta^{\pi - \theta} \frac{\sin(2\ell + 1)\psi}{\sqrt{\cos^2\theta - \cos^2\psi}} d\psi,$$

and, if m is odd,

$$\int_{\theta}^{\pi-\theta} \frac{\sin(m+1)\psi}{\sqrt{\cos^2\theta - \cos^2\psi}} d\psi = 0.$$

For $t > 0$, it is possible to justify termwise integration of the series

$$\sum_{m=0}^{\infty} (m+1)e^{-m(m+2)t} \frac{\sin(m+1)\psi}{\sin\psi}.$$

In order to do this, one establishes that

$$\int_0^{\pi} \left| \frac{\sin(m+1)\psi}{\sin\psi} \right| d\psi \sim C\ln(m).$$

One gets

$$\frac{1}{\pi} \int_{\theta}^{\pi-\theta} h^{(3)}(t;\cos\psi) \frac{\sin\psi}{\sqrt{\cos^2\theta - \cos^2\psi}} d\psi$$

$$= \sum_{\ell=0}^{\infty} (2\ell+1)e^{-4\ell(\ell+1)t} p_{\ell}^{(2)}(\cos 2\theta).$$

Finally this leads to the following integral representation for the function $h^{(2)}$:

$$h^{(2)}(t;\cos 2\theta) = \frac{1}{\pi} \int_{\theta}^{\pi-\theta} h^{(3)}\left(\frac{t}{4};\cos\psi\right) \frac{\sin\psi}{\sqrt{\cos^2\theta - \cos^2\psi}} d\psi.$$

9.10 Exercises

1. *Stereographic projection.* Let S be the unit sphere in $\mathbb{R}^n$ with centre 0. Let ϕ be the map from $S \setminus \{-e_n\}$ onto $\mathbb{R}^{n-1}$ which, to a point x in the sphere S, associates the intersection point u of the straight line joining $-e_n$ and x with the horizontal hyperplane $\simeq \mathbb{R}^{n-1}$ (with equation $x_n = 0$).
   (a) Show that

$$u_i = \frac{x_i}{1 + x_n} \quad (1 \le i \le n-1),$$

$$x_i = \frac{2u_i}{1 + \|u\|^2} \quad (1 \le i \le n-1),$$

$$x_n = \frac{1 - \|u\|^2}{1 + \|u\|^2}.$$

(b) Let ω be the differential form of degree $n-1$ defined on S by

$$\omega = \sum_{i=1}^{n} (-1)^{i-1} x_i dx_1 \wedge \cdots \wedge \widehat{dx_i} \wedge \cdots \wedge dx_n.$$

Show that

$$(\phi^{-1})^* \omega = (-1)^{n-1} 2^{n-1} \frac{du_1 \wedge \cdots \wedge du_{n-1}}{(1 + \|u\|^2)^{n-1}}.$$

Let f be an integrable function on $\mathbb{R}^n$. Show that

$$\int_S f(x) \sigma(dx) = \frac{2^{n-1}}{\varpi_n} \int_{\mathbb{R}^{n-1}} f(\phi^{-1}(u)) \frac{du_1 \ldots du_{n-1}}{(1 + \|u\|^2)^{n-1}}$$

(σ is the normalised uniform measure on S).

2. Let $B(0, R)$ be the open ball with centre 0 and radius $R > 0$ in $\mathbb{R}^3$. One defines the sequence of the functions (v_k) $(k \geq 0)$ by

$$v_k(x) = \begin{cases} \dfrac{1}{4\pi(2k+1)!} \dfrac{(R - \|x\|)^{2k+1}}{R\|x\|} & \text{if } 0 < \|x\| < R, \\ 0 & \text{if } \|x\| \geq R. \end{cases}$$

(a) Show that, for $0 < \|x\| < R$,

$$\Delta v_{k+1} = v_k.$$

(b) Let the function u be C^{2m+2} on $\overline{B(0, R)}$. Show that

$$\int_S u(Rz) \sigma(dz) = \sum_{k=0}^{m} \frac{R^{2k}}{(2k+1)!} \Delta^k u(0)$$
$$+ \int_{B(0,R)} \Delta^{m+1} u(x) v_m(x) \lambda(dx)$$

(S is the unit sphere with centre 0 in $\mathbb{R}^3$, and σ is the normalised uniform measure on S).

Hint. Observe that, on the sphere S_R with centre 0 and radius R, $v_k = 0$, and, for $k \geq 1$, $\partial v_k / \partial v = 0$.

(c) Let the function u be C^{∞} on an open set $\Omega \subset \mathbb{R}^3$ which is an eigenfunction of the Laplace operator:

$$\Delta u = \lambda u.$$

Show that u has the following mean value property: if the closed ball $\overline{B(x_0, r)}$ is contained in Ω,

$$\int_S u(x_0 + rz) \sigma(dz) = \frac{\sinh(R\sqrt{\lambda})}{R\sqrt{\lambda}} u(x_0).$$

3. Consider on $\mathbb{R}^n$ the Cauchy problem for the heat equation:

$$\frac{\partial u}{\partial t} = \Delta u, \quad u(0, x) = f(x).$$

Assume that the initial data f is a radial function:

$$f(x) = f_0(\|x\|),$$

where f_0 is a continuous function on $[0, \infty[$. Then the solution u is radial as well:

$$u(t, x) = u_0(t, \|x\|).$$

Show that

$$u_0(t, r) = \int_0^\infty H_0(t, r, \rho) f_0(\rho) \rho^{n-1} d\rho,$$

with

$$H_0(t, r, \rho) = \frac{1}{(2\sqrt{\pi t})^n} e^{-(r^2+\rho^2)/4t} \mathcal{I}_{(n-2)/2}\left(\frac{r\rho}{2t}\right),$$

where $\mathcal{I}_\nu$ is the modified Bessel function:

$$\mathcal{I}_\nu(\tau) = \mathcal{J}_\nu(i\tau).$$

In particular, for $n = 3$,

$$H_0(t, r, \rho) = \frac{1}{(2\sqrt{\pi t})^3} e^{-(r^2+\rho^2)/4t} \frac{2t}{r\rho} \sinh\left(\frac{r\rho}{2t}\right).$$

4. (a) Let ρ be a positive continuous 2π-periodic function on $\mathbb{R}$. One associates to the function ρ the set E in the plane $\mathbb{R}^2$ defined by

$$E = \{x = (r\cos\theta, r\sin\theta) \mid 0 \le r \le \rho(\theta)\}.$$

Show that the area of E is given by

$$\mathrm{area}(E) = \pi \int_0^{2\pi} \rho(\theta)^2 d\theta.$$

(b) Let ρ be a positive continuous function on the unit sphere S in $\mathbb{R}^3$. Similarly one associates to the function ρ the set Q in $\mathbb{R}^3$ defined by

$$Q = \{x = ru \mid 0 \le r \le \rho(u), \ u \in S\}.$$

For $v \in S$, let P_v denote the plane passing through 0 and orthogonal to v, and $Q \cap P_v$ is the intersection of Q by the plane P_v. The aim of this exercise is to show that the set Q is determined by the areas of its intersections with the planes passing through 0, that is by the function:

$$v \mapsto \mu_v(Q \cap P_v), \quad S \to \mathbb{R},$$

where μ_v is the Euclidean Lebesgue measure on the plane P_v.

(c) Consider the transform $\mathcal{R}$ which maps a function $f \in \mathcal{C}(S)$ onto the function $\mathcal{R}f$ defined on S by

$$(\mathcal{R}f)(v) = \int_{P_v} f(u)\mu_v(du).$$

This is a continuous operator on the space $\mathcal{C}(S)$. The image of an odd function is zero, and that of an even function is even as well. Using Schur's Lemma show that the space $\mathcal{Y}_{2k}$ is an eigenspace of $\mathcal{R}$ for the eigenvalue

$$\lambda_{2k} = 2\pi p_{2k}(0) \neq 0 \quad \left(p_{2k}(0) = (-1)^k \frac{1 \cdot 3 \cdots (2k-1)}{2 \cdot 4 \cdots 2k} \right).$$

Then show that the operator $\mathcal{R}$, acting on the space $\mathcal{C}^0(S)$ consisting of continuous and even functions on S, is injective.

(d) Conclude (b).

See: A. A. Kirillov (1976). *Elements of the Theory of Representations*. Springer (§17.1, p. 272).

10

Analysis on the spaces of symmetric and Hermitian matrices

In this chapter we consider the space $\mathcal{V} = Sym(n, \mathbb{R})$ of $n \times n$ symmetric matrices on which the orthogonal group $K = O(n)$ acts, or $\mathcal{V} = Herm(n, \mathbb{C})$ of $n \times n$ Hermitian matrices with the action of the unitary group $K = U(n)$. The group K acts on $\mathcal{V}$ by the transformations

$$x \mapsto kxk^* \quad (k \in K).$$

If a function f on $\mathcal{V}$ is K-invariant, then $f(x)$ only depends on the eigenvalues $\lambda_1, \ldots, \lambda_n$ of x,

$$f(x) = F(\lambda_1, \ldots, \lambda_n),$$

where the function F, defined on $\mathbb{R}^n$, is symmetric, that is invariant under permutation. We will see in Section 10.1 how the integral of f over $\mathcal{V}$ reduces to an integral over $\mathbb{R}^n$: this is the Weyl integration formula. The Laplace operator Δf of such an invariant function f is given by a formula involving a differential operator acting on the variables $\lambda_1, \ldots, \lambda_n$: this is the radial part of the Laplace operator, whose formula will be given in Section 10.2. In the case of $\mathcal{V} = Herm(n, \mathbb{C})$ we will see that the Fourier transform of such an invariant function is given by a formula involving the Fourier transform on $\mathbb{R}^n$.

10.1 Integration formulae

First we establish the *Weyl integration formula* for the space $Sym(n, \mathbb{R})$ of symmetric matrices. From that we will see how the Haar measure of $GL(n, \mathbb{R})$ is given in terms of the polar decomposition of matrices. We will then consider the case of $\mathcal{V} = Herm(n, \mathbb{C})$.

Every matrix $X \in \mathcal{V} = Sym(n, \mathbb{R})$ can be diagonalised in an orthogonal basis; this can be written

$$X = kak^T,$$

where k is an orthogonal matrix, $k \in K = O(n)$, and the matrix a is diagonal, $a = \mathrm{diag}(a_1, \ldots, a_n)$. The numbers $a_1, \ldots, a_n$ are the eigenvalues of X. Recall that this decomposition is not unique. Let A denote the space of diagonal matrices, and let λ be a Lebesgue measure on $\mathcal{V}$, and α a Haar measure of K.

Theorem 10.1.1 (Weyl integration formula) *There exists a constant $C > 0$ such that, if f is an integrable function on $\mathcal{V} = Sym(n, \mathbb{R})$, then*

$$\int_{\mathcal{V}} f(X)\lambda(dX) = C \int_{K \times A} f(kak^T)d\alpha(k) \prod_{i<j} |a_i - a_j| da_1 \ldots da_n.$$

If the Haar measure α of K is normalised, and if the Lebesgue measure λ is chosen so that

$$\lambda(dX) = \prod_{i \le j} dx_{ij},$$

then

$$C = c_n = \frac{\pi^{n(n+1)/4}}{n! \prod_{i=1}^{n} \Gamma\left(\frac{i}{2}\right)}$$

(see Exercise 2).

Let f be an integrable function on $\mathcal{V} = Sym(n, \mathbb{R})$ which is K-invariant:

$$f(kXk^T) = f(X) \quad (k \in K).$$

Such a function only depends on the eigenvalues of X: there exists a function F on $\mathbb{R}^n$ such that

$$f(X) = F(\lambda_1, \ldots, \lambda_n),$$

where $\lambda_1, \ldots, \lambda_n$ are the eigenvalues of X. The function F is invariant under the permutation group:

$$F\big(\lambda_{\sigma(1)}, \ldots, \lambda_{\sigma(n)}\big) = F(\lambda_1, \ldots, \lambda_n) \quad (\sigma \in \mathfrak{S}_n).$$

From Theorem 10.1.1, it follows that

$$\int_{\mathcal{V}} f(X)\lambda(dX) = C \int_{\mathbb{R}^n} F(\lambda_1, \ldots, \lambda_n) \prod_{i<j} |\lambda_i - \lambda_j| d\lambda_1 \ldots d\lambda_n.$$

Proof. (a) Let us consider the map

$$\varphi : K \times A \to \mathcal{V}, \quad (k, a) \mapsto kak^T,$$

and compute its differential. A tangent vector to K at k can be written kU where $U \in \mathfrak{k} = Skewsym(n, \mathbb{R})$:

$$(D\varphi)_{(k,a)}(kU, 0) = \left.\frac{d}{dt}\right|_{t=0} k \exp(tU)a \exp(-tU)k^T = k(Ua - aU)k^T.$$

For $Y \in A$,

$$(D\varphi)_{(k,a)}(0, Y) = \left.\frac{d}{dt}\right|_{t=0} k(a + tY)k^T = kYk^T.$$

If $U = (u_{ij})$, then $X = Ua - aU = (x_{ij})$ with

$$x_{ij} = (a_j - a_i)u_{ij}.$$

Let A' denote the set of diagonal matrices $a = \mathrm{diag}(a_1, \ldots, a_n)$ for which $a_i \neq a_j$ if $i \neq j$, and let $\mathcal{V}'$ denote the set of symmetric matrices with distinct eigenvalues. The map $\varphi : K \times A' \to \mathcal{V}'$ is a covering of order $2^n n!$. In fact in a point (k, a) of $K \times A'$ the differential of φ is invertible. Furthermore, in the diagonalisation of a symmetric matrix whose eigenvalues are distinct, the eigenvalues are determined up to order, and the orthonormal eigenvectors up to sign.

(b) Let ω be a skew q-linear form on $\mathcal{V}$ ($q = \dim \mathcal{V} = \frac{1}{2}n(n+1)$). Let $U_1, \ldots, U_p$ be p vectors in $\mathfrak{k}$ ($p = \dim \mathfrak{k} = \frac{1}{2}n(n-1)$), and $Y_1, \ldots, Y_n$ be n vectors in A:

$$\begin{aligned}
&\varphi^* \omega_{(k,a)}(kU_1, \ldots, kU_p, Y_1, \ldots, Y_n) \\
&= \omega\big((D\varphi)_{(k,a)}(kU_1), \ldots, (D\varphi)_{(k,a)}(kU_p), (D\varphi)_{(k,a)}Y_1, \ldots, (D\varphi)_{(k,a)}Y_n\big) \\
&= \omega\big(k(U_1 a - aU_1)k^T, \ldots, k(U_p a - aU_p)k^T, kY_1 k^T, \ldots, kY_n k^T\big) \\
&= \pm\omega(U_1 a - aU_1, \ldots, U_p a - aU_p, Y_1, \ldots, Y_n).
\end{aligned}$$

For the last equality we used the fact that the transformation $X \mapsto kXk^T$ ($k \in K$) has determinant ± 1.

The space $\mathcal{V}$ can be decomposed as $\mathcal{V}_0 \oplus A$, where

$$\mathcal{V}_0 = \{X = (x_{ij}) \mid x_{ii} = 0 \ (i = 1, \ldots, n)\},$$

and ω can be written $\omega = \omega_1 \otimes \omega_2$, where ω_1 is a skew p-linear form on $\mathcal{V}_0$, and ω_2 is a skew n-linear form on A. By considering the bases $\{E_{ij} - E_{ji}\}_{i<j}$ of $\mathfrak{k} = Skewsym(n, \mathbb{R})$, and $\{E_{ij} + E_{ij}\}_{i<j}$ of $\mathcal{V}_0$, one can show that there exists a constant $C \neq 0$ such that

$$\omega_1(U_1 a - aU_1, \ldots, U_p a - aU_p) = C \prod_{i<j}(a_j - a_i)\tilde{\omega}_1(U_1, \ldots, U_p),$$

where $\tilde{\omega}_1$ is a skew p-linear form on $\mathfrak{k}$.

Let κ be the differential form on K such that $\kappa_e = \tilde{\omega}_1$. From what was said above it follows that

$$\varphi^* \omega_{(k,a)}(kU_1, \ldots, kU_m, Y_1, \ldots, Y_n)$$
$$= C \prod_{i<j}(a_j - a_i)\kappa(kU_1, \ldots, kU_p)\omega_2(Y_1, \ldots, Y_n).$$

Finally, observing that $\mathcal{V} \setminus \mathcal{V}'$ has measure zero, the statement follows. $\qquad \square$

By Theorem 1.4.1 and Corollary 2.1.2 every matrix g in $G = GL(n, \mathbb{R})$ decomposes as

$$g = k \exp X,$$

with $k \in K = O(n)$, $X \in \mathcal{V} = Sym(n, \mathbb{R})$, and the map

$$\varphi : K \times \mathcal{V} \to G, \ (k, X) \mapsto k \exp X$$

is a diffeomorphism:

$$[\mathfrak{k}, \mathcal{V}] \subset \mathcal{V},$$
$$[\mathcal{V}, \mathcal{V}] \subset \mathfrak{k}.$$

Hence, for $X \in \mathcal{V}$,

$$(\text{ad } X)^k \mathcal{V} \subset \mathfrak{k}$$

if k is odd, and

$$(\text{ad } X)^k \mathcal{V} \subset \mathcal{V}$$

if k is even.

By Theorem 2.1.4, for $X, Y \in \mathfrak{g}$,

$$(D \exp)_X Y = \left.\frac{d}{dt}\right|_{t=0} \exp(X + tY) = \exp X \sum_{k=0}^{\infty} \frac{(-1)^k}{(k+1)!}(\text{ad } X)^k Y.$$

If $X, Y \in \mathcal{V}$ one can write (by transposing both sides)

$$(D \exp)_X Y = Z \exp X,$$

with

$$Z = \sum_{k=0}^{\infty} \frac{1}{(k+1)!}(\text{ad } X)^k Y = Z_1 + Z_2,$$

where

$$Z_1 = \sum_{\ell=0}^{\infty} \frac{1}{(2\ell+2)!} (\mathrm{ad}\, X)^{(2\ell+1)} Y \in \mathfrak{k},$$

$$Z_2 = \sum_{\ell=0}^{\infty} \frac{1}{(2\ell+1)!} (\mathrm{ad}\, X)^{2\ell} Y \in \mathcal{V}.$$

The element Z_2 can be written

$$Z_2 = \frac{\sinh \mathrm{ad}\, X}{\mathrm{ad}\, X} Y.$$

Put

$$B(X) = \frac{\sinh \mathrm{ad}\, X}{\mathrm{ad}\, X},$$

and let $J(X)$ denote the determinant of the restriction of $B(X)$ to $\mathcal{V}$.

Theorem 10.1.2 *Let μ be a Haar measure on G, α a Haar measure on K, and λ a Lebesgue measure on $\mathcal{V} = Sym(n, \mathbb{R})$. There exists a constant $C > 0$ such that, if f is an integrable function on G,*

$$\int_G f(g)\mu(dg) = C \int_{K \times \mathcal{V}} f(k \exp X) J(X) \alpha(dk) \lambda(dX).$$

Proof. Let ω be a differential form of degree $m = \dim G \ (= n^2)$ on G which is left and right G-invariant. Let $U_1, \ldots, U_p$ be p vectors in $\mathfrak{k}$, and $Y_1, \ldots, Y_q$ be q vectors in $\mathcal{V}$. Then

$$\varphi^* \omega_{(k,X)}(kU_1, \ldots, kU_p, Y_1, \ldots, Y_q)$$
$$= \omega_{k \exp X}\big(kU_1 \exp X, \ldots, kU_p \exp X, k(D\exp)_X Y_1, \ldots, k(D\exp)_X Y_q\big).$$

We saw that we can write

$$(D\exp)_X Y_j = \big(Z_1(X, Y_j) + Z_2(X, Y_j)\big) \exp X,$$

with $Z_1(X, Y_j) \in \mathfrak{k}$, $Z_2(X, Y_j) \in \mathcal{V}$, and that

$$Z_2(X, Y_j) = B(X)Y_j.$$

For every j the vectors $U_1, \ldots, U_p, Z_1(X, Y_j)$ are linearly dependent, therefore,

$$\varphi^* \omega_{(k,X)}(kU_1, \ldots, kU_p, Y_1, \ldots, Y_p)$$
$$= \omega_e\big(U_1, \ldots, U_p, B(X)Y_1, \ldots, B(X)Y_q\big)$$
$$= J(X)\omega_e(U_1, \ldots, U_p, Y_1, \ldots, Y_q).$$

Since $\mathfrak{g}$ is the direct sum of the subspaces $\mathfrak{k}$ and $\mathcal{V}$, the skew m-linear form ω_e on $\mathfrak{g}$ can be written

$$\omega_e = \tilde{\omega}_1 \otimes \tilde{\omega}_2,$$

where $\tilde{\omega}_1$ is a skew p-linear form on $\mathfrak{k}$, and $\tilde{\omega}_2$ is a skew q-linear form on $\mathcal{V}$. To the form $\tilde{\omega}_1$ one associates a left invariant differential form ω_1 of degree p on K, and to $\tilde{\omega}_2$ one associates a translation invariant differential form ω_2 of degree q on $\mathcal{V}$. Hence

$$\varphi^*\omega = J(X)\omega_1 \otimes \omega_2.$$

The statement follows. □

From the above theorems one deduces the following.

Corollary 10.1.3 *Let μ be a Haar measure on $G = GL(n, \mathbb{R})$, and α a Haar measure on $K = O(n)$. Recall that A denotes the space of diagonal matrices. There exists a constant $C > 0$ such that, if f is an integrable function on G,*

$$\int_G f(g)\mu(dg)$$

$$= C \int_{K \times K \times A} f(k_1 \exp t k_2)\alpha(dk_1)\alpha(dk_2) \prod_{i<j} |\operatorname{sh}(t_i - t_j)| dt_1 \ldots dt_n,$$

where $t = \operatorname{diag}(t_1, \ldots, t_n)$.

Proof. If $t = \operatorname{diag}(t_1, \ldots, t_n)$, then the eigenvalues of the restriction of $(\operatorname{ad} t)^2$ to $\mathcal{V}$ are the numbers $(t_i - t_j)^2$, and the eigenvalues of the restriction of $B(t)$ to $\mathcal{V}$ are the numbers

$$\frac{\operatorname{sh}(t_i - t_j)}{(t_i - t_j)}.$$

Therefore,

$$J(t) = \prod_{i<j} \frac{\operatorname{sh}(t_i - t_j)}{(t_i - t_j)}.$$

The statement follows. □

In the case of the space $\mathcal{V} = Herm(n, \mathbb{C})$ of Hermitian matrices one can obtain similar results using the same method. There is however a difference at the following point. In the case of $\mathcal{V} = Sym(n, \mathbb{R})$, the commutant in $K = O(n)$ of the set A of real diagonal matrices is the group of diagonal orthogonal matrices. This is a finite group isomorphic to $\{-1, 1\}^n$. But in the case of $\mathcal{V} = Herm(n, \mathbb{C})$, the commutant in $K = U(n)$ of the set A is the group T of diagonal unitary matrices which is isomorphic to $\mathbb{T}^n$. Observe that the map

$$\varphi : K/T \times A' \to \mathcal{V}', \quad (\dot{u}, a) \mapsto uau^*,$$

is a covering of order $n!$.

Theorem 10.1.4 (Weyl integration formula) *Let λ be a Lebesgue measure on $\mathcal{V} = Herm(n, \mathbb{C})$, and α a Haar measure on $K = U(n)$. There exists a constant $C > 0$ such that, if f is an integrable function on $\mathcal{V}$, then*

$$\int_{\mathcal{V}} f(x)\lambda(dx) = C \int_{K \times A} f(kak^*)\alpha(dk) \prod_{i<j}(a_i - a_j)^2 da_1 \ldots da_n,$$

where $a = \mathrm{diag}(a_1, \ldots, a_n)$.

If α is the normalised Haar measure of $K = U(n)$, and if the Lebesgue measure λ on $\mathcal{V}$ is chosen as

$$\lambda(dx) = \prod_{i=1}^n dx_{ii} \prod_{i<j} d(\mathrm{Re}\, x_{ij})d(\mathrm{Im}\, x_{ij}),$$

one can show that

$$C = c'_n = \frac{\pi^{n(n-1)/2}}{\prod_{j=1}^n j!}.$$

The proof is similar to that of Theorem 10.1.1.

As in the case of the group $GL(n, \mathbb{R})$, from Theorem 10.1.4 one obtains a formula for the Haar measure of $GL(n, \mathbb{C})$ related to the polar decomposition.

Corollary 10.1.5 *Let μ be a Haar measure of $G = GL(n, \mathbb{C})$, and α a Haar measure on $K = U(n)$. There exists a constant $C > 0$ such that, if f is an integrable function on G, then*

$$\int_G f(g)\mu(dg)$$
$$= C \int_{U \times U \times A} f(k_1 \exp(t)k_2)\alpha(dk_1)\alpha(dk_2) \prod_{i<j} \mathrm{sh}^2(t_i - t_j)dt_1 \ldots dt_n,$$

where $t = \mathrm{diag}(t_1, \ldots, t_n) \in A$.

10.2 Radial part of the Laplace operator

The vector space $Sym(n, \mathbb{R})$, is endowed with a Euclidean inner product:

$$(x|y) = \text{tr}(xy).$$

Observe that

$$\|x\|^2 = \sum_{i=1}^{n} x_{ii}^2 + 2\sum_{i<j} x_{ij}^2.$$

The associated Laplace operator is given by,

$$\Delta = \sum_{i=1}^{n} \frac{\partial^2}{\partial x_{ii}^2} + \frac{1}{2}\sum_{i<j} \frac{\partial^2}{\partial x_{ij}^2}.$$

The group $K = O(n)$ acts on $\mathcal{V}$ by the orthogonal transformations:

$$T(k) : x \mapsto k \cdot x = kxk^*.$$

The Laplace operator is K-invariant in the following sense: if f is $\mathcal{C}^2$, then

$$(\Delta f) \circ T(k) = \Delta\big(f \circ T(k)\big) \quad (k \in K).$$

Let Ω be a K-invariant open set in $\mathcal{V}$. It can be described as the set of matrices

$$k \begin{pmatrix} d_1 & & \\ & \ddots & \\ & & d_n \end{pmatrix} k^T,$$

with $k \in K$, and $d = (a_1, \ldots, a_n) \in \omega$, where ω is an open set in $\mathbb{R}^n$ which is invariant under the permutation group $\mathfrak{S}_n$. Let f be a $\mathcal{C}^2$ function which is K-invariant:

$$f(uxu^T) = f(x) \quad (k \in K).$$

Such a function can be written

$$f(x) = F(\lambda_1, \ldots, \lambda_n),$$

where $\lambda_1, \ldots, \lambda_n$ are the eigenvalues of x, and the function F is defined on ω and invariant under $\mathfrak{S}_n$. The function Δf is K-invariant as well, and therefore of the form

$$\Delta f(x) = L F(\lambda_1, \ldots, \lambda_n).$$

The operator L is called the *radial part* of the Laplace operator.

Theorem 10.2.1 *Let f be a K-invariant C^2 function. Then*

$$\Delta f(x) = LF(\lambda_1, \ldots, \lambda_n),$$

where

$$LF = \sum_{i=1}^{n} \frac{\partial^2 F}{\partial \lambda_i^2} + \sum_{i<j} \frac{1}{\lambda_i - \lambda_j} \left(\frac{\partial F}{\partial \lambda_i} - \frac{\partial F}{\partial \lambda_j} \right).$$

We use Lemma 9.2.2 once more. Recall that it says the following.

Let f be a C^2 function on an open set Ω in a finite dimensional vector space V. Let U be an endomorphism of V, and $a \in V$. Let $\varepsilon > 0$ such that, for $|t| < \varepsilon$, $\exp tU \cdot a \in \Omega$. Assume that, for $|t| < \varepsilon$,

$$f(\exp tU \cdot a) = f(a).$$

Then

$$(Df)_a(U \cdot a) = 0,$$
$$(D^2 f)_a(U \cdot a, U \cdot a) + (Df)_a(U^2 \cdot a) = 0.$$

Proof of Theorem 10.2.1. Let X be a skewsymmetric matrix. For $t \in \mathbb{R}$, $\exp tX$ is an orthogonal matrix and, for every $a \in \Omega$,

$$f(\exp tX \, a \exp tX^T) = f(a).$$

By Lemma 9.2.2, applied to the endomorphism U given by $U \cdot a = Xa + aX^T$,

$$(Df)_a(Xa + aX^T) = 0,$$
$$(D^2 f)_a(Xa + aX^T, Xa + aX^T) + (Df)_a(X^2a + 2XaX^T + a(X^T)^2) = 0.$$

Take $X = E_{ij} - E_{ji}$ ($i \neq j$), $a = \text{diag}(a_1, \ldots, a_n)$. We get

$$Xa + aX^T = (a_j - a_i)(E_{ij} + E_{ji}),$$
$$X^2 + 2XaX^T + a(X^T)^2 = 2(a_j - a_i)(E_{ii} - E_{jj}),$$

and therefore

$$(a_j - a_i)^2 (D^2 f)_a(E_{ij} + E_{ji}, E_{ij} + E_{ji}) + 2(a_j - a_i)(Df)_a(E_{ii} - E_{jj}) = 0,$$

or

$$\frac{\partial^2 f}{\partial x_{ij}^2}(a) = \frac{2}{a_i - a_j} \left(\frac{\partial f}{\partial x_{ii}}(a) - \frac{\partial f}{\partial x_{jj}}(a) \right).$$

Finally,

$$\frac{\partial^2 f}{\partial x_{ii}^2}(a) = \frac{\partial^2 F}{\partial \lambda_i^2},$$

$$\frac{1}{2}\frac{\partial^2 f}{\partial x_{ij}^2}(a) = \frac{1}{\lambda_i - \lambda_j}\left(\frac{\partial F}{\partial \lambda_i} - \frac{\partial F}{\partial \lambda_j}\right) \quad (i \neq j). \qquad \square$$

The case of $\mathcal{V} = Herm(n, \mathbb{C})$ with the Euclidean inner product $(x|y) = \mathrm{tr}(xy)$ is very similar. In that case

$$\|x\|^2 = \sum_{i=1}^{n} x_{ii}^2 + 2\sum_{i<j} |x_{ij}|^2$$

$$= \sum_{i=1}^{n} x_{ii}^2 + 2\sum_{i<j}\big((\mathrm{Re}\ x_{ij})^2 + (\mathrm{Im}\ x_{ij})^2\big),$$

and the Laplace operator can be written

$$\Delta = \sum_{i=1}^{n} \frac{\partial^2}{\partial x_{ii}^2} + \frac{1}{2}\sum_{i<j}\left(\frac{\partial^2}{\partial(\mathrm{Re}\ x_{ij})^2} + \frac{\partial^2}{\partial(\mathrm{Im}\ x_{ij})^2}\right).$$

We consider the action of the unitary group $K = U(n)$ on $\mathcal{V}$ given by

$$x \mapsto kxk^* \quad (k \in U).$$

Similarly a K-invariant open set $\Omega \subset \mathcal{V}$ consists of the matrices

$$x = k \begin{pmatrix} a_1 & & \\ & \ddots & \\ & & a_n \end{pmatrix} k^*,$$

with $k \in K$, and $a = (a_1, \dots, a_n) \in \omega$, where ω is a $\mathfrak{S}_n$-invariant open set in $\mathbb{R}^n$. Let f be a K-invariant $\mathcal{C}^2$ function on Ω,

$$f(kxk^*) = f(x) \quad (k \in K).$$

The function f can be written

$$f(x) = F(\lambda_1, \dots, \lambda_n),$$

where the function F is defined on ω and is $\mathfrak{S}_n$-invariant.

Theorem 10.2.2 (i) *If f is a K-invariant $\mathcal{C}^2$ function, then*

$$(\Delta f)(x) = (LF)(\lambda_1, \dots, \lambda_n),$$

where

$$LF = \sum_{i=1}^{n} \frac{\partial^2 F}{\partial \lambda_i^2} + 2 \sum_{i<j} \frac{1}{\lambda_i - \lambda_j} \left(\frac{\partial F}{\partial \lambda_i} - \frac{\partial F}{\partial \lambda_j} \right).$$

(ii) *The above formula can also be written*

$$LF = \frac{1}{V(\lambda)} \sum_{i=1}^{n} \frac{\partial^2}{\partial \lambda_i^2} \big(V(\lambda) F(\lambda) \big),$$

where V is the Vandermonde determinant

$$V(\lambda) = \prod_{j<k} (\lambda_j - \lambda_k) = \begin{vmatrix} \lambda_1^{n-1} & \cdots & \lambda_1 & 1 \\ \lambda_2^{n-1} & \cdots & \lambda_2 & 1 \\ \vdots & \ddots & \vdots & \vdots \\ \lambda_n^{n-1} & \cdots & \lambda_n & 1 \end{vmatrix}.$$

Proof. The proof of (i) is similar to that of Theorem 10.2.1. By taking $X = E_{ij} - E_{ji}$ $(i \neq j)$ we get

$$\frac{1}{2} \frac{\partial^2 f}{\partial (\operatorname{Re} x_{ij})^2}(a) = \frac{1}{\lambda_i - \lambda_j} \left(\frac{\partial F}{\partial \lambda_i} - \frac{\partial F}{\partial \lambda_j} \right),$$

and, for $X = i(E_{ij} + E_{ji})$,

$$\frac{1}{2} \frac{\partial^2 f}{\partial (\operatorname{Im} x_{ij})^2}(a) = \frac{1}{\lambda_i - \lambda_j} \left(\frac{\partial F}{\partial \lambda_i} - \frac{\partial F}{\partial \lambda_j} \right). \qquad \square$$

Lemma 10.2.3 *The Vandermonde polynomial V is harmonic:*

$$\sum_{i=1}^{n} \frac{\partial^2}{\partial \lambda_i^2} V = 0.$$

Proof. The polynomial

$$\sum_{i=1}^{n} \frac{\partial^2}{\partial \lambda_i^2} V$$

is skewsymmetric, hence divisible by V. Since its degree is less than the degree of V it is equal to zero. $\qquad \square$

Let us prove now part (ii) of Theorem 10.2.2. Using the formula giving the Laplace operator applied to the product of two functions on $\mathbb{R}^n$,

$$\Delta_0(\varphi \psi) = (\Delta_0 \varphi)\psi + 2(\nabla_0 \varphi | \nabla_0 \psi) + \varphi(\Delta_0 \psi),$$

($\nabla_0 \varphi$ is the gradient of φ) and the lemma above, we get

$$\frac{1}{V} L(VF) = LF + 2\frac{1}{V}(\nabla_0 V | \nabla_0 F).$$

Since

$$\frac{1}{V}\nabla_0 V = \nabla_0 \log|V| = \sum_{i<j} \frac{1}{\lambda_i - \lambda_j}(e_i - e_j),$$

where $\{e_1, \ldots, e_n\}$ is the canonical basis of $\mathbb{R}^n$, we get

$$\frac{1}{V}(\nabla_0 V | \nabla_0 F) = \sum_{i<j} \frac{1}{\lambda_i - \lambda_j}\left(\frac{\partial F}{\partial\lambda_i} - \frac{\partial F}{\partial\lambda_j}\right). \qquad \square$$

10.3 Heat equation and orbital integrals

In this section we assume that $\mathcal{V} = Herm(n, \mathbb{C})$ and $K = U(n)$. For two matrices $x, y \in \mathcal{V}$ the *orbital integral* $\mathcal{I}(x, y)$ is defined by

$$\mathcal{I}(x, y) = \int_K e^{\mathrm{tr}(xkyk^*)}\alpha(dk),$$

where α is the normalised Haar measure of K. Observe that the function $\mathcal{I}(x, y)$ is K-invariant, for K acting on x or on y:

$$\mathcal{I}(kxk^*, y) = \mathcal{I}(x, kyk^*) = \mathcal{I}(x, y) \quad (k \in K),$$

hence determined by its restriction to the subspace of diagonal matrices. We will see that solving the Cauchy problem for the heat equation, using the Weyl integration formula (Theorem 10.1.4) and the formula giving the radial part for the Laplace operator (Theorem 10.2.2) leads to evaluation of the orbital integral $\mathcal{I}(x, y)$.

The Cauchy problem for the heat equation

$$\frac{\partial u}{\partial t} = \Delta u,$$

$$u(0, x) = f(x),$$

where f is a bounded continuous function on $\mathcal{V}$, admits a unique solution which is given by

$$u(t, x) = \frac{1}{(2\sqrt{\pi t})^N} \int_{\mathcal{V}} e^{-\frac{1}{4t}\|x-y\|^2} f(y)\lambda(dy) \quad (t > 0,\ x \in \mathcal{V}),$$

where N is the dimension of $\mathcal{V}$: $N = n^2$.

We will need the following result.

Lemma 10.3.1 *Assume that f belongs to the Schwartz space $\mathcal{S}(\mathcal{V})$. Then, for every polynomial p and every $T > 0$, there exists a constant C such that*

$$|p(x)u(t, x)| \le C \quad (0 \le t \le T,\ x \in \mathcal{V}).$$

Proof. In order to establish this result one shows that, for $0 \le t \le T$,

$$\left| p\left(\frac{\partial}{\partial \xi}\right)\left(e^{-t\|\xi\|^2} \hat{f}(\xi)\right) \right| \le \gamma(\xi),$$

where γ is an integrable function on $\mathcal{V}$. $\qquad\square$

Assume that the function f is K-invariant:

$$f(kxk^*) = f(x) \quad (k \in K).$$

Then u is K-invariant as well. We can write

$$f(x) = f_0(a), \quad a = \mathrm{diag}(a_1, \ldots, a_n) \in \mathbb{R}^n,$$

the numbers a_i being the eigenvalues of x. Similarly,

$$u(t, x) = u_0(t, a).$$

Let us evaluate the integral giving $u(t, x)$ using the Weyl integration formula (Theorem 10.1.4). The Lebesgue measure λ is assumed to be chosen as

$$\lambda(dx) = \prod_{i=1}^{n} dx_{ii} \prod_{i<j} d(\mathrm{Re}\ x_{ij}) d(\mathrm{Im}\ x_{ij}).$$

We obtain

$$u_0(t, a) = \int_{\mathbb{R}^n} H_0(t, a, b) f_0(b) |V(b)|^2 db_1 \ldots db_n,$$

with

$$\begin{aligned}
H_0(t, a, b) &= c_n' \frac{1}{(2\sqrt{\pi t})^N} \int_K e^{-\frac{1}{4t}\|a - kbk^*\|^2} \alpha(dk) \\
&= c_n' \frac{1}{(2\sqrt{\pi t})^N} e^{-\frac{1}{4t}(\|a\|^2 + \|b\|^2)} \int_K e^{\frac{1}{2t} \mathrm{tr}(akbk^*)} \alpha(dk) \\
&= c_n' \frac{1}{(2\sqrt{\pi t})^N} e^{-\frac{1}{4t}(\|a\|^2 + \|b\|^2)} \mathcal{I}\left(\frac{1}{2t} a, b\right).
\end{aligned}$$

Theorem 10.3.2 (i) *If $a = \mathrm{diag}(a_1, \ldots, a_n)$ and $b = \mathrm{diag}(b_1, \ldots, b_n)$, then*

$$\mathcal{I}(a, b) = 1!2!\ldots(n-1)! \frac{1}{V(a)V(b)} \det(e^{a_i b_j})_{1 \le i,j \le n}.$$

(ii) *The kernel H_0, which is defined above, is given by*

$$H_0(t, a, b) = \frac{1}{(2\sqrt{\pi t})^n} \frac{1}{V(a)V(b)} e^{-\frac{1}{4t}(\|a\|^2 + \|b\|^2)} \frac{1}{n!} \det\left(e^{\frac{1}{2t} a_i b_j}\right)_{1 \le i,j \le n}.$$

Proof. From the formula giving the radial part of the Laplace operator (Theorem 10.2.2) it follows that the function u_0 is a solution of the equation

$$\frac{\partial u_0}{\partial t} = \frac{1}{V(a)} \sum_{i=1}^{n} \frac{\partial^2}{\partial a_i^2} \big(V(a)u_0(t,a)\big).$$

This leads to

$$v(t,a) = V(a)u_0(t,a), \quad g(a) = V(a)f_0(a).$$

The function v is a solution of the following Cauchy problem in $\mathbb{R}^n$:

$$\frac{\partial v}{\partial t} = \sum_{i=1}^{n} \frac{\partial^2 v}{\partial a_i^2},$$
$$v(0,a) = g(a).$$

Assume that the initial data f belongs to the Schwartz space $\mathcal{S}(\mathcal{V})$. From Lemma 10.3.1, for every $T > 0$, the function v is bounded on $[0,T] \times \mathbb{R}^n$, hence

$$v(t,a) = \frac{1}{(2\sqrt{\pi t})^n} \int_{\mathbb{R}^n} e^{-\frac{1}{4t}\|a-b\|^2} g(b)db_1 \ldots db_n.$$

Since the function g is skewsymmetric, this can be written

$$v(t,a) = \frac{1}{(2\sqrt{\pi t})^n} \int_{\mathbb{R}^n} e^{-\frac{1}{4t}(\|a\|^2+\|b\|^2)}$$
$$\times \frac{1}{n!} \sum_{\sigma \in \Sigma_n} \varepsilon(\sigma) e^{\frac{1}{2t}\sum_{i=1}^{n} a_i b_{\sigma(i)}} g(b)db_1 \ldots db_n.$$

This shows that, for every function $g(b) = V(b)f_0(b)$, where f_0 is a symmetric function in the Schwartz space $\mathcal{S}(\mathbb{R}^n)$,

$$\int_{\mathbb{R}^n} H_0(t,a,b)g(b)db_1 \ldots db_n$$
$$= \frac{1}{(2\sqrt{\pi t})^n} \int_{\mathbb{R}^n} e^{-\frac{1}{4t}(\|a\|^2+\|b\|^2)} \frac{1}{n!} \sum_{\sigma \in \Sigma_n} \varepsilon(\sigma) e^{\frac{1}{2t}\sum_{i=1}^{n} a_i b_{\sigma(i)}} g(b)db_1 \ldots db_n.$$

Therefore (ii) is proven, and (i) follows since

$$c_n' = \frac{\pi^{n(n-1)/2}}{\prod_{j=1}^{n} j!}.$$

$\square$

10.4 Fourier transforms of invariant functions

In this section we assume that $\mathcal{V} = Herm(n, \mathbb{C})$ and $K = U(n)$. For $x \in \mathcal{V}$ the orbital measure μ_x is defined on $\mathcal{V}$ by

$$\int_{\mathcal{V}} f(y)\mu_x(dy) = \int_K f(kxk^*)\alpha(dk),$$

where α is the normalised Haar measure of K, and f is a continuous function on $\mathcal{V}$. The Fourier transform $\widehat{\mu_x}$ of the measure μ_x is the following function:

$$\widehat{\mu_x}(\xi) = \int e^{-i\,\mathrm{tr}(y\xi)}\mu_x(dy)$$

$$= \int_G e^{-i\,\mathrm{tr}(\xi u x u^*)}\alpha(du),$$

$$= \mathcal{I}(x, -i\xi),$$

where the function $\mathcal{I}(x, y)$ is the integral orbital we introduced in Section 10.3.

Proposition 10.4.1 *If* $x = \mathrm{diag}(x_1, \ldots, x_n)$, *and* $\xi = \mathrm{diag}(\xi_1, \ldots, \xi_n)$, *then*

$$\widehat{\mu_x}(\xi) = 1!2! \cdots (n-1)! \frac{1}{V(x)V(-i\xi)} \det\left(e^{-ix_j\xi_k}\right)_{1 \le j,k \le n}.$$

As an application we will establish a formula for the Fourier transform of a K-invariant function on $\mathcal{V} = Herm(n, \mathbb{C})$. Let f be a K-invariant function in the Schwartz space $\mathcal{S}(\mathcal{V})$ and put

$$F(a_1, \ldots, a_n) = f\big(\mathrm{diag}(a_1, \ldots, a_n)\big).$$

Let $\hat{f}$ denote the Fourier transform of f on $\mathcal{V}$:

$$\hat{f}(\xi) = \int_{\mathcal{V}} e^{-i\,\mathrm{tr}(x\xi)} f(x)\lambda(dx) \quad (\xi \in \mathcal{V}),$$

and $\hat{F}$ the Fourier transform of F on $\mathbb{R}^n$:

$$\hat{F}(b) = \int_{\mathbb{R}^n} e^{-i(a|b)} F(a)da_1 \ldots da_n \quad (b \in \mathbb{R}^n).$$

The function $\hat{f}$ is K-invariant as well, hence determined by its restriction to the subspace of diagonal matrices. Put

$$\tilde{F}(b_1, \ldots, b_n) = \hat{f}\big(\mathrm{diag}(b_1, \ldots, b_n)\big).$$

Proposition 10.4.2 $\tilde{F}(b) = c_n' 1!2! \cdots n! \dfrac{1}{V(b)} V\left(\dfrac{\partial}{\partial b}\right) \hat{F}(b).$

Proof. Let us recall the Weyl integration formula for the space $\mathcal{V} = Herm(n, \mathbb{C})$ of Hermitian matrices with the action of the unitary group $K = U(n)$

(Theorem 10.1.4): if f is an integrable function on $\mathcal{V}$, then

$$\int_{\mathcal{V}} f(x)\lambda(dx) = c_n' \int_A \left(\int_K f(kak^*)d\alpha(k) \right) V(a)^2 da_1 \ldots da_n,$$

where A is the space of real diagonal matrices, and α is the normalised Haar measure of K. We use this formula to evaluate the Fourier transform of f:

$$\hat{f}(\xi) = c_n' \int_A \left(\int_K e^{-i\,\mathrm{tr}(kak^*\xi)} d\alpha(k) \right) F(a)V(a)^2 da_1 \ldots da_n$$

$$= c_n' \int_A \mathcal{I}(-i\xi, a)F(a)V(a)^2 da_1 \ldots da_n.$$

By Theorem 10.3.2,

$$\tilde{F}(b) = c_n' 1!2! \cdots (n-1)!$$

$$\times \frac{1}{V(-ib)} \int_A \det\left(\left(e^{-ia_j b_k}\right)_{1 \le j,k \le n} \right) V(a)F(a) da_1 \ldots da_n$$

$$= c_n' 1!2! \cdots (n-1)!$$

$$\times \frac{1}{V(-ib)} \sum_{\sigma \in \mathfrak{S}_n} \varepsilon(\sigma) \int_A e^{-i(a_1 b_{\sigma(1)} + \cdots + a_n b_{\sigma(n)})} V(a)F(a) da_1 \ldots da_n.$$

From the classical properties of the Fourier transform

$$G(b) := \int_A e^{-i(a_1 b_1 + \cdots + a_n b_n)} V(a)F(a) da_1 \ldots da_n$$

$$= V\left(i\frac{\partial}{\partial b} \right) \hat{F}(b).$$

Observe that the function G is skewsymmetric. Finally

$$\tilde{F}(b) = c_n' 1!2! \cdots (n-1)!n! \frac{1}{V(b)} V\left(\frac{\partial}{\partial b} \right) \hat{F}(b). \qquad \square$$

10.5 Exercises

1. The Lie algebra $\mathfrak{k}$ of $K = O(n)$ is the space $Skew(n, \mathbb{R})$ of skewsymmetric matrices. On K one considers the left and right invariant differential form κ of degree $p = \dim \mathfrak{k} = \frac{1}{2}n(n-1)$ for which the p-linear skew form κ_e on $\mathfrak{k} = Skew(n, \mathbb{R})$ is defined (up to a sign) by

$$\kappa_e = \bigwedge_{1 \le i < j \le n} dz_{ij} \quad \left(Z = (z_{ij}) \in Skew(n, \mathbb{R}) \right).$$

Let $v = |\kappa|$ be the associated Haar measure of K. Show that

$$\text{vol}(K) := v(K) = 2^n \frac{\pi^{n(n+1)/4}}{\prod_{i=1}^n \Gamma\left(\frac{i}{2}\right)}.$$

Hint. Consider on $GL(n, \mathbb{R})$ the differential form ω defined (up to a sign) by

$$\omega = |\det x|^{-n} \bigwedge_{1 \le i, j \le n} dx_{ij},$$

and the map

$$\varphi : O(n) \times T(n, \mathbb{R})_+ \to GL(n, \mathbb{R}), \quad (k, t) \mapsto kt.$$

Show that (up to a sign)

$$\varphi^* \omega = \kappa \otimes \tau,$$

where τ is the differential form on $T(n, \mathbb{R})_+$ defined (up to a sign) by

$$\tau = \prod_{i=1}^n t_{ii}^{-i} \bigwedge_{1 \le i \le j \le n} dt_{ij}.$$

(Show first that $(\varphi^* \omega)_e = \kappa_e \otimes \tau_e$.) Use Exercise 4 of Chapter 5.

2. In the statement of Theorem 10.1.1, assume that λ is the Lebesgue measure on $Sym(n, \mathbb{R})$ defined by

$$\lambda(dx) = \prod_{i \le j} dx_{ij},$$

and α the normalised Haar measure of $K = O(n)$. Show that

$$c_n = \frac{\pi^{n(n+1)/4}}{n! \prod_{i=1}^n \Gamma\left(\frac{i}{2}\right)}.$$

Hint. Follow the proof of Theorem 10.1.1 and use Exercise 1.

3. Let $\mathcal{P}_n$ denote the cone of positive definite $n \times n$ real symmetric matrices. On $\mathcal{P}_n$ consider the measure μ defined by

$$\mu(dx) = (\det x)^{-(n+1)/2} \prod_{i \le j} dx_{ij}.$$

(a) To every $g \in GL(n, \mathbb{R})$ one associates the transformation

$$T_g : \mathcal{P}_n \to \mathcal{P}_n, \quad x \mapsto gxg^T.$$

Show that the measure μ is invariant under the transformations T_g ($g \in GL(n, \mathbb{R})$).

(b) Let $T(n, \mathbb{R})_+$ denote the group of upper triangular matrices with positive diagonal entries. Show that the map

$$T(n, \mathbb{R})_+ \to \mathcal{P}_n, \quad t \mapsto tt^T,$$

is a diffeomorphism.

Show that, if f is a function on $\mathcal{P}_n$ which is integrable with respect to the measure μ,

$$\int_{\mathcal{P}_n} f(x)\mu(dx) = 2^n \int_{T(n,\mathbb{R})_+} f(tt^T) \prod_{j=1}^n t_{jj}^{j-n-1} \prod_{i \le j} dt_{ij}.$$

(c) The gamma function Γ_n of the cone $\mathcal{P}_n$ is the function of the complex variable s defined by

$$\Gamma_n(s) = \int_{\mathcal{P}_n} e^{-\operatorname{tr}(x)}(\det x)^s \mu(dx).$$

Show that the integral is well defined for $\operatorname{Re} s > \frac{n-1}{2}$, and that

$$\Gamma_n(s) = \pi^{n(n-1)/4} \prod_{j=1}^n \Gamma\left(s - \frac{j-1}{2}\right).$$

(d) Show that there is a constant $d_n > 0$ such that, if the function f on $\mathcal{P}_n$ is μ-integrable, then

$$\int_{\mathcal{P}_n} f(x)\mu(dx)$$

$$= d_n \int_{K \times A} f(k \exp tk^T)\alpha(dk) \prod_{i<j} \left| \operatorname{sh} \frac{t_i - t_j}{2} \right| d_1 \ldots dt_n,$$

where $t = \operatorname{diag}(t_1, \ldots t_n)$, $K = O(n)$, and α is the normalised Haar measure on K. Show that

$$d_n = \frac{2^{n(n-1)/2} \pi^{n(n+1)/4}}{n! \prod_{i=1}^n \Gamma\left(\frac{i}{2}\right)}.$$

4. By taking the special function

$$f(x) = e^{-\operatorname{tr}(x^2)/2},$$

determine the constants $C = c_n$ (Theorem 10.1.1) and $C = c_n'$ (Theorem 10.1.4) using the Mehta integral:

$$\int_{\mathbb{R}^n} e^{-\frac{1}{2}\sum_{j=1}^n t_j^2} | \prod_{i<j}(t_i - t_j)|^{2\gamma} dt_1 \ldots dt_n = (2\pi)^{n/2} \prod_{j=1}^n \frac{\Gamma(1 + j\gamma)}{\Gamma(1 + \gamma)}.$$

11

Irreducible representations
of the unitary group

The unitary group $G = U(n)$ is a connected and compact linear Lie group. To every irreducible representation of G one associates a highest weight and this gives a parameterisation of the set $\hat{G}$ of equivalence classes of irreducible representations of G.

We establish the Weyl formula for the character of an irreducible representation. Its restriction to the subgroup T of diagonal unitary matrices is a Schur function.

The Lie algebra $\mathfrak{g} = Lie(G)$ consists of skewHermitian matrices:

$$\mathfrak{g} = i\,Herm(n, \mathbb{C}).$$

The complex Lie algebra $\mathfrak{g}_{\mathbb{C}} = \mathfrak{g} + i\mathfrak{g}$ is the Lie algebra $\mathfrak{gl}(n, \mathbb{C}) = M(n, \mathbb{C})$ of the group $GL(n, \mathbb{C})$. The subgroup T of G consists of unitary diagonal matrices t:

$$t = \begin{pmatrix} t_1 & & \\ & \ddots & \\ & & t_n \end{pmatrix}, \quad t_j \in \mathbb{C}, \ |t_j| = 1.$$

Its Lie algebra $\mathfrak{t}$ consists of diagonal matrices whose diagonal entries are pure imaginary. We will denote by $\mathfrak{h}$ its complexification, which consists of diagonal matrices with complex coefficients, and by $\mathfrak{n}$ the nilpotent subalgebra of $\mathfrak{g}_{\mathbb{C}}$ consisting of upper triangular matrices with diagonal entries equal to zero.

11.1 Highest weight theorem

A continuous character γ of the group T can be written

$$\gamma(t) = t_1^{\mu_1} \dots t_n^{\mu_n}, \quad \mu_j \in \mathbb{Z}.$$

The associate linear form μ on $\mathfrak{h}$,

$$\mu(H) = \sum_{j=1}^{n} \mu_j h_j, \quad H = \mathrm{diag}(h_1, \ldots, h_n) \in \mathfrak{h},$$

is called a *weight*. The following holds: $\gamma(\exp H) = e^{\mu(H)}$. The set P of all weights is a lattice in $i\mathfrak{t}^*$; hence $P \simeq \mathbb{Z}^n$.

Let $(\pi, \mathcal{V})$ be a finite dimensional representation of G. We consider on $\mathcal{V}$ a Hermitian inner product for which the representation π is unitary. (By Proposition 6.1.1 we know that such an inner product exists.) The derived representation $d\pi$ extends as a $\mathbb{C}$-linear representaton of the Lie algebra $\mathfrak{g}_\mathbb{C}$. Observe that

$$d\pi(X)^* = d\pi(X^*).$$

A linear form μ on $\mathfrak{h}$ is called a *weight of the representation* π if there exists a non-zero vector $v \in \mathcal{V}$ such that

$$d\pi(H)v = \mu(H)v, \quad H \in \mathfrak{h}.$$

This means that the subspace

$$\mathcal{V}_\mu = \{v \in \mathcal{V} \mid \forall H \in \mathfrak{h}, \ d\pi(H)v = \mu(H)v\}$$

does not reduce to $\{0\}$. Since

$$\pi(\exp H)v = e^{\mu(H)}v,$$

μ is a weight: $\mu \in P$. Let $P(\pi) \subset P$ denote the set of weights of the representation π. The dimension m_μ of $\mathcal{V}_\mu$ is called the multiplicity of the weight μ.

The operators $\pi(t)$ $(t \in T)$ are unitary, hence normal, and commute with each other. Therefore they can be diagonalised simultaneously. Hence,

$$\mathcal{V} = \bigoplus_{\mu \in P(\pi)} \mathcal{V}_\mu.$$

A non-zero vector $v \in \mathcal{V}$ is called a *highest weight vector* if there exists a weight $\lambda \in P(\pi)$ such that

$$d\pi(H)v = \lambda(H)v, \quad H \in \mathfrak{h},$$
$$d\pi(X)v = 0, \quad X \in \mathfrak{n}.$$

Theorem 11.1.1 (Highest weight theorem) *Let $(\pi, \mathcal{V})$ be a finite dimensional representation of G.*

(i) *There exists a highest weight vector.*

(ii) *Let v_0 be a highest weight vector. The representation π is irreducible if and only if every highest weight vector is proportional to v_0. In that case the corresponding weight λ is called the highest weight of the representation π.*

Proof. (a) Fix $H_0 \in i\mathfrak{t}$:

$$H_0 = \begin{pmatrix} h_1 & & \\ & \ddots & \\ & & h_n \end{pmatrix}, \quad h_j \in \mathbb{R},$$

such that $h_1 > h_2 > \cdots > h_n$. The eigenvalues of $d\pi(H_0)$ are the real numbers $\mu(H_0)$ with $\mu \in P(\pi)$. Let $\lambda \in P(\pi)$ be such that

$$\forall \mu \in P(\pi), \quad \mu(H_0) \le \lambda(H_0),$$

and $v \in \mathcal{V}_\lambda$, $v \ne 0$. We will show that v is a highest weight vector. Clearly

$$d\pi(H)v = \lambda(H)v, \quad H \in \mathfrak{h}.$$

Let $X = E_{ij}$, with $i < j$ ($\{E_{ij}\}$ denotes the canonical basis of $M(n, \mathbb{C})$). Since $[H_0, X] = (h_i - h_j)X$, we get

$$d\pi(H_0)d\pi(X)v = d\pi(X)d\pi(H_0)v + d\pi([H_0, X])v$$
$$= \big(\lambda(H_0) + (h_i - h_j)\big)d\pi(X)v,$$

hence $d\pi(X)v = 0$ since $\lambda(H_0) + (h_i - h_j) > \lambda(H_0)$.

(b) Assume π to be irreducible and let v_0 be a highest weight vector. Let v be a vector such that $(v_0|v) = 0$, and such that

$$d\pi(X)v = 0, \quad X \in \mathfrak{n}.$$

If $H \in \mathfrak{h}$, then

$$\big(d\pi(H)v_0|v\big) = \lambda(H)(v_0|v) = 0.$$

If $X \in \mathfrak{n}^*$, $d\pi(X)^* = d\pi(X^*)$, and $X^* \in \mathfrak{n}$, hence

$$\big(d\pi(X)v_0|v\big) = \big(v_0|d\pi(X^*)v\big) = 0.$$

Since $\mathfrak{g}_\mathbb{C} = \mathfrak{h} + \mathfrak{n} + \mathfrak{n}^*$, for every $X \in \mathfrak{g}_\mathbb{C}$, $\big(d\pi(X)v_0|v\big) = 0$. Since the representation is irreducible, it follows that $v = 0$.

(c) Let v_0 be a highest weight vector and assume that every highest weight vector is proportional to v_0. Let $\mathcal{W} \ne \{0\}$ be an invariant subspace. This subspace contains a highest weight vector, hence $v_0 \in \mathcal{W}$. The subspace $\mathcal{W}^\perp$ is invariant as well. If it were not reduced to $\{0\}$ it would contain v_0, and this is not possible; hence $\mathcal{W}^\perp = \{0\}$ and $\mathcal{W} = \mathcal{V}$. $\qquad\square$

A weight μ is said to be *dominant* if $\mu_1 \geq \mu_2 \geq \cdots \geq \mu_n$, and *strongly dominant* if $\mu_1 > \mu_2 > \cdots > \mu_n$. One denotes by P^+ the set of dominant weights and by P^{++} the set of strongly dominant weights.

Proposition 11.1.2 *Let (π, V) be an irreducible representation of G. The highest weight λ of π is a dominant weight.*

Proof. Fix i, $1 \leq i < n$, and consider the representation π_0 of $G_0 = SU(2)$ on V which is defined by $\pi_0(g_0) = \pi(g)$ if

$$g_0 = \begin{pmatrix} \alpha & \beta \\ -\bar{\beta} & \bar{\alpha} \end{pmatrix},$$

and

$$g = \alpha E_{ii} + \beta E_{i,i+1} - \bar{\beta} E_{i+1,i} + \bar{\alpha} E_{i+1,i+1}$$

$$= \begin{pmatrix} 1 & & & & & & \\ & \ddots & & & & & \\ & & 1 & & & & \\ & & & \alpha & \beta & & \\ & & & -\bar{\beta} & \bar{\alpha} & & \\ & & & & & 1 & \\ & & & & & & \ddots & \\ & & & & & & & 1 \end{pmatrix}.$$

The space V can be decomposed as a direct sum of irreducible invariant subspaces for π_0 (Corollary 6.1.2)

$$V = V_1 \oplus \cdots \oplus V_N.$$

The restriction of the representation π_0 to V_k is equivalent to one of the representations π_m (Theorem 7.5.3),

$$\pi_0\big|_{V_k} \simeq \pi_{m_k}.$$

Let v be a highest weight vector for the representaton π. It decomposes as

$$v = v_1 + \cdots + v_N, \quad v_k \in V_k.$$

The vector v_k verifies

$$d\pi_0\left(\begin{pmatrix} 1 & 0 \\ 0 & -1 \end{pmatrix}\right) v_k = d\pi(E_{ii} - E_{i+1,i+1})v_k = (\lambda_i - \lambda_{i+1})v_k,$$

$$d\pi_0\left(\begin{pmatrix} 0 & 1 \\ 0 & 0 \end{pmatrix}\right) v_k = d\pi(E_{i,i+1})v_k = 0.$$

From Section 7.5, it follows that

$$\lambda_i - \lambda_{i+1} = m_k \geq 0. \qquad \qquad \square$$

Hence, to every irreducible representation π of G one associates its highest weight $\lambda \in P^+$. It only depends on the equivalence class of π. Therefore this defines a map:

$$\hat{G} \to P^+.$$

We will see in the next section that this map is a bijection.

11.2 Weyl formulae

We will establish three formulae due to Hermann Weyl: an integration formula, a formula for the character of an irreducible representation, and lastly a formula for the dimension of an irreducible representation. Using the character formula we will show that the map, which associates its highest weight to an equivalence class of irreducible representations, is a bijection from $\hat{G}$ onto the set P^+ of dominant weights.

A central function on G is determined by its restriction to the subgroup T of unitary diagonal matrices,

$$t = \begin{pmatrix} t_1 & & \\ & \ddots & \\ & & t_n \end{pmatrix}, \quad t_j \in \mathbb{C}, \ |t_j| = 1,$$

and this restriction is a symmetric function of the numbers $t_1, \ldots, t_n$. By the Weyl integration formula, the integral of a central function on G reduces to an integral on T. Let μ denote the normalised Haar measure of G, and ν the normalised Haar measure of T,

$$\nu(dt) = \frac{1}{(2\pi)^n} d\theta_1 \ldots d\theta_n, \quad t_j = e^{i\theta_j}.$$

Let V denote the *Vandermonde polynomial*,

$$V(t) = \prod_{j<k}(t_j - t_k) = \begin{vmatrix} t_1^{n-1} & \ldots & t_1 & 1 \\ t_2^{n-1} & \ldots & t_2 & 1 \\ \vdots & \ddots & \vdots & \vdots \\ t_n^{n-1} & \ldots & t_n & 1 \end{vmatrix}.$$

Theorem 11.2.1 (Weyl integration formula) *For an integrable function f on G,*

$$\int_G f(x)\mu(dx) = \frac{1}{n!}\int_T \left(\int_G f(gtg^{-1})\mu(dg)\right)|V(t)|^2 v(dt).$$

In particular, if f is central, then

$$\int_G f(x)\mu(dx) = \frac{1}{n!}\int_T f(t)|V(t)|^2 v(dt).$$

Observe that

$$|V(t)|^2 = \prod_{j<k} 4\sin^2\frac{\theta_j - \theta_k}{2}, \qquad t_j = e^{i\theta_j}.$$

Proof. (a) For $g \in G, t \in T$, put

$$\varphi(g,t) = gtg^{-1}.$$

Since $\varphi(gs,t) = \varphi(g,t)$ $(s \in T)$, $\varphi(g,t)$ only depends on the class gT, and φ can be seen as a map which is defined on $G/T \times T$:

$$\varphi : G/T \times T \to G.$$

On the set $G/T \times \{t \in T \mid V(t) \neq 0\}$ the map φ is a covering of order $n!$. Let $\mathfrak{m}$ denote the subspace consisting of matrices in $\mathfrak{g} = i Herm(n, \mathbb{C})$ whose diagonal entries are zero. The tangent space to G/T at eT can be identified with $\mathfrak{m}$ and that to gT can be identified with $g\mathfrak{m}$.

Let us compute the differential of φ. For $X \in \mathfrak{m}$,

$$\begin{aligned}
(D\varphi)_{(gT,t)}(gX, 0) &= \frac{d}{ds}\varphi(g\exp sX, t)\Big|_{s=0} \\
&= \frac{d}{ds}g\exp sXt\exp -sXg^{-1}\Big|_{s=0} \\
&= gt\frac{d}{ds}\left(\exp s\big(Ad(t^{-1})X\big)\exp -sX\right)\Big|_{s=0}g^{-1} \\
&= gt\big((Ad(t^{-1}) - I)X\big)g^{-1}.
\end{aligned}$$

And for $Y \in \mathfrak{t}$,

$$\begin{aligned}
(D\varphi)_{(gT,t)}(0, tY) &= \frac{d}{ds}\varphi(g, t\exp sY)\Big|_{s=0} \\
&= \frac{d}{ds}gt\exp sYg^{-1}\Big|_{s=0} = gtYg^{-1}.
\end{aligned}$$

Finally, for $X \in \mathfrak{m}$, $Y \in \mathfrak{t}$,

$$(D\varphi)_{(gT,t)}(gX, tY) = gt\Big(\big(\mathrm{Ad}(t^{-1}) - I\big)X + Y\Big)g^{-1}.$$

(b) Let ω be a biinvariant differential form on G of degree $\dim G = n^2$. Let $m = \dim G/T = \dim \mathfrak{m} = n^2 - n$, and $X_1, \dots, X_m \in \mathfrak{m}$, $Y_1, \dots, Y_n \in \mathfrak{t}$:

$$(\varphi^*\omega)_{(gT,t)}(gX_1, \dots, gX_m, tY_1, \dots, tY_n)$$
$$= \omega_{gtg^{-1}}\big((D\varphi)_{(gT,t)}(gX_1), \dots, (D\varphi)_{(gT,t)}(gX_m),$$
$$(D\varphi)_{(gT,t)}(tY_1), \dots, (D\varphi)_{(gT,t)}(tY_n)\big)$$
$$= \omega_{gtg^{-1}}\Big(gt\big(\mathrm{Ad}(t^{-1}) - I\big)X_1g^{-1}, \dots, gt$$
$$\times \big(\mathrm{Ad}(t^{-1}) - I\big)g^{-1}, gtY_1g^{-1}, \dots, gtY_ng^{-1}\Big),$$

and, since the form ω is biinvariant, this is equal to

$$\omega_e\Big(\big(\mathrm{Ad}(t^{-1}) - I\big)X_1, \dots, \big(\mathrm{Ad}(t^{-1}) - I\big)X_m, Y_1, \dots, Y_n\Big)$$
$$= \det\Big(\big(\mathrm{Ad}(t^{-1}) - I\big)\big|_{\mathfrak{m}}\Big)\omega_e(X_1, \dots X_m, Y_1, \dots, Y_n).$$

Consider first the case $n = 2$,

$$\big(\mathrm{Ad}(t^{-1}) - I\big)\begin{pmatrix} 0 & z \\ -\bar{z} & 0 \end{pmatrix} = \begin{pmatrix} 0 & (\bar{t}_1 t_2 - 1)z \\ -(t_1\bar{t}_2 - 1)\bar{z} & 0 \end{pmatrix},$$

and

$$\det\big(\mathrm{Ad}(t^{-1}) - I\big) = |t_1 - t_2|^2.$$

It follows that, for $n \geq 2$,

$$\det\big(\mathrm{Ad}(t^{-1}) - I\big) = |V(t)|^2.$$

This computation shows that

$$\varphi^*\omega = |V(t)|^2\omega_1 \otimes \omega_2$$

where ω_1 is a left G-invariant differential form of degree m on G/T and ω_2 a T-invariant differential form of degree n on T. It follows that, for an integrable function f on G,

$$\int_U f(x)\mu(dx) = c\int_T \left(\int_U f(gtg^{-1})\mu(dg)\right)|V(t)|^2\nu(dt).$$

In fact, the set $\{t \in T \mid V(t) = 0\}$ has measure zero (for ν), and the image by φ of $G/T \times \{t \in T \mid V(t) = 0\}$ is of measure zero (for μ).

(c) In order to compute the constant c let us take $f = 1$,

$$\frac{1}{c} = \int_T |V(t)|^2 v(dt),$$

and we evaluate below this integral using the Parseval formula for multiple Fourier series. □

For $\alpha = (\alpha_1, \ldots, \alpha_n) \in \mathbb{Z}^n$ let t^α denote the monomial

$$t^\alpha = t_1^{\alpha_1} \ldots t_n^{\alpha_n}.$$

Let $\mathcal{F}(T)$ denote the space *trigonometric polynomials*, that is functions of the form

$$p(t) = \sum_{\alpha \in \mathbb{Z}^n} a_\alpha t^\alpha,$$

whose coefficients a_α are complex numbers, which are equal to zero but a finite number of them. The polynomial p is said to be *symmetric* if, for every permutation $\sigma \in \mathfrak{S}_n$,

$$p(\sigma \cdot t) = p(t),$$

where $\sigma \cdot t = (t_{\sigma(1)}, \ldots, t_{\sigma(n)})$, and *skewsymmetric* if

$$p(\sigma \cdot t) = \varepsilon(\sigma) p(t),$$

where $\varepsilon(\sigma)$ is the signature of the permutation σ. Let $\mathcal{F}_0(T)$ denote the space of symmetric trigonometric polynomials and $\mathcal{F}_1(T)$ the space of those which are skewsymmetric. For $\alpha \in P^{++}$, that is if $\alpha_1 > \cdots > \alpha_n$, the polynomial

$$A_\alpha(t) = \begin{vmatrix} t_1^{\alpha_1} & \cdots & t_1^{\alpha_n} \\ \vdots & & \vdots \\ t_n^{\alpha_1} & \cdots & t_n^{\alpha_n} \end{vmatrix} = \sum_{\sigma \in \mathfrak{S}_n} \varepsilon(\sigma) t^{\sigma(\alpha)},$$

where $\sigma(\alpha) = (\alpha_{\sigma(1)}, \ldots, \alpha_{\sigma(n)})$, is skewsymmetric. By the Parseval formula

$$\int_T |A_\alpha(t)|^2 v(dt) = \#(\mathfrak{S}_n) = n!.$$

In particular, for $\alpha = \delta := (n-1, n-2, \ldots, 0)$, $A_\delta = V$, and

$$\int_T |V(t)|^2 v(dt) = n!,$$

hence

$$c = \frac{1}{n!}.$$

Proposition 11.2.2 *The polynomials A_α ($\alpha \in P^{++}$) form a basis of $\mathcal{F}_1(T)$.*

Proof. The polynomials A_α are linearly independent. In fact assume that

$$\sum_{\alpha \in P^{++}} a_\alpha A_\alpha = 0.$$

For $\alpha_1 > \cdots > \alpha_n$ the coefficient of t^α in this sum is equal to a_α, hence $a_\alpha = 0$.
   Let p be a skewsymmetric polynomial

$$p(t) = \sum_{\alpha \in P} a_\alpha t^\alpha.$$

Let $\alpha^0 = (\alpha_1^0, \ldots, \alpha_n^0)$ be such that $\alpha_i^0 = \alpha_j^0$ ($i \neq j$), and let τ be the transposition which exchanges i and j. Then $\tau(\alpha^0) = \alpha^0$, and since

$$\sum_{\alpha \in P} a_{\tau(\alpha)} t^\alpha = -\sum_{\alpha \in P} a_\alpha t^\alpha,$$

then $a_{\alpha^0} = 0$. Hence one can write

$$p(t) = \sum_{\alpha \in P^{++}} \sum_{\sigma \in \mathfrak{S}_n} a_{\alpha,\sigma} t^{\sigma(\alpha)},$$

and

$$a_{\alpha,\sigma} = \varepsilon(\sigma) a_{\alpha,e}.$$

Therefore

$$p(t) = \sum_{\alpha \in P^{++}} a_{\alpha,e} A_\alpha(t). \qquad \square$$

For $\alpha \in P^+$ the *Schur function* s_α is defined as

$$s_\alpha(t) = \frac{A_{\alpha+\delta}(t)}{V(t)}.$$

The function s_α is a symmetric trigonometric polynomial. In fact the polynomial $A_{\alpha+\delta}$ vanishes if $t_i = t_j$ ($i \neq j$), hence is divisible by $t_i - t_j$. Since the factors $t_i - t_j$ are mutually prime, it is divisible by their product. As a quotient of two skewsymmetric functions, the function s_α is symmetric.

Proposition 11.2.3 (Character formula) *Let π be an irreducible representation of G, λ its highest weight, and χ_π its character. For $t \in T$.*

$$\chi_\pi(t) = s_\lambda(t).$$

Proof. The restriction to T of the character χ_π is a trigonometric polynomial

$$\chi_\pi(t) = \sum_{\mu \in P(\pi)} m_\mu t^\mu,$$

which is symmetric. The coefficients m_μ are positive integers, and $m_\lambda = 1$. The polynomial

$$V(t)\chi_\pi(t) = \sum_{\sigma \in \mathfrak{S}_n} \sum_{\mu \in P(\pi)} \varepsilon(\sigma) m_\mu t^{\sigma(\delta)+\mu}$$

is skewsymmetric and its coefficients are integers. This can be expanded in the basis $\{A_\alpha\}$:

$$V\chi_\pi = \sum_{\alpha \in P^{++}} a_\alpha A_\alpha.$$

The coefficients a_α are integers and, for $\alpha = \lambda + \delta$, $a_\alpha = 1$. From the Parseval formula it follows that

$$\int_T |V(t)\chi_\pi(t)|^2 \nu(dt) = n! \sum_\alpha |a_\alpha|^2.$$

On the other hand, since χ_π is the character of an irreducible representation,

$$\int_U |\chi_\pi(g)|^2 \mu(dg) = 1,$$

by Proposition 6.5.1. This can be written, using the Weyl integration formula (Theorem 11.2.1),

$$\frac{1}{n!} \int_T |\chi_\pi(t)|^2 |V(t)|^2 \nu(dt) = 1.$$

It follows that $a_\alpha = 1$ in the case when $\alpha = \lambda + \delta$, and $a_\alpha = 0$ otherwise. □

Observe that the map $\lambda \mapsto \lambda + \delta$ is a bijection from P^+ onto P^{++}.

Corollary 11.2.4 (i) *If two irreducible representations of G have the same highest weight, then they are equivalent.*

(ii) *For every $\lambda \in P^+$ there is an irreducible representation of G with highest weight λ.*

Hence, the map which associates to an irreducible representation its highest weight induces a bijection from $\hat{G}$ onto P^+.

Proof. (i) If two irreducible representations have the same highest weight, then their characters are equal by the character formula. Hence they are equivalent by Proposition 6.5.1.

(ii) Let $\lambda \in P^+$. Assume that there is no irreducible representation of G with highest weight λ. The trigonometric polynomial s_λ is symmetric and therefore is the restriction to T of a continuous central function $\tilde{s}_\lambda$ on G. Let π be an irreducible representation of G, with highest weight μ, and character χ_π. Then

$$\int_G \tilde{s}_\lambda(g)\overline{\chi_\pi(g)}\mu(dg) = \frac{1}{n!}\int_T A_{\lambda+\delta}(t)\overline{A_{\mu+\delta}(t)}\nu(dt) = 0,$$

since $\lambda \neq \mu$. But this is impossible because the characters of the irreducible representations of G form a Hilbert basis of the space of (classes of) square integrable central functions (Proposition 6.5.3). $\qquad\square$

Let π be an irreducible representation of G with highest weight λ, and character χ_π, and let d_λ be the dimension of the representation space. Then

$$d_\lambda = \chi_\pi(e).$$

We cannot evaluate the value χ_λ at $t = e$ directly using the character formula because it appears as the quotient of two functions vanishing at $t = e$. We obtain a formula for d_λ by a limit procedure.

Corollary 11.2.5 (Dimension formula)

$$d_\lambda = \frac{V(\lambda+\delta)}{V(\delta)} = \frac{\prod_{j<\ell}(\alpha_j - \alpha_\ell)}{\prod_{j<\ell}(\ell - j)},$$

where $\alpha = \lambda + \delta$, that is $\alpha_j = \lambda_j + n - j$.

From this formula we get the following estimate which will be useful when studying the convergence of Fourier series on G:

$$d_\lambda \leq C(1 + \|\lambda\|)^{n(n-1)/2}.$$

Proof. Let $v = (v_1, \ldots, v_n)$. If p is a trigonometric polynomial, we put

$$(H_v p)(\eta) = p(e^{iv_1\eta}, \ldots, e^{iv_n\eta}) \quad (\eta \in \mathbb{R}).$$

Observe that $H_v A_\alpha = H_\alpha A_v$, in fact

$$\begin{aligned}(H_v A_\alpha)(\eta) &= \sum_{w \in \mathfrak{S}_n} \varepsilon(w)e^{i\langle v, w(\alpha)\rangle\eta} \\ &= \sum_{w \in \mathfrak{S}_n} \varepsilon(w)e^{i\langle w(v), \alpha\rangle\eta} = (H_\alpha A_v)(\eta).\end{aligned}$$

For $\nu = \delta = (n-1, n-2, \ldots, 0)$, $A_\nu(t) = V(t)$. Furthermore

$$A_\alpha(e^{i\nu_1\eta}, \ldots, e^{i\nu_n\eta}) = V(e^{i\alpha_1\eta}, \ldots, e^{i\alpha_n\eta})$$
$$= \prod_{j<\ell}\left(e^{i\alpha_j\eta} - e^{i\alpha_\ell\eta}\right) \sim (i\eta)^{\frac{n(n-1)}{2}} \prod_{j<\ell}(\alpha_j - \alpha_\ell) \quad (\eta \to 0).$$

The statement follows. □

11.3 Holomorphic representations

Let Ω be a domain in $\mathbb{C}^N$, and let the complex valued $\mathcal{C}^1$ function f be defined on Ω. The differential Df of f at z can be written

$$(Df)_z(w) = \frac{d}{dt}\Big|_{t=0} f(z + tw) \qquad (w \in \mathbb{C}^N, t \in \mathbb{R}).$$

We introduce the differential operators

$$\frac{\partial}{\partial z_j} = \frac{1}{2}\left(\frac{\partial}{\partial x_j} - i\frac{\partial}{\partial y_j}\right), \qquad \frac{\partial}{\partial \bar{z}_j} = \frac{1}{2}\left(\frac{\partial}{\partial x_j} + i\frac{\partial}{\partial y_j}\right),$$

With this notation, if $w_j = u_j + iv_j$ ($u_j, v_j \in \mathbb{R}$),

$$(Df)(w) = \sum_{j=1}^N \frac{\partial f}{\partial x_j}u_j + \sum_{j=1}^N \frac{\partial f}{\partial y_j}v_j$$
$$= \sum_{j=1}^N \frac{\partial f}{\partial z_j}w_j + \sum_{j=1}^N \frac{\partial f}{\partial \bar{z}_j}\bar{w}_j.$$

The function f is said to be *holomorphic* in Ω if, for every $z \in \Omega$, the differential $(Df)_z$ of f is $\mathbb{C}$-linear. This can be written as

$$\frac{\partial f}{\partial \bar{z}_j} = 0 \qquad (j = 1, \ldots, N).$$

Since the group $GL(n, \mathbb{C})$ is an open set in $M(n, \mathbb{C}) \simeq \mathbb{C}^N$, with $N = n^2$, it makes sense for a function defined on an open set in $GL(n, \mathbb{C})$ to be holomorphic.

For $R > 1$ we denote by Ω_R the open set given by

$$\Omega_R = \{g \in GL(n, \mathbb{C}) \mid \|g\| < R, \|g^{-1}\| < R\}.$$

It is a neighbourhood of $G = U(n)$.

Proposition 11.3.1 *Let the function f be holomorphic on Ω_R. If f vanishes on $G = U(n)$, then f vanishes identically.*

Proof. Fix $X \in Herm(n, \mathbb{C})$ with $\|X\| < \log R$, and $u \in G$. For $\tau \in \mathbb{C}$ put

$$\varphi(\tau) = f(u \exp i\tau X).$$

The function φ is defined and holomorphic for $|\Im \tau| < \log R/\|X\|$. It vanishes on $\mathbb{R}$, since, for real τ, $u \exp i\tau X \in G$. Hence φ vanishes identically. In particular, $\varphi(-i) = f(u \exp X) = 0$. On the other hand every $g \in \Omega_R$ decomposes as $g = u \exp X$ with $u \in G$, $X \in Herm(n, \mathbb{C})$, $\|X\| < \log R$. Hence f vanishes identically on Ω_R. □

Let π be a representation of $GL(n, \mathbb{C})$ on a finite dimensional complex vector space $\mathcal{V}$, and let $d\pi$ be the derived representation. The representation π is said to be *holomorphic* if π is a holomorphic function on $GL(n, \mathbb{C})$ with values in $End(\mathcal{V})$, that is if its coefficients $g \mapsto \langle \pi(g)u, v \rangle$ ($u \in \mathcal{V}$, $v \in \mathcal{V}^*$) are holomorphic functions on $GL(n, \mathbb{C})$.

Proposition 11.3.2 *The representation π is holomorphic if and only if the derived representation*

$$d\pi : \mathfrak{g}_{\mathbb{C}} = M(n, \mathbb{C}) \to End(\mathcal{V})$$

is $\mathbb{C}$-linear.

Proof. (a) Assume the representation π is holomorphic. For every $X \in M(n, \mathbb{C})$ the function

$$\varphi(\tau) = \pi(\exp \tau X)$$

is holomorphic on $\mathbb{C}$, and

$$d\pi(\tau X) = \frac{d}{dt}\varphi(t\tau)\Big|_{t=0} = \varphi'(0)\tau = \tau d\pi(X),$$

hence $d\pi$ is $\mathbb{C}$-linear.

(b) Assume $d\pi$ is $\mathbb{C}$-linear. For $g \in GL(n, \mathbb{C})$, $X \in \mathfrak{g}$,

$$\frac{d}{dt}(g \exp tX)\Big|_{t=0} = gX,$$

hence

$$(D\pi)_g(gX) = \frac{d}{dt}\pi(g \exp tX)\Big|_{t=0} = \pi(g)\frac{d}{dt}\pi(\exp tX)\Big|_{t=0}$$

$$= \pi(g)d\pi(X).$$

It follows that $(D\pi)_g$ is $\mathbb{C}$-linear, and this means that π is holomorphic. □

A function f on $GL(n, \mathbb{C})$ is said to be *regular* if it can be written

$$f(x) = p(x_{11}, \ldots, x_{ij}, \ldots, x_{nn}, (\det x)^{-1}),$$

where p is a polynomial in $n^2 + 1$ variables with complex coefficients. We denote by $\mathcal{R}$ the algebra of regular functions on $GL(n, \mathbb{C})$. Observe that $\mathcal{R}$ is invariant under $GL(n, \mathbb{C})$ acting on left and right sides, that is under the operators $L(g)$ and $R(g)$ $(g \in G)$ defined by

$$(L(g)f)(x) = f(g^{-1}x), \qquad (R(g)f)(x) = f(xg).$$

Also, let $\mathcal{R}^0$ denote the algebra of the restrictions to $G = U(n)$ of the functions in $\mathcal{R}$. By Proposition 11.3.1 the restriction map $\mathcal{R} \to \mathcal{R}^0$ is a bijection.

Proposition 11.3.3 *The algebra $\mathcal{R}^0$ is dense in the algebra $C(G)$ of continuous functions on G.*

Proof. We will apply the Stone–Weierstrass Theorem stated in Chapter 6 (Theorem 6.4.3). The constant functions belong clearly to $\mathcal{R}^0$, and $\mathcal{R}^0$ separates points in G. We will show that, for $f \in \mathcal{R}^0$, the function, which is defined on G by $u \mapsto \overline{f(u)}$, belongs to $\mathcal{R}^0$ as well. From Cramer's rule, it follows that, if $f \in \mathcal{R}$, the function $\check{f}$ given by

$$\check{f}(x) = f(x^{-1})$$

belongs to $\mathcal{R}$ as well. Similarly the function f^T given by

$$f^T(x) = f(x^T),$$

and the function $\bar{f}$ given by

$$\bar{f}(x) = \overline{f(\bar{x})}$$

also belongs to $\mathcal{R}$. It follows that the function f^* given by

$$f^*(x) = \bar{f}((x^T)^{-1})$$

belongs to $\mathcal{R}$; on the other hand, for $u \in G$,

$$f^*(u) = \overline{f(u)},$$

since $(u^T)^{-1} = \bar{u}$. □

Recall that $\mathcal{M}$ denotes the space of finite linear combinations of coefficients of finite dimensional representations of G (see Section 6.4).

Theorem 11.3.4

$$\mathcal{R}^0 = \mathcal{M}.$$

Proof. (a) Let $\mathcal{R}_{k\ell}$ denote the subspace of $\mathcal{R}$ consisting of functions of the form

$$f(x) = (\det x)^k p(x_{ij}),$$

with $k \in \mathbb{Z}$, and where p is a polynomial in the n^2 variables x_{ij} of degree $\leq \ell$. The space $\mathcal{R}_{k\ell}$ is finite dimensional and G-biinvariant. It decomposes as a direct sum of irreducible subspaces for the action of $G \times G$. But the finite dimensional subspaces of $L^2(G)$ which are irreducible for the action of $G \times G$ are the subspaces $\mathcal{M}_\lambda$ ($\lambda \in \hat{G}$). Therefore

$$\mathcal{R}_{k\ell}^0 = \bigoplus_{\lambda \in \Sigma(k,\ell)} \mathcal{M}_\lambda,$$

where

$$\Sigma(k, \ell) = \{\lambda \in \hat{G} \mid \mathcal{M}_\lambda \subset \mathcal{R}_{k\ell}^0\}.$$

(b) To show that

$$\mathcal{R}^0 = \mathcal{M} := \bigoplus_{\lambda \in \hat{G}} \mathcal{M}_\lambda,$$

we argue by contradiction. Assume that there exists $\lambda \in \hat{G}$ such that $\mathcal{M}_\lambda \not\subset \mathcal{R}^0$. This implies that

$$\forall (k, \ell), \quad \mathcal{M}_\lambda \not\subset \mathcal{R}_{k\ell}^0.$$

Therefore,

$$\forall \mu \in \Sigma(k, \ell), \quad \mathcal{M}_\lambda \perp \mathcal{M}_\mu,$$

and $\mathcal{M}_\lambda \perp \mathcal{R}_{k\ell}^0$; since

$$\mathcal{R}^0 = \bigcup_{k,\ell} \mathcal{R}_{k\ell}^0,$$

it follows that $\mathcal{M}_\lambda \perp \mathcal{R}^0$. But we know that $\mathcal{R}^0$ is dense in $\mathcal{C}(G)$ (Proposition 11.3.3). Therefore this is impossible. $\qquad \square$

Corollary 11.3.5 *Every finite dimensional representation π of $G = U(n)$ extends uniquely as a holomorphic representaton $\tilde{\pi}$ of $GL(n, \mathbb{C})$.*

Proof. Let π be a representation of G on a finite dimensional complex vector space $\mathcal{V}$. Fix a basis $\{e_1, \dots, e_N\}$ in $\mathcal{V}$. By Theorem 11.3.4 the functions

$$u \mapsto \pi_{ij}(u) = \big(\pi(u)e_j | e_i\big)$$

extend as functions $g \mapsto \tilde{\pi}_{ij}(g)$ in $\mathcal{R}$. Let $\tilde{\pi}(g)$ denote the endomorphism of $\mathcal{V}$ with matrix entries $\big(\tilde{\pi}_{ij}(g)\big)$. The relation

$$\tilde{\pi}(g_1 g_2) = \tilde{\pi}(g_1)\tilde{\pi}(g_2)$$

is satisfied if $g_1, g_2 \in G$, and both sides are holomorphic functions in g_1 and g_2. The equality holds for $g \in GL(n, \mathbb{C})$ by Proposition 11.3.1. The uniqueness of the extension follows from that proposition as well.   □

11.4 Polynomial representations

Let π be a finite dimensional holomorphic representation of $GL(n, \mathbb{C})$. The representation π is said to be *polynomial* if the coefficients of π are restrictions to $GL(n, \mathbb{C})$ of polynomials in the n^2 variables x_{ij}.

Let λ be a dominant weight

$$\lambda = (\lambda_1, \dots, \lambda_n), \quad \lambda_i \in \mathbb{Z}, \quad \lambda_1 \geq \cdots \geq \lambda_n,$$

and π_λ an irreducible representation of $G = U(n)$ on a complex vector space $\mathcal{V}$ with highest weight λ. We will consider on $\mathcal{V}$ a Hermitian inner product for which the representation is unitary. By Corollary 11.3.5 this extends as a holomorphic representation of $GL(n, \mathbb{C})$ which will also be denoted by π_λ.

Theorem 11.4.1 *The representation π_λ is polynomial if and only if $\lambda_n \geq 0$.*

Proof. Since the representation of $G \times G$ on $\mathcal{M}_\lambda$ is irreducible, for π_λ to be polynomial it is necessary and sufficient that one of its coefficients is the restriction to $GL(n, \mathbb{C})$ of a polynomial.

Let v be a normalised highest weight vector. We will see that the coefficient

$$f(g) = (\pi_\lambda(g)v|v)$$

is equal to

$$\Delta_\lambda(g) := \Delta_1(g)^{\lambda_1 - \lambda_2} \Delta_2(g)^{\lambda_2 - \lambda_3} \dots \Delta_n(g)^{\lambda_n},$$

where $\Delta_k(g)$ denotes the principal minor determinant of order k:

$$\Delta_k(g) = \det\big((g_{ij})_{1 \leq i, j \leq k}\big).$$

In fact

$$f(d) = d_1^{\lambda_1} \dots d_n^{\lambda_n},$$

for a diagonal matrix $d = \mathrm{diag}(d_1, \dots, d_n)$, and

$$f(n_2^* g n_1) = f(g),$$

if n_1, n_2 are upper triangular matrices with diagonal entries equal to one. Hence,

$$f(n_2^* d n_1) = d_1^{\lambda_1} \dots d_n^{\lambda_n}.$$

This holds also for the function Δ_λ:

$$\Delta_\lambda(n_2^* d n_1) = d_1^{\lambda_1} \dots d_n^{\lambda_n}.$$

Let N denote the subgroup of upper triangular matrices with diagonal entries equal to one, and D the subgroup of diagonal matrices. Since the set $N^* D N$ is dense in $GL(n, \mathbb{C})$ (see Exercise 13 of Chapter 1), the functions f and Δ_λ are equal.

On the other hand Δ_λ is a polynomial if and only if $\lambda_n \geq 0$. This proves the statement. $\square$

If the coefficients of π_λ extend as holomorphic functions on $M(n, \mathbb{C})$, then the representation π_λ is polynomial. In fact, for the regular function Δ_λ to be the restriction to $GL(n, \mathbb{C})$ of a holomorphic function on $M(n, \mathbb{C})$, it is necessary and sufficient that $\lambda_n \geq 0$.

In particular, the character of a polynomial representation is a polynomial. Recall that the restriction to the subspace of diagonal matrices of the character χ_λ of π_λ is equal to the Schur function s_λ. Recall also that the Schur function s_α is given by

$$s_\alpha(t) = \frac{A_{\alpha+\delta}(t)}{V(t)}, \qquad t = (t_1, \dots, t_n) \in (\mathbb{C}^*)^n,$$

where

$$A_\alpha(t) = \begin{vmatrix} t_1^{\alpha_1} & \cdots & t_1^{\alpha_n} \\ \vdots & \vdots & \vdots \\ t_n^{\alpha_1} & \cdots & t_n^{\alpha_n} \end{vmatrix},$$

and $\delta = (n-1, n-2, \dots, 1, 0)$. The denominator $V(t) = A_\delta(t)$ is the Vandermonde polynomial.

Example 1. Consider the space $\Lambda^k(\mathbb{C}^n)$ of skew k-linear forms on $\mathbb{C}^n$ ($1 \leq k \leq n$). An element f in $\Lambda^k(\mathbb{C}^n)$ is a k-linear map

$$f : (\mathbb{C}^n)^k \to \mathbb{C},$$
$$(x_1, \dots, x_k) \mapsto f(x_1, \dots, x_k) \quad (x_i \in \mathbb{C}^n),$$

such that

$$f\big(x_{\sigma(1)}, \dots, x_{\sigma(k)}\big) = \varepsilon(\sigma) f(x_1, \dots, x_k) \quad (\sigma \in \mathfrak{S}_k).$$

Let π be the representation of $GL(n, \mathbb{C})$ on $\Lambda^k(\mathbb{C}^n)$ defined by

$$(\pi(g)f)(x_1, \dots, x_k) = f(x_1 g, \dots, x_k g).$$

The form

$$f_0(x_1, \ldots, x_k) = \det\left(\left(x_i^j\right)_{1 \le i,j \le k}\right)$$

is a highest weight vector. In fact, for a diagonal matrix $d = \mathrm{diag}(d_1, \ldots, d_n)$,

$$\pi(d)f_0 = d_1 \ldots d_k f_0,$$

and one can check that $\pi(n)f_0 = f_0$ for every $n \in N$. One can show also that every highest weight vector is proportional to f_0. Therefore the representation π is irreducible, and its highest weight is equal to

$$\lambda = 1^k := (1, \ldots, 1, 0, \ldots, 0)$$

(1 occurs k times and 0 occurs $n - k$ times).

To every sequence $J = (j_1, \ldots, j_k)$ such that $1 \le j_1 < \cdots < j_k \le n$ one associates the form

$$f_J(x_1, \ldots, x_k) = \det\left(\left(x_i^{j_\ell}\right)_{1 \le i,\ell \le k}\right).$$

The forms f_J constitute a basis of $\Lambda^k(\mathbb{C}^n)$ and, if $g = \mathrm{diag}(d_1, \ldots, d_n)$, then

$$\pi(g)f_J = d_{j_1} \ldots d_{j_k} f_J.$$

It follows that

$$\chi_\pi(g) = \sum_J d_{j_1} \ldots d_{j_k} = e_k(a_1, \ldots, a_n),$$

where e_k denotes the *elementary symmetric function*:

$$e_k(t) = \sum_{1 \le j_1 < j_2 < \cdots < j_k \le n} t_{j_1} t_{j_2} \ldots t_{j_k}, \qquad t = (t_1, \ldots, t_n) \in \mathbb{C}^n.$$

This shows that, if $\alpha = 1^k$, then

$$s_\alpha(t) = e_k(t).$$

This fact can be established directly. Observe that the generating function of the elementary symmetric functions e_k is given by

$$E(z; t) := \sum_{k=0}^n e_k(t)z^k = \prod_{j=1}^n (1 + zt_j), \qquad z \in \mathbb{C}, t = (t_1, \ldots, t_n) \in \mathbb{C}^n.$$

Consider the Vandermonde polynomial in $n + 1$ variables:

$$V(z, t_1, \ldots, t_n).$$

This can be written

$$V(z, t_1, \ldots, t_n)$$

$$= \prod_{i=1}^{n} (z - t_i) \prod_{i<j} (t_i - t_j)$$

$$= V(t_1, \ldots, t_n)\left(z^n - e_1(t)z^{n-1} + \cdots + (-1)^k e_k(t)z^{n-k} + \cdots + (-1)^n e_n(t)\right).$$

We can also write it as a determinant:

$$V(z, t_1, \ldots, t_n) = \begin{vmatrix} z^n & z^{n-1} & \cdots & z & 1 \\ t_1^n & t_1^{n-1} & \cdots & t_1 & 1 \\ \vdots & \vdots & \vdots & \vdots & \vdots \\ t_n^n & t_n^{n-1} & \cdots & t_n & 1 \end{vmatrix}.$$

We expand it with respect to the entries of the first row:

$$V(z, t_1, \ldots, t_n) = \sum_{k=0}^{n} (-1)^k z^{n-k} A_{1^k + \delta}(t).$$

It follows that

$$A_{1^k + \delta}(t) = V(t)e_k(t),$$

or

$$s_{1^k}(t) = e_k(t).$$

Example 2. Let π be the representation of $GL(n, \mathbb{C})$ on the space $S^m(\mathbb{C}^n)$ of homogeneous polynomials of degree m defined by

$$\left(\pi(g)p\right)(x) = f(xg).$$

The polynomial p_0,

$$p_0(x) = x_1^m,$$

is a highest weight vector. In fact, for a diagonal matrix $d = \operatorname{diag}(d_1, \ldots, d_n)$,

$$\pi(d)p_0 = d_1^m p_0,$$

and $\pi(n)p_0 = p_0$ for every $n \in N$. One can show also that every highest weight vector is proportional to it. Therefore the representation π is irreducible, and its highest weight vector is equal to

$$\lambda = [m] := (m, 0, \ldots, 0).$$

The monomials

$$p_\alpha(x) = x^\alpha, \quad (|\alpha| = m)$$

form a basis of $S^m(\mathbb{C}^n)$, and for a diagonal matrix $d = \mathrm{diag}(d_1, \ldots, d_n)$,

$$\pi(d)p_\alpha = d_1^{\alpha_1} \ldots d_n^{\alpha_n} p_\alpha.$$

It follows that

$$\chi_\pi(d) = h_m(d_1, \ldots, d_n),$$

where h_m is the *complete symmetric function*

$$h_m(t_1, \ldots, t_n) = \sum_{|\alpha|=m} t_1^{\alpha_1} \ldots t_n^{\alpha_n}.$$

This shows that, if $\alpha = [m]$, then

$$s_\alpha(t) = h_m(t).$$

This fact can be established directly as well. Let us show first that the generating function of the functions h_m is given by

$$H(z;t) := \sum_{m=0}^\infty h_m(t)z^m = \prod_{j=1}^n \frac{1}{1 - zt_j},$$

for $z \in \mathbb{C}$ with small enough modulus. In fact

$$\sum_{\alpha \in \mathbb{N}^n} (tx)^\alpha = \sum_{m=0}^\infty \left(\sum_{|\alpha|=m} x^\alpha\right) t^m = \sum_{m=0}^\infty h_m(x)t^m$$

$$= \prod_{j=1}^n \sum_{\alpha_j=0}^\infty (tx_j)^{\alpha_j} = \prod_{j=1}^n \frac{1}{1 - tx_j}.$$

Let us compute the sum of the following series which converges for $z \in \mathbb{C}$, $|z| < 1$,

$$\sum_{m=0}^\infty z^m A_{[m]+\delta}(t) = \begin{vmatrix} \sum_{m=0}^\infty z^m t_1^{m+n-1} & t_1^{n-2} & \cdots & t_1 & 1 \\ \sum_{m=0}^\infty z^m t_2^{m+n-1} & t_2^{n-2} & \cdots & t_2 & 1 \\ \vdots & \vdots & \cdots & \vdots & \vdots \\ \sum_{m=0}^\infty z^m t_n^{m+n-1} & t_n^{n-2} & \cdots & t_n & 1 \end{vmatrix}$$

$$= \begin{vmatrix} \frac{t_1^{n-1}}{1-zt_1} & t_1^{n-2} & \cdots & t_1 & 1 \\ \frac{t_2^{n-1}}{1-zt_2} & t_1^{n-2} & \cdots & t_2 & 1 \\ \vdots & \vdots & \cdots & \vdots & \vdots \\ \frac{t_n^{n-1}}{1-zt_n} & t_n^{n-2} & \cdots & t_n & 1 \end{vmatrix}.$$

Since

$$t_i^k + z\frac{t_i^{k+1}}{1 - zt_i} = \frac{t_i^k}{1 - zt_i},$$

this determinant is equal to

$$\begin{vmatrix} \frac{t_1^{n-1}}{1-zt_1} & \frac{t_1^{n-2}}{1-zt_1} & \cdots & \frac{t_1}{1-zt_1} & \frac{1}{1-zt_1} \\ \frac{t_2^{n-1}}{1-zt_2} & \frac{t_2^{n-2}}{1-zt_2} & \cdots & \frac{t_2}{1-zt_2} & \frac{1}{1-zt_2} \\ \vdots & \vdots & \cdots & \vdots & \vdots \\ \frac{t_n^{n-1}}{1-zt_n} & \frac{t_n^{n-2}}{1-zt_n} & \cdots & \frac{t_n}{1-zt_n} & \frac{1}{1-zt_n} \end{vmatrix} = V(t)\prod_{i=1}^{n}\frac{1}{1 - zt_i} = V(t)\sum_{m=0}^{\infty}z^m h_m(t).$$

Therefore

$$A_{[m]+\delta}(t) = V(t)h_m(t),$$

or

$$s_{[m]}(t) = h_m(t).$$

Observe that $h_m(1, \ldots, 1)$ is equal to the dimension of the space $S^m(\mathbb{C}^n)$ of m-homogeneous polynomials in n variables. Hence

$$\sum_{m=0}^{\infty}z^m \dim S^m(\mathbb{C}^n) = \frac{1}{(1 - z)^n}.$$

It follows that

$$\dim S^m(\mathbb{C}^n) = \frac{(m + n - 1)!}{m!(n - 1)!}.$$

11.5 Exercises

1. Let π be a representation of $U(2)$, with highest weight $\lambda = (\lambda_1, \lambda_2)$. Show that the restriction of π to $SU(2)$ is equivalent to the representation π_m as introduced in Section 7.5, with $m = \lambda_1 - \lambda_2$.
2. Let $\pi = \mathrm{Ad}$ be the adjoint representaton of $GL(n, \mathbb{C})$ on $\mathfrak{g} \simeq M(n, \mathbb{C})$. Determine the highest weight vectors of π. Is the representation π irreducible? Determine the weights of π.
3. Let π be an irreducible unitary representation of $U(n)$ with highest weight $\lambda = (\lambda_1, \ldots, \lambda_n)$. Show that the highest weight of the conjugate representation $\bar{\pi}$ is $\lambda' = (-\lambda_n, \ldots, -\lambda_1)$.

Hint. Put

$$w_0 = \begin{pmatrix} & & 1 \\ & \cdot^{\displaystyle\cdot^{\displaystyle\cdot}} & \\ 1 & & \end{pmatrix}.$$

Show that $\mathrm{Ad}\, w_0(N) = N^*$. Let v be a highest weight vector for the representation π. Show that $v' = \pi(w_0)v$ is a highest weight vector for the conjugate representation $\bar{\pi}$.

4. In order to be specific about the number of variables let us denote by $s_\alpha^{(n)}$ the Schur function in n variables $\big(\alpha = (\alpha_1, \ldots, \alpha_n)\big)$. Show that, for $\alpha_n > 0$,

$$s_\alpha^{(n)}(t_1, \ldots, t_{n-1}, 0) = 0,$$

and, for $\alpha_n = 0$,

$$s_\alpha^{(n)}(t_1, \ldots, t_{n-1}, 0) = s_{\alpha'}^{(n-1)}(t_1, \ldots, t_{n-1}),$$

with $\alpha' = (\alpha_1, \ldots, \alpha_{n-1})$.

5. For $\lambda \in P^+$, and $g \in GL(n, \mathbb{C})$, show that

$$|\chi_\lambda(g)| \le d_\lambda |\det g|^{\lambda_n} \|g\|^\alpha,$$

where

$$\alpha = \lambda_1 + \cdots + \lambda_{n-1} - (n-1)\lambda_n.$$

Hint. Use

$$|\operatorname{tr} \pi_\lambda(g)| \le d_\lambda \|\pi_\lambda(g)\|,$$

and the following decomposition of g,

$$g = u_1 d u_2,$$

where $k_1, k_2 \in U(n)$, and d is a real diagonal matrix.

6. Let π be the representation of the group $GL(n, \mathbb{C})$ on the space $V = M(n, \mathbb{C})$, consisting of $n \times n$ square complex matrices, defined by

$$\pi(g)x = gxg^T.$$

The subspaces $V_1 = Sym(n, \mathbb{C})$ and $V_2 = Skew(n, \mathbb{C})$ are invariant. Let π_1 and π_2 be the restrictions of π to the subspaces V_1 and V_2.

   (a) Compute the restrictions to the subgroup of invertible diagonal matrices of the characters χ, χ_1, χ_2 of the representations π, π_1, π_2.
   (b) Let N be the subgroup of G consisting of upper triangular matrices whose diagonal elements are equal to 1. Let W be the subspace of V

consisting of the matrices x which are invariant under N:

$$\mathcal{W} = \{x \in V \mid \forall n \in N, \ \pi(n)x = x\}.$$

Show that

$$\mathcal{W} = \left\{ \begin{pmatrix} \alpha & -\beta & 0 & \cdots & 0 \\ \beta & 0 & 0 & \cdots & 0 \\ 0 & 0 & 0 & \cdots & 0 \\ \vdots & \vdots & \vdots & \cdots & \vdots \\ 0 & 0 & 0 & \cdots & 0 \end{pmatrix} \ \middle| \ \alpha, \beta \in \mathbb{C} \right\}.$$

Hint. Show first that, if the matrix

$$x = \begin{pmatrix} x_1 & z^T \\ y & x_0 \end{pmatrix}, \quad x_1 \in \mathbb{C}, \ y, z \in \mathbb{C}^{n-1}, \quad x_0 \in M(n-1, \mathbb{C}),$$

satisfies $\pi(n)x = x$ for every matrix n of the form

$$n = \begin{pmatrix} 1 & u^T \\ 0 & I_{n-1} \end{pmatrix}, \quad u \in \mathbb{C}^{n-1},$$

then $x_0 = 0$, $z = -y$. Then show that, if furthermore $\pi(n)x = x$, for every matrix n of the form

$$n = \begin{pmatrix} 1 & 0 \\ 0 & n_0 \end{pmatrix},$$

where n_0 is a $(n-1) \times (n-1)$ upper triangular matrix with diagonal elements equal to 1, then $y^T = (\beta, 0, \ldots, 0)$.

(c) Consider now the restrictions of the representations π, π_1 and π_2 to the unitary group $U(n)$, which will also be denoted by π, π_1 and π_2. Show that the representations π_1 and π_2 are irreducible. Determine the highest weights of the representations π_1 and π_2.

(d) Evaluate the following integral:

$$\int_{U(n)} |\chi(g)|^2 d\mu(g),$$

where μ denotes the normalised Haar measure of the group $U(n)$. Evaluate the following integral:

$$\int_0^{2\pi} \cdots \int_0^{2\pi} \left| \sum_{j=1}^n e^{i\theta_j} \right|^4 \prod_{j<k} |e^{i\theta_j} - e^{i\theta_k}|^2 d\theta_1 \ldots d\theta_n.$$

7. Let π be a unitary representation of $U(n)$ on a complex Euclidean vector space $\mathcal{H}$ (finite dimensional). Let N be the subgroup of $GL(n, \mathbb{C})$ consisting

of upper triangular matrices with diagonal entries equal to 1, and let $\mathcal{H}^N$ be the subspace of $\mathcal{H}$ consisting of N-invariant vectors:

$$\mathcal{H}^N = \{v \in \mathcal{H} \mid \forall n \in N, \ \pi(n)v = v\}.$$

(a) Show that $\mathcal{H}^N$ is invariant under the operators $\pi(d)$ where $d \in GL(n, \mathbb{C})$ is diagonal.

(b) Let μ be a weight
Define

$$\mathcal{H}_\mu^N = \{v \in \mathcal{H}^N \mid \forall H \in \mathfrak{h}, \ d\pi(H)v = \mu(H)v\}.$$

Assume that, for every μ,

$$\dim \mathcal{H}_\mu^N = 0 \text{ or } 1,$$

and let M be the set of weights μ such that $\dim \mathcal{H}_\mu^N = 1$. Show that

$$\mathcal{H}^N = \bigoplus_{\mu \in M} \mathcal{H}_\mu^N.$$

(c) Consider a decomposition of $\mathcal{H}$ as a sum of irreducible subspaces:

$$\mathcal{H} = \bigoplus_{\alpha \in A} \mathcal{H}_\alpha.$$

Let $\mu(\alpha)$ denote the highest weight of the irreducible subspace $\mathcal{H}_\alpha$. Show that $\mu(\alpha) \in M$. Hence one defines a map:

$$\alpha \mapsto \mu(\alpha), \ A \to M.$$

(d) Show that the map $\alpha \mapsto \mu(\alpha)$ is injective.

(e) Show that the map $\alpha \mapsto \mu(\alpha)$ is surjective.
    Hint. Let $\mu \in M$ and $v \in \mathcal{H}_\mu^N$, $v \neq 0$. Show that the subspace $\mathcal{H}_\mu$ of $\mathcal{H}$ which is generated by the vectors

$$\{\pi(g)v \mid g \in GL(n, \mathbb{C})\}$$

is irreducible.

(f) Show that

$$\mathcal{H} = \bigoplus_{\mu \in M} \mathcal{H}_\mu.$$

8. In this exercise we propose to apply the results of the preceding excercise to the following representation: $\mathcal{H}$ is the space $\mathcal{P}^k\big(Sym(n, \mathbb{C})\big)$ of k-homogeneous polynomial functions on the vector space $Sym(n, \mathbb{C})$ of $n \times n$

complex symmetric matrices, and π is the representation of $GL(n, \mathbb{C})$ on $\mathcal{H}$ defined by

$$\big(\pi(g)p\big)(x) = p(g^T x g).$$

(a) Let $\mu \in M$ and $p \in \mathcal{H}_\mu^M$, $p \not\equiv 0$. Show that $p(I_n) \neq 0$.
   Hint. Similar to Exercise 13 of Chapter 1, show that the set

$$\{x = n^T d^2 n \mid n \in N, d \in D\}$$

is a dense open set in $Sym(n, \mathbb{C})$.
   Using the relation

$$p(n^T d^2 n) = e^{\mu(H)} p(I_n) \quad (d = \exp H),$$

show that $\mu = (2m_1, \ldots, 2m_n)$ with $m_i \in \mathbb{N}$, $m_1 + \cdots + m_n = 2k$.
(b) Let $\Delta_1(x), \ldots, \Delta_n(x)$ denote the principal minors of the matrix x. Let $m_1, \ldots, m_n \in \mathbb{N}$ be such that $m_1 \geq \cdots \geq m_n$, $m_1 + \cdots + m_n = k$. Define

$$\Delta_{\mathbf{m}}(x) = \Delta_1(x)^{m_1 - m_2} \Delta_2(x)^{m_2 - m_3} \ldots \Delta_n(x)^{m_n}.$$

Show that the polynomial $\Delta_{\mathbf{m}}$ belongs to $\mathcal{H}_\mu^N$ with $\mu = (2m_1, \ldots, 2m_n)$.
(c) Let $\mathcal{P}_{\mathbf{m}}$ denote the subspace in $\mathcal{H}$ generated by

$$\{\pi(g)\Delta_{\mathbf{m}} \mid g \in GL(n, \mathbb{C})\}.$$

Show that $\mathcal{P}_{\mathbf{m}}$ is an irreducible subspace with highest weight $\mu = (2m_1, \ldots, 2m_n)$, and that

$$\mathcal{H} = \bigoplus \mathcal{P}_{\mathbf{m}},$$

the sum being over the multi-indices $\mathbf{m} \in \mathbb{N}^n$ such that $m_1 \geq \cdots \geq m_n$, $m_1 + \cdots + m_n = k$.

12

Analysis on the unitary group

In this last chapter we present several applications of representation theory for the unitary group to harmonic analysis. The Laplace operator will be useful for studying convergence of Fourier series. In particular we consider Fourier series of central functions which can be written in terms of Schur function series. We will also see the analogue of the Taylor series for a holomorphic function of a matrix variable. We will determine the radial part of the Laplace operator, and finally study the heat kernel on the unitary group.

12.1 Laplace operator

We consider on the Lie algebra $\mathfrak{g} = i\,Herm(n, \mathbb{C})$ of the unitary group $G = U(n)$ the inner product

$$(X|Y) = \mathrm{tr}(XY^*) = -\,\mathrm{tr}(XY).$$

Let ρ be a representation of $\mathfrak{g}$ on a complex vector space $\mathcal{V}$. Recall that the Casimir operator associated to ρ is defined by

$$\Omega_\rho = \sum_{i=1}^{N} \rho(X_i)^2,$$

where $\{X_1, \ldots, X_N\}$ is an orthonormal basis of $\mathfrak{g}$ ($N = \dim \mathfrak{g} = n^2$). The representation ρ extends as a $\mathbb{C}$-linear representation of the complex Lie algebra $\mathfrak{g}_\mathbb{C} = \mathfrak{g} + i\mathfrak{g} = M(n, \mathbb{C})$.

Proposition 12.1.1

$$-\Omega_\rho = \sum_{j=1}^{n} \rho(E_{jj})^2 + \sum_{j=1}^{n} (n - 2j + 1)\rho(E_{jj}) + 2\sum_{j<k} \rho(E_{kj})\rho(E_{jk}).$$

Proof. Let us consider the orthonormal basis of $\mathfrak{g}$ consisting of the matrices

$$i E_{jj}, \quad X_{jk} = \frac{1}{\sqrt{2}}(E_{jk} - E_{kj}), \quad Y_{jk} = \frac{i}{\sqrt{2}}(E_{jk} + E_{kj}) \quad (j < k).$$

From the relations

$$\rho(X_{jk})^2 = \tfrac{1}{2}\left(\rho(E_{jk})^2 + \rho(E_{kj})^2 - \rho(E_{jk})\rho(E_{kj}) - \rho(E_{kj})\rho(E_{jk})\right),$$
$$\rho(Y_{jk})^2 = \tfrac{1}{2}\left(-\rho(E_{jk})^2 - \rho(E_{kj})^2 - \rho(E_{jk})\rho(E_{kj}) - \rho(E_{kj})\rho(E_{jk})\right),$$

it follows that

$$-\Omega_\rho = \sum_{j=1}^{n} \rho(E_{jj})^2 + \sum_{j\neq k} \rho(E_{jk})\rho(E_{kj}).$$

Using the relation

$$[E_{jk}, E_{kj}] = E_{jj} - E_{kk},$$

we get

$$\rho(E_{jk})\rho(E_{kj}) = \rho(E_{kj})\rho(E_{jk}) + \rho(E_{jj} - \rho(E_{kk}),$$

and finally

$$-\Omega_\rho = \sum_{j=1}^{n} \rho(E_{jj})^2 + \sum_{j=1}^{n}(n - 2j + 1)\rho(E_{jj}) + 2\sum_{j<k} \rho(E_{kj})\rho(E_{jk}). \quad \square$$

Let $(\pi, \mathcal{V})$ be an irreducible representation of G on a finite dimensional complex vector space $\mathcal{V}$. Let Ω_π denote the Casimir operator of the derived representation $d\pi$ of $\mathfrak{g}$. There exists a number κ_π such that $\Omega_\pi = -\kappa_\pi I$ (Corollary 6.7.2). If λ is the highest weight of the representation π, we will write $\kappa_\pi = \kappa_\lambda$.

Proposition 12.1.2

$$\kappa_\lambda = \sum_{j=1}^{n} \lambda_j^2 + \sum_{j=1}^{n}(n - 2j + 1)\lambda_j.$$

Proof. Let $v \in \mathcal{V}$ be a highest weight vector:

$$\rho(E_{jj})v = \lambda_j v,$$
$$\rho(E_{jk})v = 0 \quad \text{if } j < k.$$

By Proposition 12.1.1 it follows that

$$-\Omega_\pi v = \sum_{j=1}^{n} \lambda_j^2 v + \sum_{j=1}^{n}(n - 2j + 1)\lambda_j v. \quad \square$$

Recall that, for a function $f \in C^1(G)$, we wrote

$$\big(\rho(X)f\big)(g) = \frac{d}{dt}\bigg|_{t=0} f(g \exp tX).$$

The Laplace operator Δ of the group G is given, for a function $f \in C^2(G)$, by

$$\Delta f(g) = \sum_{j=1}^{N} \frac{d^2}{dt^2}\bigg|_{t=0} f(gtX_i),$$

that is

$$\Delta = \sum_{j=1}^{N} \rho(X_j)^2,$$

where $\{X_1, \ldots, X_N\}$ is an orthonormal basis of $\mathfrak{g}$.

Recall that $\mathcal{M}_\lambda$ denotes the subspace of $C(G)$ generated by the coefficients of an irreducible representation of G with highest weight λ. A function $f \in \mathcal{M}_\lambda$ is an eigenfunction of the Laplace operator:

$$\Delta f = -\kappa_\lambda f$$

(Proposition 8.2.1).

12.2 Uniform convergence of Fourier series on the unitary group

For every dominant weight $\lambda \in P^+$, we consider an irreducible representation π_λ with highest weight λ on a finite dimensional complex vector space $\mathcal{V}_\lambda$ with dimension d_λ. The space $\mathcal{V}_\lambda$ will be endowed with a Hermitian inner product for which the representation π_λ is unitary. The Fourier coefficient $\hat{f}(\lambda)$ ($\lambda \in P^+$) of an integrable function f on G is the endomorphism of $\mathcal{V}_\lambda$ defined by

$$\hat{f}(\lambda) = \int_G f(x)\pi_\lambda(x^{-1})\mu(dx)$$

(μ denotes the normalised Haar measure of G). If f is square integrable, then the Fourier series of f converges to f in the $L^2(G)$ sense,

$$f(x) = \sum_{\lambda \in P^+} d_\lambda \operatorname{tr}\big(\hat{f}(\lambda)\pi_\lambda(x)\big)$$

(Theorem 6.4.2). We will study the uniform convergence of this Fourier series using the Laplace operator Δ of G.

Proposition 12.2.1 *For a function $f \in C^2(G)$,*

$$\widehat{\Delta f}(\lambda) = -\kappa_\lambda \hat{f}(\lambda).$$

One can show this relation as in the case of the group $SU(2)$ (Proposition 8.4.1).

Theorem 12.2.2 *If $f \in C^{2k}(G)$ with $2k > \frac{1}{2} \dim G = \frac{1}{2} n^2$, then*

$$\sum_{\lambda \in P^+} d_\lambda^{3/2} \||\hat{f}(\lambda)\|| < \infty,$$

and

$$f(x) = \sum_{\lambda \in P^+} d_\lambda \operatorname{tr}\big(\hat{f}(\lambda)\pi_\lambda(x)\big);$$

the convergence is absolute and uniform.

Lemma 12.2.3 *For $2k > n^2/2$,*

$$\sum_{\lambda \in P^+, \lambda \neq 0} \frac{d_\lambda^2}{\kappa_\lambda^{2k}} < \infty.$$

Proof. By the dimension formula (Corollary 11.2.5) there is a constant $C > 0$, which depends only on n, such that, if $\lambda \neq 0$, then

$$d_\lambda \leq C \|\lambda\|^{n(n-1)/2}.$$

On the other hand,

$$\kappa_\lambda \geq \|\lambda\|^2.$$

Therefore,

$$\sum_{\lambda \in P^+, \lambda \neq 0} \frac{d_\lambda^2}{\kappa_\lambda^{2k}} \leq \|\lambda\|^{n(n-1)-4k}.$$

One uses the fact that the Epstein series

$$\sum_{\mathbf{m} \in \mathbb{Z}^n \setminus \{0\}} \|\mathbf{m}\|^{-s}$$

converges for $s > n$. $\qquad\qquad\square$

Proof of Theorem 12.2.2. Using Proposition 12.2.1 we obtain

$$\sum_{\lambda \neq 0} d_\lambda^{3/2} \||\hat{f}(\lambda)\|| = \sum_{\lambda \neq 0} d_\lambda^{3/2} \||\widehat{\Delta^k f}(\lambda)\||$$

$$\leq \left(\sum_{\lambda \neq 0} \frac{d_\lambda^2}{\kappa_\lambda^{2k}} \right)^{1/2} \left(\sum_{\lambda \neq 0} d_\lambda \||\widehat{\Delta^k f}(\lambda)\||^2 \right)^{1/2} < \infty.$$

This follows from Lemma 12.2.3 and from the relation

$$\sum_{\lambda \in P^+} d_\lambda \||\widehat{\Delta^k f}(\lambda)\||^2 = \int_G |\Delta^k f(x)|^2 \mu(dx).$$

The statement follows (see Proposition 6.6.1). □

As in the case of the group $SU(2)$, it is possible to characterise the Fourier coefficients of a C^∞ function.

Theorem 12.2.4 *Let f be a continuous function on G. The function f is C^∞ on G if and only if*

$$\forall k > 0, \quad \sup_{\lambda \in P^+} \|\lambda\|^k \||\hat{f}(\lambda)\|| < \infty.$$

The proof is similar to that of Theorem 8.4.3. Lemma 8.4.4 should be modified as follows.

Lemma 12.2.5

$$\||d\pi_\lambda(X)\|| \le \sqrt{d_\lambda} \|\lambda\| \|X\|.$$

Proof. For $X = iH = \mathrm{diag}(ih_1, \dots, ih_n)$ $(h_j \in \mathbb{R})$, the eigenvalues of $d\pi_\lambda(X)$ are the numbers $\mu(X) = i\mu(H)$, where μ is a weight of the representation π_λ. If $h_1 \ge \cdots \ge h_n$, then $-\lambda(H) \le \mu(H) \le \lambda(H)$. Hence

$$\|d\pi_\lambda(X)\| \le |\lambda(X)| \le \|\lambda\| \|X\|.$$

Every $X \in \mathfrak{g}$ can be written $x = \mathrm{Ad}(g)X_0$, with $g \in G$, $X_0 = \mathrm{diag}(ih_1, \dots, ih_n)$ $(h_1 \ge \cdots \ge h_n)$; therefore

$$\|d\pi_\lambda(X)\| \le \|\lambda\| \|X\|,$$

and

$$\||d\pi_\lambda(X)\|| \le \sqrt{d_\lambda} \|d\pi_\lambda(X)\| \le \sqrt{d_\lambda} \|\lambda\| \|X\|. \qquad \square$$

12.3 Series expansions of central functions

We saw in Section 6.5 that the characters of irreducible representations form a Hilbert basis of the subspace of $L^2(G)$ consisting of square integrable central functions (Proposition 6.5.3). This can be stated equivalently as follows: the Schur functions s_α form a Hilbert basis of the subspace of $L^2(\mathbb{T}^n)$ consisting of symmetric functions which are square integrable with respect to the measure

$$\frac{1}{n!} |V(t)|^2 \nu(dt).$$

In this section we develop a method for computing Schur function expansions explicitly.

Consider n power series

$$f_i(z) = \sum_{m=0}^{\infty} c_m^{(i)} z^m,$$

which converge for $|z| < r$. Then, for $|z_1| < r, \ldots, |z_n| < r$,

$$\det\big(f_i(z_j)\big)_{1 \le i,j \le n}$$

is a skewsymmetric analytic function in n variables. Hence it is divisible by the Vandermonde polynomial $V(z)$,

$$V(z) = \prod_{j<k}(z_j - z_k),$$

and the quotient is a symmetric function which admits an expansion as a Schur function series.

Proposition 12.3.1

$$\frac{\det\big(f_i(z_j)\big)_{1 \le i,j \le n}}{V(z)} = \sum_{m_1 \ge \cdots \ge m_n \ge 0} a_\mathbf{m} s_\mathbf{m}(z),$$

with

$$a_\mathbf{m} = \det\big(c_{m_j+\delta_j}^{(i)}\big)_{1 \le i,j \le n}.$$

Proof. By a simple computation,

$$\det\big(f_i(z_j)\big) = \sum_{\alpha_1 > \cdots > \alpha_n \ge 0} \det\big(c_{\alpha_j}^{(i)}\big) A_\alpha(z).$$

By putting $\alpha_j = m_j + \delta_j$, and dividing both sides by $V(z)$ one gets the statement. □

Corollary 12.3.2

$$\lim_{z \to a, \ldots, z_n \to a} \frac{\det\big(f_i(z_j)\big)_{1 \le i,j \le n}}{V(z)} = \frac{(-1)^{n(n-1)/2}}{\delta!} \det\big(f_i^{(j-1)}(a)\big).$$

For $\mathbf{m} = (m_1, \ldots, m_n) \in \mathbb{N}^n$, one defines

$$\mathbf{m}! = m_1! \ldots m_n!$$

In particular, for $\delta = (n-1, n-2, \ldots, 0)$,

$$\delta! = 1!2! \ldots (n-1)!.$$

Proof. By replacing z_i by $z_i - a$ if necessary, we may assume that $a = 0$. By the expansion in Proposition 12.3.1,

$$\lim_{z_1 \to 0, \ldots, z_n \to 0} \frac{\det(f_i(z_j))}{V(z)} = a_0 = \det\left(c_{\delta_j}^{(i)}\right).$$

Since

$$c_m^{(i)} = \frac{1}{m!} f_i^{(m)}(0),$$

we get

$$\begin{aligned}
a_0 &= \det\left(\frac{1}{(n-j)!} f_i^{(n-j)}(0)\right) \\
&= \frac{1}{(n-1)!} \cdots \frac{1}{2!} \det\left(f_i^{(n-j)}(0)\right) \\
&= \frac{(-1)^{n(n-1)/2}}{\delta!} \det\left(f_i^{(j-1)}(0)\right). \qquad \square
\end{aligned}$$

To a power series

$$f(z) = \sum_{m=0}^{\infty} c_m z^m,$$

which converges for $|z| < r$, we associate the function in the $2n$ variables $x_1, \ldots, x_n, y_1, \ldots, y_n$ defined by

$$\frac{\det(f(x_i y_j))_{1 \le i, j \le n}}{V(x)V(y)}.$$

This is a symmetric function in the variables x_i and also in the variables y_j.

Proposition 12.3.3 *For $|x_i y_j| < r$,*

$$\frac{\det(f(x_i y_j))_{1 \le i, j \le n}}{V(x)V(y)} = \sum_{m_1 \ge \ldots \ge m_n \ge 0} a_{\mathbf{m}} s_{\mathbf{m}}(x) s_{\mathbf{m}}(y),$$

with

$$a_{\mathbf{m}} = c_{m_1+\delta_1} \cdots c_{m_n+\delta_n}.$$

Proof. To the numbers $x_1, \ldots, x_n$ we associate the n power series

$$f_i(z) = f(x_i z) = \sum_{m=0}^{\infty} c_m x_i^m z^m.$$

In the present case $c_m^{(i)} = c_m x_i^m$, and

$$\det(c_{\alpha_j}^{(i)}) = \det(c_{\alpha_j} x_i^{\alpha_j}) = c_{\alpha_1} \cdots c_{\alpha_n} A_\alpha(x).$$

By Proposition 12.3.1 the statement follows. $\qquad \square$

We will look at two important examples.

First example. For f we take the exponential function

$$f(z) = e^z = \sum_{m=0}^{\infty} \frac{z^m}{m!}.$$

Corollary 12.3.4

(i)
$$\frac{\det\left(e^{x_i y_j}\right)_{1 \le i, j \le n}}{V(x)V(y)} = \sum_{m_1 \ge \ldots \ge m_n \ge 0} \frac{1}{(\mathbf{m} + \delta)!} s_{\mathbf{m}}(x) s_{\mathbf{m}}(y).$$

(ii)
$$e^{x_1 + \cdots + x_n} = \delta! \sum_{m_1 \ge \cdots m_n \ge 0} \frac{1}{(\mathbf{m} + \delta)!} s_{\mathbf{m}}(\underline{1}) s_{\mathbf{m}}(x).$$

We have used the notation $\underline{1} = (1, \ldots, 1)$.

Proof. One gets (ii) from (i) by passing to the limit as $y_1 \to 1, \ldots, y_n \to 1$. In fact one applies Corollary 12.3.2 to the functions $f_i(z) = e^{x_i z}$. Since

$$f_i^{(j-1)}(1) = x_i^{j-1} e^{x_i},$$

we get

$$\lim_{y \to 1, \ldots, y_n \to 1} \frac{\det(e^{x_i y_j})}{V(y)} = \det\left(x_i^{j-1} e^{x_i}\right) = (-1)^{n(n-1)/2} V(x) e^{x_1 + \cdots + x_n},$$

and (ii) follows. $\qquad\qquad\qquad\qquad\qquad\qquad\qquad\qquad\qquad\qquad\qquad\square$

Second example. Take

$$f(z) = \frac{1}{1-z} = \sum_{m=0}^{\infty} z^m \quad (|z| < 1).$$

Corollary 12.3.5 *For $|x_i y_j| < 1$,*

$$\prod_{i,j}^{n} \frac{1}{1 - x_i y_j} = \sum_{m_1 \ge \ldots \ge m_n \ge 0} s_{\mathbf{m}}(x) s_{\mathbf{m}}(y).$$

In particular, for $|x_i| < 1$, $y_j = 1$,

$$\prod_{i=1}^{n} \frac{1}{(1 - x_i)^n} = \sum_{m_1 \ge \ldots \ge m_n \ge 0} s_{\mathbf{m}}(\underline{1}) s_{\mathbf{m}}(x).$$

Proof. The formula in Proposition 12.3.3 can be written in the present case

$$\det\left(\frac{1}{1-x_iy_j}\right) = V(x)V(y)\sum_{m_1\geq\ldots\geq m_n\geq 0} s_{\mathbf{m}}(x)s_{\mathbf{m}}(y).$$

It is possible to evaluate this determinant. It is essentially the Cauchy determinant:

$$\det\left(\frac{1}{1-x_iy_j}\right) = V(x)V(y)\prod_{i,j}^{n}\frac{1}{1-x_iy_j}$$

(see Exercise 1). $\square$

Let f be a central C^∞ function on $G = U(n)$. It admits a uniformly convergent Fourier expansion:

$$f(g) = \sum_{\lambda\in P^+} a_\lambda \chi_\lambda(g),$$

with

$$a_\lambda = \int_{U(n)} f(y)\overline{\chi_\lambda(g)}\mu(dg).$$

Consider the function f on G defined by

$$f(g) = e^{\alpha\,\mathrm{tr}(g)} \quad (\alpha \in \mathbb{C}).$$

From Corollary 12.3.4 (ii) it follows that

$$e^{\alpha\,\mathrm{tr}(g)} = \sum_{\lambda\in P^+, \lambda_n\geq 0} d_\lambda \frac{\delta!}{(\lambda+\delta)!}\alpha^{|\lambda|}\chi_\lambda(g),$$

where $|\lambda| = \lambda_1 + \cdots + \lambda_n$, and also that

$$\int_G e^{\alpha\,\mathrm{tr}(g)}\overline{\chi_\lambda(g)}\mu(dg) = \frac{\delta!}{(\lambda+\delta)!}\alpha^{|\lambda|},$$

if $\lambda_n \geq 0$, and is equal to 0 otherwise. The expansion converges even for $g \in M(n,\mathbb{C})$ (in fact one can see this using Exercise 5 in Chapter 11), and equality still holds by Proposition 11.3.1.

Corollary 12.3.4 provides an alternative for evaluating the orbital integral,

$$\mathcal{I}(x,y) = \int_{U(n)} e^{\mathrm{tr}(xkyk^*)}\mu(dk) \quad (x,y \in Herm(n,\mathbb{C})),$$

we considered in Section 10.3.

Proposition 12.3.6 *For $x = \mathrm{diag}(x_1, \ldots, x_n)$ and $y = \mathrm{diag}(y_1, \ldots, y_n)$,*

$$\mathcal{I}(x, y) = \delta! \sum_{m_1 \geq \cdots \geq m_n \geq 0} \frac{1}{(\mathbf{m} + \delta)!} s_{\mathbf{m}}(x) s_{\mathbf{m}}(y)$$

$$= \delta! \frac{1}{V(x)V(y)} \det\left(e^{x_j y_k}\right)_{1 \leq j, k \leq n}.$$

Proof. We have seen that, for $z \in M(n, \mathbb{C})$,

$$e^{\mathrm{tr}(z)} = \sum_{\lambda \in P_+, \lambda_n \geq 0} d_\lambda a_\lambda \chi_\lambda(z),$$

with

$$a_\lambda = \frac{\delta!}{(\lambda + \delta)!},$$

and this series converges uniformly on every compact set. Therefore

$$\mathcal{I}(x, y) = \int_{U(n)} e^{\mathrm{tr}(xuyu^*)} \alpha(du)$$

$$= \sum_{\lambda \in P_+, \lambda_n \geq 0} d_\lambda a_\lambda \int_{U(n)} \chi_\lambda(xuyu^*) \alpha(du).$$

Since (Proposition 6.5.2)

$$\int_{U(n)} \chi_\lambda(xuyu^*) \alpha(du) = \frac{1}{d_\lambda} \chi_\lambda(x) \chi_\lambda(y),$$

it follows that

$$\mathcal{I}(x, y) = \sum_{\lambda \in P^+, \lambda_n \geq 0} a_\lambda \chi_\lambda(x) \chi_\lambda(y).$$

For $x = \mathrm{diag}(x_1, \ldots, x_n)$, $y = \mathrm{diag}(y_1, \ldots, y_n)$,

$$\mathcal{I}(x, y) = \delta! \frac{1}{V(x)V(y)} \det\left(e^{x_i y_j}\right). \qquad \square$$

Consider now the function f on $G = U(n)$ defined by

$$f(g) = \det(I - \alpha g)^{-n} \quad (|\alpha| < 1).$$

By Corollary 12.3.5 it follows that

$$\det(I - \alpha g)^{-n} = \sum_{\lambda \in P^+, \lambda_n \geq 0} d_\lambda \alpha^{|\lambda|} \chi_\lambda(g),$$

and that

$$\int_{U(n)} \det(I - \alpha g)^{-n} \overline{\chi_\lambda(g)} \mu(dg) = \alpha^{|\lambda|},$$

if $\lambda_n \geq 0$, and is equal to 0 otherwise. In fact, for $g \in M(n, \mathbb{C})$, $\|g\| < 1$,

$$\det(I - g)^{-n} = \sum_{\lambda \in P^+, \lambda_n \geq 0} d_\lambda \chi_\lambda(g)$$

12.4 Generalised Taylor series

Let $\mathcal{D}_R$ be the ball with centre 0 and radius R in $M(n, \mathbb{C})$,

$$\mathcal{D}_R = \{x \in M(n, \mathbb{C}) \mid \|x\| < R\}.$$

Let the function f be holomorphic on $\mathcal{D}_R$. For $\lambda \in P^+$, $g \in GL(n, \mathbb{C})$, $\|g\| < R$, put

$$A_\lambda(g) = \int_{U(n)} f(gk)\pi_\lambda(gk)^{-1}\mu(dk).$$

The function A_λ, with values in $\mathrm{End}(\mathcal{V}_\lambda)$ is defined and holomorphic in $\mathcal{D}_R$. By the invariance of the Haar measure μ we get, for $h \in U(n)$,

$$A_\lambda(gk) = \int_{U(n)} f(ghk)\pi_\lambda(ghk)^{-1}\mu(dk)$$

$$= \int_{U(n)} f(gk')\pi_\lambda(gk')^{-1}\mu(dk') = A_\lambda(g).$$

By Proposition 11.3.1, it follows that $A_\lambda(g)$ does not depend on g. We will write A_λ instead of $A_\lambda(g)$. We can write

$$A_\lambda \pi_\lambda(g) = \int_{U(n)} f(gk)\pi_\lambda(k^{-1})\mu(dk).$$

The right-hand side is the restriction to $\{g \in GL(n, \mathbb{C}) \mid \|g\| < R\}$ of a holomorphic function in $\mathcal{D}_R$. Therefore, if $A_\lambda \neq 0$, then there is a non-zero coefficient of the representation π_λ which is holomorphic on $\mathcal{D}_R$, hence π_λ is polynomial, and this implies $\lambda_n \geq 0$. The endomorphisms A_λ are called *generalised Taylor coefficients* of the function f.

Define, for $0 \leq r < R$,

$$M(r) = \sup_{k \in U(n)} |f(rk)|.$$

Lemma 12.4.1 (Cauchy inequalities) *For* $0 < r < R$,

$$\|A_\lambda\| \leq r^{-|\lambda|}M(r),$$

where $|\lambda| = \lambda_1 + \cdots + \lambda_n$.

Proof. Take $g = rI$. Then

$$\pi_\lambda(g) = r^{|\lambda|} Id,$$

and

$$r^{|\lambda|} \|A_\lambda\| \le \int_U |f(rk)| \mu(dk) \le M(r). \qquad \square$$

Theorem 12.4.2 *Let the function f be defined and holomorphic on $\mathcal{D}_R$. Then*

$$f(x) = \sum_{\lambda \in P^+, \lambda_n \ge 0} d_\lambda \operatorname{tr}\big(A_\lambda \pi_\lambda(x)\big);$$

the series converges absolutely and uniformly on every compact set in $\mathcal{D}_R$.

Proof. Fix $g \in GL(n, \mathbb{C})$, $\|g\| < R$, and define the function φ on $U(n)$ by

$$\varphi(k) = f(gk).$$

The generalised Fourier coefficients of φ are given by

$$\int_{U(n)} \varphi(k) \pi_\lambda(k^{-1}) \mu(dk) = A_\lambda \pi_\lambda(g).$$

Since the function φ is C^∞, it is equal to the sum of its Fourier series (Theorem 12.2.2):

$$\varphi(k) = \sum_{\lambda \in P^+} d_\lambda \operatorname{tr}\big(A_\lambda \pi_\lambda(gk)\big),$$

and, for $k = e$,

$$f(g) = \sum_{\lambda \in P^+} d_\lambda \operatorname{tr}\big(A_\lambda \pi_\lambda(g)\big).$$

Let us prove that the series converges uniformly on

$$Q_r = \{x \in M(n, \mathbb{C}) \mid \|x\| \le r\},$$

for every $r < R$. Let $r < r_1 < R$. By Lemma 12.4.1,

$$\|A_\lambda\| \le r_1^{-|\lambda|} M(r_1),$$

and, if $x \in Q_r$,

$$|d_\lambda \operatorname{tr}\big(A_\lambda \pi_\lambda(g)\big)| \le d_\lambda^2 \|A_\lambda\| \|\pi_\lambda(g)\| \le d_\lambda^2 M(r_1) \left(\frac{r}{r_1}\right)^{|\lambda|},$$

since

$$|\operatorname{tr}(A_\lambda \pi_\lambda(g)| \le \|\!|A_\lambda|\!\| \|\!|\pi_\lambda(g)|\!\|,$$

$$\|\!|A_\lambda|\!\| \le \sqrt{d_\lambda} \|A_\lambda\|, \quad \|\!|\pi_\lambda(g)|\!\| \le \sqrt{d_\lambda} \|g\|^{|\lambda|}.$$

On the other hand if, in the second formula of Corollary 12.3.5, we take $x_i = t$, with $0 \leq t < 1$, we get

$$\sum_{\lambda \in P^+, \lambda_n \geq 0} d_\lambda^2 t^{|\lambda|} = (1 - t)^{-n^2}. \qquad \square$$

Let $\mathcal{D} = \mathcal{D}_1$ be the unit ball in $M(n, \mathbb{C})$. We denote by $A(\mathcal{D})$ the space of continuous functions on $\overline{\mathcal{D}}$, holomorphic on $\mathcal{D}$.

Theorem 12.4.3 (Cauchy–Bochner formula) *Let* $f \in A(\mathcal{D})$ *and* $x \in \mathcal{D}$.

$$f(x) = \int_{U(n)} \det(I - k^{-1}x)^{-n} f(k)\mu(dk).$$

Proof. By Theorem 12.4.2,

$$f(x) = \sum_{\lambda \in P^+, \lambda \geq 0} d_\lambda \operatorname{tr}\big(A_\lambda \pi_\lambda(x)\big),$$

and

$$A_\lambda = \int_{U(n)} f(k)\pi_\lambda(k^{-1})\mu(dk).$$

Therefore

$$f(g) = \sum_{\lambda \in P^+, \lambda_n \geq 0} d_\lambda \int_{U(n)} f(k)\chi_\lambda(k^{-1}x)\mu(dk).$$

By Corollary 12.3.5, if $\|y\| < 1$,

$$\sum_{\lambda \in P^+, \lambda_n \geq 0} d_\lambda \chi_\lambda(y) = \det(I - y)^{-n};$$

the convergence is uniform on every compact set in $\mathcal{D}$. Hence we can permute integration and summation:

$$f(x) = \int_{U(n)} f(k) \left(\sum_{\lambda \in P^+, \lambda_n \geq 0} d_\lambda \chi_\lambda(k^{-1}x) \right) \mu(dk)$$

$$= \int_{U(n)} f(k) \det(I - k^{-1}x)^{-n} d\mu(k). \qquad \square$$

By the maximum principle, the maximum of $|f|$ for $f \in A(\mathcal{D})$ is reached at a point in the topological boundary $\partial \mathcal{D}$. It even turns out that the maximum is reached at a point in $U(n) \subset \partial \mathcal{D}$. Furthermore $U(n)$ is the smallest closed set in $\partial \mathcal{D}$ with this property. One says that $U(n)$ is the *Shilov boundary* of $\mathcal{D}$. This follows from the next two propositions.

Proposition 12.4.4 *Let $f \in A(\mathcal{D})$. Then*

$$\max_{x \in \overline{\mathcal{D}}} |f(x)| = \max_{u \in U(n)} |f(u)|.$$

Proof. Fix $k_1, k_2 \in U(n)$, and put

$$F(z_1, \ldots, z_n) = f\big(k_1 \mathrm{diag}(z_1, \ldots, z_n)k_2\big).$$

The function F is continuous on the closure $\overline{D}^n$ of the polydisc

$$D^n = \{(z_1, \ldots, z_n) \in \mathbb{C}^n \mid |z_j| < 1\},$$

and holomorphic on D^n. The maximum of $|F|$ is reached at a point in

$$\mathbb{T}^n = \{(u_1, \ldots, u_n) \in \mathbb{C}^n \mid |u_j| = 1\}.$$

Since every $x \in \overline{\mathcal{D}}$ can be written

$$x = k_1 \mathrm{diag}(z_1, \ldots, z_n)k_2,$$

with $k_1, k_2 \in U(n)$, $|z_j| \leq 1$, the statement follows. $\square$

Proposition 12.4.5 *For $k \in U(n)$, define*

$$f(z) = e^{\mathrm{tr}(zk^{-1})}.$$

The maximum of $|f|$ on $\overline{\mathcal{D}}$ is reached at $z = k$, and uniquely at this point.

This follows from the next lemma. Let $S(0, r)$ denote the Euclidean sphere with centre 0 and radius r in $M(n, \mathbb{C})$:

$$S(0, r) = \{z \in M(n, \mathbb{C}) \mid |||z||| = r\}.$$

Lemma 12.4.6

$$\overline{\mathcal{D}} \cap S(0, \sqrt{n}) = U(n).$$

Proof. We have to show that, for $\|Z\| \leq 1$ with $|||Z||| = \sqrt{n}$, then $z \in U(n)$. Let $Z_1, \ldots, Z_n$ be the columns of the matrix Z. Since $\|Z\| \leq 1$,

$$\|Z_j\| \leq 1 \quad (j = 1, \ldots, n),$$

and since $|||Z||| = \sqrt{n}$,

$$\|Z_1\|^2 + \cdots + \|Z_n\|^2 = n.$$

Therefore $\|Z_j\| = 1$ $(j = 1, \ldots, n)$. For $\xi \in \mathbb{C}^n$

$$Z\xi = \sum_{j=1}^{n} \xi_j Z_j,$$

and

$$\|Z\xi\|^2 = \sum_{j=1}^{n} \|Z_j\|^2 |\xi_j|^2 + \sum_{j \neq k} (Z_j | Z_k) \xi_j \overline{\xi_k}$$

$$\leq \sum_{j=1}^{n} |\xi_j|^2.$$

Therefore, for every $\xi \in \mathbb{C}^n$,

$$\sum_{j \neq k} (Z_j | Z_k) \xi_j \overline{\xi_k} \leq 0.$$

It follows that $(Z_j | Z_k) = 0 \; (j \neq k)$. Hence we have shown that the matrix Z is unitary. $\qquad\qquad\qquad\qquad\qquad\qquad\qquad\qquad\qquad\qquad\qquad\qquad$ $\square$

12.5 Radial part of the Laplace operator on the unitary group

We consider on the Lie algebra $\mathfrak{g} = \mathfrak{u}(n)$ of the unitary group $G = U(n)$ the Euclidean inner product

$$(X|Y) = \mathrm{tr}(XY^*) = -\mathrm{tr}(XY).$$

The Laplace operator Δ acts on $\mathcal{C}^2$ functions on G. It is defined by

$$\Delta = \sum_{i=1}^{N} \rho(X_i)^2 \quad (N = \dim G = n^2),$$

where $\{X_1, \ldots, X_n\}$ is an orthonormal basis in $\mathfrak{g}$. This definition does not depend on the choice of orthonormal basis. The operator Δ commutes with the operators $L(g)$ et $R(g)$. It is symmetric: if $f, \varphi \in \mathcal{C}^2(G)$, then

$$\int_G \Delta f(x) \overline{\varphi(x)} \mu(dx) = \int_G f(x) \overline{\Delta \varphi(x)} \mu(dx).$$

And $-\Delta$ is positive:

$$-\int_G \Delta f(x) \overline{f(x)} \mu(dx) = \int_G \sum_{i=1}^{N} |\rho(X_i) f(x)|^2 \mu(dx) \geq 0.$$

If the function f is central, then the function Δf is central as well. A central function is determined by its restriction to the subgroup of unitary diagonal matrices

$$a(\theta) = \begin{pmatrix} e^{i\theta_1} & & \\ & \ddots & \\ & & e^{i\theta_n} \end{pmatrix}.$$

We put

$$f_0(\theta_1, \ldots, \theta_n) = f\big(a(\theta)\big).$$

Proposition 12.5.1 (i) *If f is a central function then*

$$(\Delta f)_0 = L f_0,$$

where L is the following differential operator, the radial part of the Laplace operator,

$$L = \sum_{j=1}^{n} \frac{\partial^2}{\partial\theta_j^2} + \sum_{j<k} \cot \frac{\theta_j - \theta_k}{2} \left(\frac{\partial}{\partial\theta_j} - \frac{\partial}{\partial\theta_k} \right).$$

(ii) *This operator can also be written*

$$L f_0 = \frac{1}{\Pi(\theta)} \left(\sum_{j=1}^{n} \frac{\partial^2}{\partial\theta_j^2} + \gamma \right) \big(\Pi(\theta) f_0\big),$$

with

$$\Pi(\theta) = \prod_{j<k} \sin \frac{\theta_j - \theta_k}{2}, \quad \gamma = \sum_{j=1}^{n} \left(\delta_j - \frac{n-1}{2} \right)^2.$$

Proof. Consider the orthonormal basis of $\mathfrak{g}$ consisting of the following matrices

$$i E_{jj}, \quad X_{jk} = \frac{1}{\sqrt{2}}(E_{jk} - E_{kj}), \quad Y_{jk} = \frac{i}{\sqrt{2}}(E_{jk} + E_{kj}).$$

We follow the method of the second proof of Proposition 8.3.3. First,

$$\rho(i E_{jj})^2 f\big(a(\theta)\big) = \frac{\partial^2}{\partial\theta_j^2} f\big(a(\theta)\big).$$

Fix j and k ($j \neq k$), and put, for $z \in \mathbb{C}$,

$$T(z) = z E_{jk} - \bar{z} E_{kj}.$$

We apply relation (d) in Section 8.2

$$\frac{d^2}{dt^2} f(g \exp tX \exp tY) \bigg|_{t=0} = \big(\rho(X+Y)^2 f\big)(g) + \big(\rho([X,Y])f\big)(g),$$

by taking

$$X = \text{Ad}(a(-\theta))T(z) = T(e^{-i(\theta_j - \theta_k)}z),$$
$$Y = -T(z),$$

for which

$$X + Y = T((e^{-i(\theta_j - \theta_k)} - 1)z),$$
$$[X, Y] = -2\sin(\theta_j - \theta_k)|z|^2(E_{jj} - E_{kk}).$$

If

$$z = ie^{i(\theta_j - \theta_k)/2},$$

then

$$X + Y = 2\sqrt{2}\sin\frac{\theta_j - \theta_k}{2}X_{jk}.$$

The relation (d) can be written

$$8\sin^2\frac{\theta_j - \theta_k}{2}\rho(X_{jk})^2 f(a(\theta)) - 2\sin(\theta_j - \theta_k)\rho(E_{jj} - E_{kk})f(a(\theta)) = 0,$$

or

$$\rho(X_{jk})^2 f(a(\theta)) = \frac{1}{2}\cot\frac{\theta_j - \theta_k}{2}\rho(E_{jj} - E_{kk})f(a(\theta)).$$

If

$$z = e^{i(\theta_j - \theta_k)/2},$$

then

$$X + Y = -2\sin\frac{\theta_j - \theta_k}{2}Y_{jk},$$

and

$$\rho(Y_{jk})^2 f(a(\theta)) = \frac{1}{2}\cot\frac{\theta_j - \theta_k}{2}\rho(E_{jj} - E_{kk})f(a(\theta)).$$

This proves (i) since

$$\Delta = \sum_{j=1}^{n}\rho(iE_{jj})^2 + \sum_{j<k}(\rho(X_{jk})^2 + \rho(Y_{jk})^2). \qquad \square$$

Lemma 12.5.2

$$\sum_{j=1}^{n}\frac{\partial^2}{\partial\theta_j^2}\Pi(\theta) = -\gamma\,\Pi(\theta),$$

where

$$\gamma = \sum_{j=1}^{n} \left(\delta_j - \frac{n-1}{2} \right)^2 .$$

Proof. From the relation

$$e^{i\theta_j} - e^{i\theta_k} = 2i e^{i(\theta_j + \theta_k)/2} \sin \frac{\theta_j - \theta_k}{2}$$

it follows that

$$V(e^{i\theta_1}, \dots, e^{i\theta_k}) = \prod_{j<k}(e^{i\theta_j} - e^{i\theta_k})$$
$$= (2i)^{n(n-1)/2} \exp\left(i^{(n-1)/2} \sum_{j=1}^{n} \theta_j \right) \Pi(\theta),$$

hence

$$\Pi(\theta) = (2i)^{-n(n-1/2)} \sum_{\sigma \in \mathfrak{S}_n} \exp\left(i \sum_{j=1}^{n} \left(\sigma(\delta_j) - \frac{n-1}{2} \right) \theta_j \right).$$

The statement follows. □

Let us prove part (ii) in the proposition. For a C^2 function f on $\mathbb{R}^n$,

$$\Delta_0(\Pi f) = \Pi \Delta_0 f + 2(\nabla_0 \Pi | \nabla_0 f) + f \Delta_0 \Pi,$$

where Δ_0 is the ordinary Laplace operator on $\mathbb{R}^n$, and ∇_0 is the associated gradient. A simple computation gives

$$\nabla_0 \Pi = \Pi \nabla_0 \log \Pi$$
$$= \Pi \nabla_0 \sum_{j<k} \log \sin \frac{\theta_j - \theta_k}{2}$$
$$= \tfrac{1}{2} \Pi \sum_{j<k} \cot \frac{\theta_j - \theta_k}{2} (e_j - e_k),$$

where $\{e_1, \dots, e_n\}$ is the canonical basis in $\mathbb{R}^n$. Hence

$$\Delta_0(\Pi f) = \Pi \left(\Delta_0 f + \sum_{j<k} \cot \frac{\theta_j - \theta_k}{2} \left(\frac{\partial f}{\partial \theta_j} - \frac{\partial f}{\partial \theta_k} \right) - \gamma f \right),$$

and this is the stated formula. □

12.6 Heat equation on the unitary group

In this section we study the Cauchy problem for the *heat equation* on the unitary group $G = U(n)$,

$$\frac{\partial u}{\partial t} = \Delta u,$$
$$u(0, x) = f(x),$$

where f is a continuous function on G. We follow the same method used in Section 8.6 where we studied the heat equation on the group $SU(2)$. Propositions 8.6.1 and 8.6.2 together with their proofs hold for the group $G = U(n)$. Each of them implies uniqueness of the solution of the Cauchy problem. To establish existence one uses the Fourier method. Assume first that $f \in C^{2k}$ with $2k > n^2/2$. Then the Fourier coefficients of f satisfy

$$\sum_{\lambda \in P^+} d_\lambda^{3/2} \|\|\hat{f}(\lambda)\|\| < \infty,$$

and the Fourier series of f,

$$\sum_{\lambda \in P^+} d_\lambda \, \mathrm{tr}\big(\hat{f}(\lambda)\pi_\lambda(x)\big),$$

converges uniformly to f (Theorem 12.2.2). In this case the solution of the Cauchy problem is given by

$$u(t, x) = \sum_{\lambda \in P^+} e^{-\kappa_\lambda t} \, \mathrm{tr}\big(\hat{f}(\lambda)\pi_\lambda(x)\big),$$

with

$$\kappa_\lambda = \sum_{j=1}^{n} \lambda_j^2 + \sum_{j=1}^{n}(n - 2j + 1)\lambda_j.$$

The *heat kernel* is defined on $]0, \infty[\times G$ by

$$H(t, x) = \sum_{\lambda \in P^+} d_\lambda \, e^{-\kappa_\lambda t} \chi_\lambda(x).$$

For $t \geq t_0 > 0$, this series converges absolutely and uniformly. In fact,

$$|\chi_\lambda(x)| \leq \chi_\lambda(e) = d_\lambda,$$

and

$$\sum_{\lambda \in P^+} d_\lambda^2 \, e^{-\kappa_\lambda t_0} < \infty,$$

since

$$d_\lambda \le C(1 + \|\lambda\|)^{n(n-1)/2},$$
$$\kappa_\lambda \ge \|\lambda\|^2.$$

The solution of the Cauchy problem can be written

$$u(t, x) = \int_G H(t, xy^{-1}) f(y) \mu(dy).$$

One can show the following, as in Section 8.6.

Proposition 12.6.1 *The heat kernel H has the following properties:*

(i) $H(t, x) \ge 0$,
(ii) $\int_G H(t, x) \mu(dx) = 1$,
(iii) *for every neighbourhood V of e,*

$$\lim_{t \to 0} \int_V H(t, x) \mu(dx) = 1.$$

The proof is the same as that of Proposition 8.6.3. For the proof of (iii) one uses a C^{2k} positive function, with $k > n^2/2$, whose support is contained in V. Similarly one can deduce the following.

Theorem 12.6.2 *Let f be a continuous function on G. The Cauchy problem admits a unique solution which is given, for $t > 0$, by*

$$u(t, x) = \int_G H(t, xy^{-1}) f(y) \mu(dy).$$

Let $H_0(t, \theta)$ denote the restriction of the kernel $H(t, x)$ to the subgroup of diagonal matrices:

$$H_0(t, \theta) = H(t, a(\theta)).$$

By Proposition 12.5.1 (ii) the function

$$H_1(t, \theta) = \Pi(\theta) H_0(t, \theta)$$

should be a solution of

$$\frac{\partial H_1}{\partial t} = \sum_{j=1}^n \frac{\partial^2 H_1}{\partial \theta_j^2} + \gamma H_1.$$

In fact this can be checked:

$$H_1(t, \theta) = (2i)^{-n(n-1)/2} \exp\left(-i\frac{n-1}{2}\sum_{j=1}^n \theta_j\right)$$
$$\times \sum_{m_1 \ge \cdots \ge m_n} d_m e^{-\kappa_m t} A_{m+\delta}(e^{i\theta_1}, \ldots, e^{i\theta_n}),$$

Analysis on U(n)

and

$$\kappa_{\mathbf{m}} = \sum_{j=1}^{n} m_j^2 + \sum_{j=1}^{n} (n - 2j + 1)m_j$$

$$= \sum_{j=1}^{n} \left(m_j + \delta_j - \frac{n-1}{2} \right)^2 - \sum_{j=1}^{n} \left(\delta_j - \frac{n-1}{2} \right)^2.$$

Finally we establish a formula which is analogous to Proposition 8.6.6.

Proposition 12.6.3

$$H_0(t, \theta) = C_n \frac{e^{\gamma t}}{t^{n^2/2}} \sum_{k \in \mathbb{Z}^n} \frac{V(\theta - 2k\pi)}{\Pi(\theta)} e^{-\|\theta - 2k\pi\|^2/4t}.$$

Proof. As at the end of Section 8.5 we use the Poisson summation formula. □

Lemma 12.6.4 (Poisson summation formula) *Let* $f \in \mathcal{S}(\mathbb{R}^n)$. *Put*

$$\check{f}(x) = \int_{\mathbb{R}^n} e^{i(x|\xi)} f(\xi) d\xi.$$

Then

$$\sum_{\mathbf{m} \in \mathbb{Z}^n} f(\mathbf{m}) e^{i(\mathbf{m}|x)} = \sum_{k \in \mathbb{Z}^n} \check{f}(x - 2k\pi).$$

If

$$f(\xi) = e^{-t\|\xi - a\|^2} \quad (a \in \mathbb{R}^n),$$

then

$$\check{f}(x) = e^{i(a|x)} \left(\frac{\pi}{t} \right)^{\frac{n}{2}} e^{-\|x\|^2/4t},$$

and we get

$$\sum_{\mathbf{m} \in \mathbb{Z}^n} e^{-t\|\mathbf{m} - a\|^2} e^{i(\mathbf{m}|x)} = e^{i(a|x)} \left(\frac{\pi}{t} \right)^{n/2} \sum_{k \in \mathbb{Z}^n} e^{-\|x - 2k\pi\|^2/4t}.$$

Both series converge uniformly and can be differentiated termwise. We apply to both sides the differential operator $V(\partial/\partial x)$. Let us compute the left-hand side. Put

$$F(t, x) := V\left(\frac{\partial}{\partial x}\right)\left(e^{-i(a|x)}\sum_{\mathbf{m}\in\mathbb{Z}^n} e^{-t\|\mathbf{m}-a\|^2} e^{i(\mathbf{m}|x)}\right)$$

$$= e^{-i(a|x)} V\left(\frac{\partial}{\partial x}\right)\left(\sum_{\mathbf{m}\in\mathbb{Z}^n} e^{-t\|\mathbf{m}-a\|^2} e^{i(\mathbf{m}|x)}\right)$$

$$= i^p e^{-i(a|x)}\sum_{\mathbf{m}\in\mathbb{Z}^n} e^{-t\|\mathbf{m}-a\|^2} V(\mathbf{m})e^{i(\mathbf{m}|x)},$$

where $p = n(n-1)/2$. Since the polynomial V is skewsymmetric, we get

$$F(t, x) = i^p e^{-i(a|x)}\sum_{m_1 > \cdots > m_n} e^{-t\|\mathbf{m}-a\|^2} V(\mathbf{m})A_{\mathbf{m}}(e^{ix_1}, \ldots, e^{ix_n})$$

$$= i^p e^{-i(a|x)}\sum_{m_1 \geq \cdots \geq m_n} e^{-t\|\mathbf{m}+\delta-a\|^2} V(\mathbf{m}+\delta)A_{\mathbf{m}+\delta}(e^{ix_1}, \ldots, e^{ix_n}).$$

By the dimension formula (Corollary 11.2.5)

$$d_{\mathbf{m}} = \frac{V(\mathbf{m}+\delta)}{V(\delta)},$$

and we saw that

$$\kappa_{\mathbf{m}} = \|\mathbf{m}+\delta-a\|^2 - \gamma,$$

with

$$a = \left(\frac{n-1}{2}, \ldots, \frac{n-1}{2}\right).$$

We get finally

$$\frac{1}{\Pi(x)} F(t, x) = C_n' e^{-\gamma t}\sum_{m_1 \geq \cdots \geq m_n} d_{\mathbf{m}} e^{-\kappa_{\mathbf{m}} t} s_{\mathbf{m}}(e^{ix_1}, \ldots, e^{ix_n})$$

$$= c_n' e^{-\gamma t} H_0(t, x).$$

To compute the right-hand side we use the following lemma.

Lemma 12.6.5 *Let the function f be C^p ($p = n(n-1)/2$) on an open ball with centre 0 in $\mathbb{R}^n$ and radial:*

$$f(x) = F(\|x\|).$$

Then

$$\left(V\left(\frac{\partial}{\partial x}\right)f\right)(x) = \left(\left(\frac{1}{r}\frac{d}{dr}\right)^p F\right)(\|x\|) \cdot V(x).$$

In particular, if

$$f(x) = e^{s\|x\|^2/2} \quad (s \in \mathbb{R}),$$

we get

$$\left(V\left(\frac{\partial}{\partial x} \right) f \right)(x) = s^p f(x) V(x).$$

Proof. We will show first that, for a m-homogeneous polynomial P,

$$P\left(\frac{\partial}{\partial x} \right) f(x) = \left(\frac{1}{r}\frac{d}{dr} \right)^m F(\|x\|) P(x) + \sum_{k<m} \left(\frac{1}{r}\frac{d}{dr} \right)^k F(\|x\|) P_k(x),$$

where P_k, if non-zero, is a k-homogeneous polynomial. Note first that

$$\frac{\partial}{\partial x_i} f(x) = \frac{1}{r}\frac{d}{dr} F(\|x\|) x_i.$$

We will prove the formula recursively with respect to m. Assume that it holds for m:

$$\frac{\partial}{\partial x_i} P\left(\frac{\partial}{\partial x} \right) f(x) = \left(\frac{1}{r}\frac{d}{dr} \right)^{m+1} F(\|x\|) x_i P(x) + \left(\frac{1}{r}\frac{d}{dr} \right)^m F(\|x\|) \frac{\partial P}{\partial x_i}(x)$$

$$+ \sum_{k<m} \left(\frac{1}{r}\frac{d}{dr} \right)^{k+1} F(\|x\|) x_i P_k(x)$$

$$+ \sum_{k<m} \left(\frac{1}{r}\frac{d}{dr} \right)^k F(\|x\|) \frac{\partial P_k}{\partial x_i}(x).$$

The Vandermonde polynomial is skewsymmetric. Therefore, for $P = V$, ($m = p$), the polynomials P_k are skewsymmetric, hence divisible by V. Since $\deg P_k < \deg V = p$, necessarily the polynomials P_k are zero. $\qquad\square$

We get finally

$$V\left(\frac{\partial}{\partial x} \right) e^{-\|x-2\mathbf{k}\pi\|^2/4t} = \left(-\frac{1}{2t} \right)^p V(x - 2\mathbf{k}\pi) e^{-\|x-2\mathbf{k}\pi\|^2/4t}.$$

By observing that $p + n/2 = n(n-1)/2 + n/2 = n^2/2$, this finishes the proof. $\qquad\square$

In the series in Proposition 12.6.3, the dominant term in a neighbourhood of the identity element is that corresponding to $\mathbf{k} = 0$. This can be said more precisely as follows. Put

$$H(t, x) = \tilde{H}(t, x) + R(t, x),$$

where $\tilde{H}(t, x)$ is the central function in x for which

$$\tilde{H}(t, a(\theta)) = C_n \frac{e^{\gamma t}}{t^{n^2/2}} \frac{V(\theta)}{H(\theta)} e^{-\|\theta\|^2/4t}.$$

There exists a constant $c > 0$ such that

$$\int_G |R(t, x)| \mu(dx) = O(e^{-c/t}).$$

12.7 Exercises

1. *Cauchy determinant.* Show that

$$\det \left(\frac{1}{1 - x_i y_j} \right)_{1 \le i, j \le n} = V(x) V(y) \prod_{i,j=1}^{n} \frac{1}{1 - x_i y_j}.$$

Hint. Subtract the last row from the $n - 1$ first rows in such a way that the following factor appears:

$$\prod_{i=1}^{n-1} (x_i - x_n) \prod_{i=1}^{n-1} \frac{1}{1 - x_n y_i}.$$

Then subtract the last column from the $n - 1$ first columns.

2. To a function F on the torus $\mathbb{T}$, whose Fourier series is absolutely convergent,

$$F(t) = \sum_{m=-\infty}^{\infty} c_m t^m, \qquad \sum_{m=-\infty}^{\infty} |c_m| < \infty,$$

one associates the following function f on the unitary group $U(n)$:

$$f(g) = \det F(g).$$

This means that f is a central function and

$$f(\text{diag}(t_1, \ldots, t_n)) = F(t_1) \times \cdots \times F(t_n).$$

The aim of this exercise is to determine the Fourier series expansion of the function f:

$$f(g) = \sum_{\lambda \in P^+} a_\lambda \chi_\lambda(g),$$

by showing that

$$a_\lambda = \det((c_{\lambda_i - i + j})_{1 \le i, j \le n}).$$

Observe that this amounts to showing that

$$\prod_{j=1}^{n} F(t_j) = \sum_{m_1 \geq \cdots \geq m_n} a_{\mathbf{m}} s_{\mathbf{m}}(t_1, \ldots, t_n).$$

Two methods are proposed.

(a) *First method.* Determine, in the power series expansion of $V(t)F(t_1) \times \cdots \times F(t_n)$, the coefficient of the monomial $t_1^{m_1+\delta_1} \times \cdots \times t_n^{m_n+\delta_n}$.

(b) *Second method.* Show that

$$a_{\mathbf{m}} = \frac{1}{n!} \int_{\mathbb{T}^n} \prod_{i=1}^{n} F(t_i) \overline{A_{\mathbf{m}+\delta}(t)} V(t) \nu(dt),$$

and compute this integral by expanding the trigonometric polynomial $A_{\mathbf{m}+\delta}(t)\overline{V(t)}$.

Bibliography

N. Bourbaki (1971). *Groupes et Algèbres de Lie.* Masson.

Th. Bröcker, T. Tom Dieck (1985). *Representation Theory of Compact Lie Groups.* Springer.

R. Carter, G. Segal, I. Macdonald (1995). *Lectures on Lie Groups and Lie Algebras.* London Mathematical Society.

C. Chevalley (1946). *Theory of Lie Groups.* Princeton University Press.

P. Cohn (1968). *Lie Groups.* Cambridge University Press.

A. Deitmar (2005). *A First Course in Harmonic Analysis.* Springer.

J. J. Duistermaat, J. A. C. Kolk (2000). *Lie Groups.* Universitext, Springer.

W. Fulton, J. Harris (1991). *Representation Theory. A First Course.* Springer.

H. Freudenthal, H. de Vries (1965). *Linear Lie Groups.* Academic Press.

R. Godement (1982). *Introduction à la Théorie des Groupes de Lie.* Publications mathématiques de l'Université Paris VII.

R. Goodman, N. R. Wallach (1998). *Representations and Invariants of the Classical Groups.* Cambridge University Press.

B. C. Hall (2003). *Lie Groups, Lie Algebras, and Representations. An Elementary Introduction.* Springer.

J. Hilgert, K.-H. Neeb (1991). *Lie-Gruppen und Lie-Algebren.* Vieweg.

J. E. Humphreys (1972). *Introduction to Lie Algebras and Representation Theory.* Springer.

A. Kirillov (1976). *Elements of the Theory of Representations.* Springer.

R. Mneimné, F. Testard (1986). *Introduction à la Théorie des Groupes de Lie Classiques.* Hermann.

G. Pichon (1973). *Groupes de Lie, Représentations et Applications.* Hermann.

M. Postnikov (1986). *Lectures in Geometry, Lie Groups and Lie Algebras.* Mir Publishers.

W. Rossmann (2002). *Lie Groups, an Introduction Through Linear Groups.* Oxford University Press.

A. A. Sagle, R. W. Walde (1973). *Introduction to Lie Groups and Lie Algebras.* Academic Press.

J. P. Serre (1965). *Lie Algebras and Lie Groups.* Benjamin.

B. Simon (1996). *Representations of Finite and Compact Groups*. American Mathematical Society.

M. Sugiura (1975). *Unitary Representations and Harmonic Analysis*. North-Holland/ Kodansha.

V. S. Varadarajan (1984). *Lie Groups, Lie Algebras, and their Representations*. Springer.

E. B. Vinberg (1989). *Linear Representations of Groups*. Birkhäuser.

D. P. Želobenko (1973). *Compact Lie Groups and their Representations*. American Mathematical Society.

Index

adjoint operator, 99
adjoint representation, 53
automorphism (of a Lie algebra), 51

Bessel function, 209
Bochner–Hecke relations, 211

Campbell–Hausdorff formula, 46
Cartan's criterion, 65
Casimir operator, 120
Cauchy determinant, 282
Cauchy inequality, 284
Cayley transform, 93
centre (of a Lie algebra), 54
character, 97, 110, 115
character formula, 257
Clebsh–Gordan coefficients, 112, 151
compact operator, 100
complex type (representation of), 124
conjugate (representation), 113
contragredient (representation), 112
covering, 54

decomposition (Gauss), 16
decomposition (Gram), 13
decomposition (polar), 8
derivation, 51
derived group, 56
derived ideal, 57
derived representation, 53
derived series, 57
descending central series, 57
differential form, 82, 83
dimension formula, 259
Dirichlet problem, 215

disjoint representations, 124
divergence, 213

Engel's Theorem, 60
equivalent representations, 52, 96
Euler angles, 135
exponential of a matrix, 18

Fourier coefficient, 109, 167
Fourier series, 158, 168
Funk–Hecke Theorem, 205

Gauss' Theorem, 213
Gegenbauer polynomial, 220
Green's formula, 214
Green's kernel, 215

Haar measure, 74
harmonic, 195, 212
heat equation, 172, 176, 225, 292
heat kernel, 173, 178, 225, 292
Heisenberg Lie algebra, 57
highest weight, 251
Hilbert–Schmidt (norm of), 109
holomorphic representations, 261

ideal, 52
identity component, 1
intertwinning operator, 52, 96
invariant subspace, 52, 95

Jacobi identity, 38, 50

Killing form, 63

Laplace equation, 195, 212
Laplace operator, 162, 193